CIRCULATING

The Encyclopedia of Women's Travel and Exploration

The Encyclopedia of Women's Travel and Exploration

Patricia D. Netzley

Oryx Press
2001

The rare Arabian Oryx is believed to have inspired the myth of the unicorn. This desert antelope became virtually extinct in the early 1960s. At that time, several groups of international conservationists arranged to have nine animals sent to the Phoenix Zoo to be the nucleus of a captive breeding herd. Today, the Oryx population is over 1,000, and over 500 have been returned to the Middle East.

Published by The Oryx Press
An imprint of Greenwood Publishing Group, Inc.
88 Post Road West
Westport, CT 06881-5007
(203) 226-3571, (800) 225-5800
www.oryxpress.com

Published simultaneously in Canada
Printed and bound in the United States of America

∞ The paper used in this publication meets the minimum requirements of American National Standard for Information Science—Permanence of Paper for Printed Library Materials, ANSI Z39.48, 1984.

Library of Congress Cataloging-in-Publication Data

Netzley, Patricia D.
The encyclopedia of women's travel and exploration / by Patricia D. Netzley.
p. cm.
Includes bibliographical references and index.
ISBN 1-57356-238-6 (alk. paper)
1. Voyages and travels—Encyclopedias. 2. Women travelers—Encyclopedias. I. Title.
G465 .N47 2000
910'.82—dc21

00-010720
CIP

For Sarah Netzley, a future explorer

CONTENTS

PREFACE

Women have traveled the world throughout history, but their contributions to the field of exploration have largely been ignored. Few schoolchildren learn about Victorian explorer Mary Kingsley, for example, although her adventures traveling through Africa equaled or perhaps even exceeded those of her male counterpart, David Livingston. Similarly, many people know that Sir Samuel White Baker discovered the source of the Nile River in 1864, but few realize that his wife Florence was standing beside him at the time.

Even among their peers, women like Mary Kingsley and Florence Baker failed to receive the support for their efforts that they deserved. In fact, prior to the mid-twentieth century, women were discouraged from traveling at all, unless they were planning a trip through safe, civilized territory—and even then, they were typically chaperoned by male relatives. Women were also expected to behave in a "ladylike" manner at all times, while avoiding strenuous physical exertion. Consequently clothing and gear for such travel-related activities as hiking, mountain climbing, and horseback riding were originally designed only for male bodies. Male travelers and adventurers have had other advantages over women as well; society has long encouraged their exploits, governments and major organizations have offered them financial support, and their achievements have often brought them fame.

In recent years, however, women have made great strides in gaining financial and emotional support for their own efforts in the field of travel and exploration, and authors have begun to take note of their accomplishments. *The Encyclopedia of Women's Travel and Exploration* is an attempt to add to this body of literature, providing biographical information on a wide variety of women explorers, adventurers, and travelers throughout history. Arranged alphabetically from A to Z, the *Encyclopedia* also offers entries on types of travel, such as migration, as well as on categories of women travelers, which include anthropologists, archaeologists, aviators, and sailors. Through extensive cross-references, these entries direct the reader to individual travelers and explorers. In addition, they provide insights into why women travel and the ways in which their reasons for traveling have changed over time. Whereas men have historically traveled for adventure, enjoyment, physical challenge, financial reward, and perhaps fame, women have primarily traveled for self-discovery, enlightenment, and education. Today, however, there is less distinction between genders in regard to travel motivations.

To further illustrate women's travel motivations and experiences, *The Encyclopedia of Women's Travel and Exploration* includes discussions of women's travel literature. Biographical entries on female travel writers provide information on the best known of their works—usually one per author—and there are separate entries on women's travel anthologies as well. In addition, travel writers are discussed in entries on each major world region (e.g., North America, Latin America, Africa, etc.)—as are explorers and adventurers, in order to help readers recognize patterns and connections among women from specific geographical regions and cultural backgrounds. There are also entries on certain types of transportation, including air travel, ground transportation, and the travel industry, in order to show how women have typically chosen to travel. Only space exploration has been excluded in this volume.

Throughout *The Encyclopedia of Women's Travel and Exploration,* which has 315 entries in all, the emphasis is on women who have been the first to accomplish a travel- or exploration-related feat or whose exploits have been extensive enough to warrant fame. Of course, determining which travelers deserve recognition is highly subjective, but in most cases the deciding factor was whether or not the reader could be directed to other works written either about or by a particular biographical subject. In other words, the *Encyclopedia* is intended to be the beginning of the reader's journey of exploration into these women's lives, rather than the sole source of information on them. Consequently individual entries include references to books and Internet Web sites for further reading, and the *Encyclopedia* also includes an extensive bibliography. A subject index concludes the book.

Accommodations

When women travel they are typically attracted to different types of accommodations than are men. Health spas and resorts, for example, are more popular with women than with men, as are bed-and-breakfast establishments, which provide sleeping accommodations in a private home. Although women are often drawn to a homey atmosphere or the type of pampering offered at a resort, they usually choose to stay in a hotel or motel. There are currently about 10 million hotel rooms in the world, with approximately three million in the United States alone.

Even when choosing the simplest accommodations, women typically rate cleanliness as the most important feature of a room. To help women travelers determine whether a particular hotel will meet their needs in this regard, automobile clubs, innkeepers' associations, and other organizations rate accommodations according to the services they offer. In the United States, these rating systems emphasize cleanliness and safety; in Europe, they emphasize the number of amenities a facility provides. Thus, a hotel in the United States might earn a high rating simply for being clean, affordable, and well located, whereas in Europe the top-rated hotels are those that provide guests with such extras as dry cleaning services, shops, private meeting rooms, and assistance with theater and restaurant reservations.

Many guidebooks are also available to help women choose appropriate accommodations, whether they want to stay in a hotel, motel, resort, spa, or retreat. These guidebooks typically not only rate various establishments but also offer detailed information on amenities, location, and pricing. Regarding the last, most guidebooks differentiate between places that price accommodations separately and ones that are "all-inclusive," a term that means that the total vacation experience, including food, accommodations, and recreational activities, has a predetermined price.

Further Reading: Bain-Dror International Travel. *Spas: The International Spa Guide: An International Passport to Beauty, Fitness, and Well-Being*. Flushing, NY: B.D.I.T., 1999; Bowler, Gail Hellund. *Artists and Writers' Colonies: Retreats, Residencies, and Respites for the Creative Mind*. Hillsboro, OR: Blue Heron Publications, 1995; Dyson, Katharine. *100 Best Romantic Resorts of the World*. Guilford, CT: Globe Pequot Press, 2000; Holms, Karin Baji. *101 Vacations to Change Your Life: A Guide to Wellness Centers, Spiritual Retreats, and Spas*. Secaucus, NJ: Carol Publishing Group, 1999; McConnell, Amy, ed. *Fodor's Healthy Escapes*. Fodor's Travel Publications, 1999; Miller, Jenifer. *Healing Centers & Retreats: Healthy Getaways for Every Body and Budget*. Santa Fe, NM: John Muir Publications, 1998; Paris, Jay, and Carmi Zona-Paris. *100 Best All-Inclusive Resorts of the World*. Guilford: CT: Globe Pequot Press, 1999; Short, Linda. *The Complete Idiot's Guide to Self-Healing with Spas and Retreats*. Indianapolis, IN: Alpha Books, 1999.

Across New Worlds

Published in 1990, *Across New Worlds: Nineteenth-Century Women Travellers and Their Writings* by Shirley Foster is an important reference work on women's travel literature of the nineteenth century. It discusses this literature in general before focusing on three regions: Italy, North America, and the Far East, including Tibet. The work uses brief but numerous quotes to illustrate various points about women's travel writing. For example, in discussing women's travel experiences, Foster says:

> As a result of their undertakings, nineteenth-century Englishwomen acquired a reputation for their intrepitude and energy as travellers, noted by their own countrymen and foreigners alike. . . . [Such] energetic and admirable women were of course assisted by the ever-widening opportunities which became available to them throughout the century, the result both of improved communications and modes of transportation, and of the gradual loosening of restrictions on their movements which made it easier for them to broaden their activities. . . . But in the same way that professionally—as writers, educationalists and doctors—they found it hard to gain recognition, so as travellers they often encountered if not outright hostility at least patronising ridicule. . . . the eccentric lady traveller, like the old maid . . . took her place in society's collection of caricatures. There may have been some truth in the exaggerations. With somewhat regrettable disloyalty to her own sex, [travel writer Fanny] Kemble herself laughs at the way English female tourists are immediately recognisable, as she observes four veiled women on board a ship at Marseilles in the 1840s, "who began stumping up and down the deck, . . . betraying in the very hang of their multitudinous shawls, the English creature." (p. 6)

Foster concludes that while women travelers were sometimes mocked, their writings offered new and important insights on foreign cultures because they focused more on social experiences and family life than did male travelers of the same period. Moreover, Foster believes that nineteenth-century women travel writers were more accepting of foreign cultures. She states, "Though for these women, conscious of their own marginalisation, travel abroad could exacerbate their sense of unease about their equivocal gender and social roles, it also inspired a willingness to embrace 'difference.' It is this openness, combined with so much evident enjoyment of . . . foreign travel, which makes them so fascinating as adventurers and writers and which gives them value today as interpreters of a significant area of Victorian female experience." (pp. 174–175). *See also* TRAVEL WRITERS.

Further Reading: Foster, Shirley. *Across New Worlds: Nineteenth-Century Women Explorers and Their Writings.* New York: Harvester Wheatsheaf, 1990.

Adams, Harriet Chalmers (1875–1937)

Born in 1875, Harriet Chalmers Adams helped found the Society of Women Geographers, an organization dedicated to encouraging women's exploration, in 1925. She was also the society's first president, serving until 1933. Her main interest was the history of Spanish exploration, and in her own travels she concentrated on countries that were or had been controlled by Spain or Portugal.

Adams was raised in central California, an area that had once been under the control of Spain. When she was a girl, she and her father embarked on a pleasure trip on horseback throughout the state, spending much of their time in the Sierra Nevadas. Adams later attributed her love of travel to this adventure. In 1899 she married an electrical engineer, Franklin Pierce Anderson, who also loved to travel; in 1900 she accompanied him on a job assignment in Mexico. The two enjoyed the trip so much that from 1903 to 1906 they traveled together throughout South America, venturing into jungles to view ancient ruins and indigenous cultures. Adams wrote many articles about her travels, which were published in newspapers or in *National Geographic.* In the September 1908 issue of *National Geographic*, she wrote about the Peruvian Andes:

> Those were long days in the saddle, with little food and less water. We knew the river water to be impure . . . and the brooks are also contaminated as they pass through the villages. At night we slept on the ground, wrapped in our blankets, at times finding shelter in a ruined temple. . . .We met no travelers save the highland Indians, and picked up a few words of their tongue. I felt that we had left civilization far behind. . . . To know a country and a people, one must leave the highway and live near to Nature. We traveled much in the saddle on this great elevated plateau . . . and gradually my standpoint changed. I started out as an outsider. . . . In time I grew, through study and observation, but more through sharing the life, half-Andean myself, and find, in looking back over years of travel in South America—years in which we visited every country—that my greatest heart interest lies in the highlands of Peru and Bolivia. (Tinling, p. 4)

In 1908 Adams's husband took an editing job that made taking long vacations difficult. Nonetheless, Adams continued to travel even when she had to go alone, and she became increasingly self-confident about her ability to deal with all kinds of adventures. As historian Marion Tinling reports, people who saw Adams during her visits to the United States "were amazed that this petite woman, elegantly dressed, could have roughed it in wild countries." (p. 6)

In 1916, Adams accepted a job working for *Harper's* magazine as a war correspondent in France. Later she visited Spain and Morocco to study South America's historical connection to those countries. In the 1930s she toured North Africa and Asia Minor, despite the fact that she was recovering from a serious back injury. Adams documented many of her experiences with her own photographs, which she published along with her articles, and she lectured about her adventures throughout the United States. In her later speeches, Adams argued that human beings originated in Central Asia, perhaps in India, and migrated into North America and then South America via a land bridge from the Asian continent near Siberia. In 1933, she retired with her husband to Europe, intending to write a book on Peru. However, she never finished it and died in 1937. *See also* EUROPE, CONTINENTAL; ORGANIZATIONS AND ASSOCIATIONS.

Further Reading: Anema, Durlynn. *Harriet Chalmers Adams: Explorer and Adventurer*. Greensboro, NC: Morgan Reynolds, 1997; Tinling, Marion. *Women Into the Unknown: A Sourcebook on Women Travelers*. Westport, CT: Greenwood Press, 1989.

Adamson, Joy (1910–1980)

Born Friederike Victoria Gessner in 1910 in Austria, Joy Adamson helped popularize African wildlife vacations during the 1960s by publishing three bestselling books on her experiences with lions. *Born Free* (1960), *Living Free* (1961), and *Forever Free* (1963) inspired several film projects, including the movie *Born Free* in 1966, the movie *Living Free* in 1971, the documentary *The Lions Are Free* in 1970, and the television series *Born Free* in 1975.

Adamson began living on a wildlife preserve in Kenya, Africa, in 1942, after marrying a naturalist who took a job there as a game warden. In 1949, after she was hired by the British government to paint portraits of representatives from 20 of Kenya's indigenous tribes, she traveled throughout the region sketching and photographing various subjects.

She also established a nonprofit organization, the Elsa Wild Animal Appeal, to support African wildlife conservation and continued to write books on Africa and its wildlife. Her later works include *The People of Kenya* (1967), *The Spotted Sphinx* (1969), *Pippa: The Cheetah and Her Cubs* (1970), and *Joy Adamson's Africa* (1972), which reproduced many of her paintings and sketches, as well as an autobiography entitled *The Searching Spirit* (1979).

Adamson and her husband eventually separated, but both continued to live in the African wilderness. While living alone, Adamson became the victim of repeated thefts, and in January 1980 she was discovered dead. At first police thought she had been attacked by a lion, but later they determined

Joy Adamson in Africa in 1971. © *Hulton-Deutsch Collection/CORBIS.*

that she had been murdered; eventually, a disgruntled former employee confessed to the crime. *See also* AFRICA; PHOTOGRAPHERS AND ARTISTS; SPOUSES.

Further Reading: Adamson, George. *My Pride and Joy*. New York: Simon & Schuster, 1987; Adamson, Joy. *Born Free: A Lioness of Two Worlds*. New York: Random House, 1960. ———. *Forever Free*. New York: Harcourt, Brace, and World, 1963; ———. *Living Free*. New York: Harcourt, Brace, and World, 1961; ———. *The Searching Spirit: An Autobiography*. With a foreword by Elspeth Huxley. London: Collins and Harvill Press, 1979; Cass, Caroline. *Joy Adamson: Behind the Mask*. London: Weidenfeld and Nicolson, 1992.

Adventure Travel

Adventure travel for women is one of the fastest growing segments of the travel industry. This type of travel experience involves physical challenge—rafting down a turbulent river, climbing a difficult peak, getting close to wild animals on safari, or hiking into a nearly inaccessible wilderness region. Until fairly recently, many men—and, in some cases, many women as well—believed that women were not up to such challenges. Today, however, women of all ages and skill levels participate in adventure travel.

The first women-specific adventure travel companies appeared in the late 1970s and early 1980s, largely because of the women's movement. During this period women were encouraged to find ways to support themselves financially, and many chose to establish businesses that furthered women's growing desire for independence. Of the many adventure-travel businesses founded at this time, one of the longest-running is Adventurewomen, Inc. (http://www.rainbowadventures.com), which was established in 1982. Other groups specialize according to activity and/or geographical region. For example, Women in Good Company (http://mariposa.yosemite.net/wigc/) provides river rafting trips for women of all ages, particularly along the Colorado River. One of the first women to promote adventure tourism in America was a river rafter. Georgie White Clark introduced large-scale river tourism to the American Southwest; during the 1970s and 1980s she led dozens of tourists in boats down the Colorado River through the Grand Canyon, encouraging an appreciation of nature and an interest in whitewater rafting.

White's business offered adventure travel experiences for both men and women. So do many other companies, including the Adventure Center (http://www.adventurecenter.com) and Outward Bound (http://www.outwardbound.com). Founded in Wales in 1941 by educator Kurt Hahn, Outward Bound focuses on teaching wilderness skills in an effort to bolster self-esteem and environmental awareness. Many women have gained confidence in their physical and mental abilities through such programs, in part because each Outward Bound course requires students to spend time alone, not as a survival exercise, but as a time for introspection. Outward Bound has five wilderness schools in the United States and outreach programs offered in local communities.

Women interested in participating in adventure travel have many books to turn to. One of the best guidebooks written specifically for women is *Adventures in Good Company: The Complete Guide to Women's Tours and Outdoor Trips* (1994) by Thalia Zepatos. It offers essays by women travelers along with advice and information related to experiences such as scuba diving, rock climbing, bicycling, and dogsledding. Adventure travelers of both genders often turn to the "Adventuring in" series published by the Sierra Club, with each volume focusing on a particular geographical region. Parents wanting to participate in adventure travel with their children can consult *Adventuring with Children: The Complete Manual for Family Adventure Travel* (1990) by Nan Jeffrey.

In addition to guidebooks, adventure travelers can read about the experiences of others in such works as Patricia McCairen's *Canyon Solitude: A Woman's Solo River Journey Through the Grand Canyon* (1998), which describes the author's solo journey down the Colorado River from northern Arizona to Lake Mead. A more strenuous example of river rafting is offered in Tracy Johnston's book *Shooting the Boh: A Woman's Voyage Down the Wildest River in Borneo* (1992), which describes the author's journey down a river that had never before been fully navigated.

Another notable adventure traveler who wrote about her experiences is Eleanor Clark, whose *Tamrart: Thirteen Days in the Sahara* (1984) describes her trek across the Sahara on a camel, after which she climbed an Algerian mountain. Similarly, Robyn Davidson's *Tracks* (1980) describes her adventure travel experience going across the Australian bush alone except for four camels and a dog. Mountaineer Judy Fracher wrote about her climbing vacations in *Hey Lady! How Did You Get Way Up Here?: Climbing the 4,000 Footers of New Hampshire* (1996). Fracher spent the years 1980 to 1992 methodically hiking up every 4,000-foot mountain in New Hampshire.

Women who participate in adventure travel should know, however, that the industry has come under much criticism lately about its safety practices because of a few highly publicized disasters. One of the most notable was a 1996 tragedy in which several climbers died when a storm hit during their descent from Mount Everest's summit. Yasuko Nanba, a Japanese woman, who at age 47 had just become the oldest woman to reach the top of the mountain, was one of those killed. There had been a total of 39 climbers ascending to the summit on the day of the disaster, including two other women—Susan Allen of Australia and Nancy Hutchison of Canada, both of whom survived.

A similar tragedy occurred in July 1999 near Interlaken, Switzerland, where 17 men and four women drowned while participating in an activity called canyoning. Canyoners travel through a deep canyon or gorge with a fast-flowing river at the bottom, scrambling along rocks and swimming and sliding in the water. Canyoning has been growing in popularity in Europe, with approximately 30,000 people participating in Switzerland each year. However, the Swiss government suspended all canyoning activities in the Interlaken area after the 1999 tragedy, Switzerland's worst whitewater river accident since the 1993 deaths of 17 adventure tourists in a rafting expedition.

All adventure travel experiences entail risk; in fact, some would say that it is the element of danger that makes such experiences attractive. Adventure tours do vary widely in their level of danger and physical challenge, thus making it possible for women to find relatively safe yet exciting adventure travel activities. Those who want to pursue such activities can find travel agencies and guidebooks devoted exclusively to adventure travel. *See also* ADVENTURES IN GOOD COMPANY; CLARK, ELEANOR; DAVIDSON, ROBYN; GUIDEBOOKS, TRAVEL; MCCAIREN, PATRICIA.

Further Reading: Clark, Eleanor. *Tamrart: Thirteen Days in the Sahara.* Winston-Salem, NC: S. Wright, 1984; Clark, Georgie White, and Duane Newcomb. *Georgie Clark: Thirty Years of River Running.* San Francisco: Chronicle Books, 1977; Davidson, Robyn. *Tracks.* New York: Pantheon Books, 1980; Fracher, Judy. *Hey, Lady! How Did You Get Way Up Here?: Climbing the 4,000 Footers of New Hampshire.* Etna, NH: Durand Press, 1996; Jeffrey, Nan, with Kevin Jeffrey. *Adventuring with Children: The Complete Manual for Family Adventure Travel.* Marstons Mills, MA: Avalon House, 1990; Johnston, Tracy. *Shooting the Boh: A Woman's Voyage Down the Wildest River in Borneo.* New York: Vintage Books, 1992; McCairen, Patricia. *Canyon Solitude: A Woman's Solo River Journey Through the Grand Canyon.* Seattle, WA: Seal Press, 1998; Teal, Louise. *Breaking into the Current: Boatwomen of the Grand Canyon.* Tucson: University of Arizona Press, 1994; Westwood, Dick, and Richard E. Westwood. *Woman of the River: Georgie White Clark, White Water Pioneer.* Logan: Utah State University Press, 1997; Zepatos, Thalia. *Adventures in Good Company: The Complete Guide to Women's Tours and Outdoor Trips.* Portland, OR: Eighth Mountain Press, 1994.

Adventures in Good Company

Published in 1994, *Adventures in Good Company: The Complete Guide to Women's Tours and Outdoor Trips* by travel author Thalia Zepatos is representative of guidebooks devoted to women's travel experiences. Much of its information relates to adventure travel; for example, the book offers resources for women interested in such activities as rock climbing, scuba diving, dog sledding, skiing, bicycling, canoeing, fishing, surfing, and sailing. Zepatos also provides information on group travel, spa vacations, spiritual retreats, and tourism opportunities for lesbians, mothers, disabled women, and older women. Interspersed with Zepatos's informative text are first-person essays by women travelers. In addition, the book includes profiles on more than 100 travel-related companies worldwide. *See also* ADVENTURE TRAVEL.

Further Reading: Zepatos, Thalia. *Adventures in Good Company: The Complete Guide to Women's Tours and Outdoor Trips.* Portland, OR: Eighth Mountain Press, 1994.

Aebi, Tania (1969–)

Born in 1969, Tania Aebi was widely publicized as being the first American woman and the youngest person to circumnavigate the globe solo. However, a year after her 1987 adventure, her claim was denied because she had a friend on board for 80 miles of sailing in the South Pacific. Record books therefore indicate that Karen Thorndike is the first and only American woman to date to have circumnavigated solo. (She made her journey in 1996–1998.)

A college dropout, Aebi undertook her journey after her father offered to buy her a boat (a 26-foot sloop) if she would use it to circumnavigate solo; he thought that this challenge would help his daughter mature. Once she agreed to his terms, he contacted the news media, and Aebi was interviewed on several talk shows before setting sail from New York. She was also interviewed upon her return. During her journey, she sent articles for publication to *Cruising World* magazine, and, in 1989, she published a book entitled *Maiden Voyage* to capitalize on her claim of solo circumnavigation—a claim that she admits was false, as written in an epilogue to the 1996 paperback edition of *Maiden Voyage.*

Despite the controversy regarding Aebi's solo status, *Maiden Voyage* offers an interesting glimpse of an around-the-world sail. Aebi faced many difficulties during her trip, including a fall on deck that sprained her wrist, problems with lice, the death of the cat that accompanied her, and several major storms. She was at sea for two and one-half years, during which she traveled more than 27,000 miles and visited several ports. While docked on an island in the South Pacific, she met a 33-year-old Swiss solo sailor called Olivier and fell in love with him. She then met with him at several prearranged locations along her route. Today Aebi is married to Olivier and living in New York. *See also* CIRCUMNAVIGATORS AND ROUND-THE-WORLD TRAVELERS; SEA TRAVEL.

Further Reading: Aebi, Tania, with Bernadette Brennan. *Maiden Voyage*. New York: Ballantine, 1996.

Africa

The continent of Africa attracted some of the most significant women explorers and hunters of the late nineteenth and early twentieth centuries; in modern times it has continued to attract female naturalists, scientists, photographers, and adventure travelers. It has also been the subject of several classic works of travel literature, largely because of British involvement in African settlement.

Human beings have been present in Africa for millions of years; in fact, most scientists believe that humankind originated in Africa. But for most of its history, civilization on the continent was concentrated in the north, where Egypt became a powerful empire in approximately 3000 B.C.E. and remained influential for roughly 3000 years. In the fifth century, trading empires began to develop in western Africa, and by the thirteenth century they had developed in eastern Africa as well.

Most people, however, lived along the coast; the interior of Africa remained largely unexplored until the nineteenth century. At that time, European interest in the continent led prominent British explorers such as Sir Richard Burton, John Hanning Speke, David Livingstone, and Henry Morton Stanley to travel into the region and report on their discoveries. Most of these men explored solely for the joy of discovery. Some, however, had additional reasons for venturing into the African interior. For example, Livingstone was both an explorer and a Christian missionary, who combined his quest for information with a quest to spread his religious beliefs.

But regardless of male explorers' reasons for traveling through Africa, their accounts caught the attention of several women who decided to match the men's efforts.

One of these women was Dutchwoman "Alexine" Tinne, who went to Egypt in 1856 with her mother as part of a tour of Europe and the Middle East. While in Egypt she took a boat a short distance up the Nile and became fascinated with the region. She returned to Egypt in 1861 to travel up the Nile in search of the river's source, but after five months on the river she became ill and returned to Khartoum in the Sudan. When she recovered she set out again, this time with some European scientists and soldiers on a yearlong expedition to explore the African interior. After many difficulties, Tinne returned to Khartoum in March 1864 without ever finding the source of the Nile. However, she did collect valuable botanical specimens, and her adventures inspired other women to explore Africa. In 1869 Tinne was killed by a hostile tribe in the Sahara while attempting to become the first European woman to cross this desert.

Many women were also influenced by an 1897 book, *Travels in West Africa,* by Englishwoman Mary Kingsley. Kingsley traveled through West Africa on two expeditions, one in 1893 and the other in 1894–1895. Friends advised her that the journey was too dangerous for a woman, but she ignored their warnings and traveled down the West African coast to Angola, then headed into the continent's interior. Her book describes the customs of the tribes she encountered as well as scenery and animals, and details a variety of interesting adventures. To further promote interest in African travel, Kingsley gave lectures on the subject when she returned to England.

Another significant British explorer was Florence Baker, with accomplishments made in concert with her husband, explorer Sir Samuel White Baker. In 1864 the two set out on a major expedition to find the source of the Nile but were unsuccessful. From 1870 to 1873 they led another African expedition, this time to fight slave trading along the Nile at the request of the Egyptian government.

In 1873, Englishwoman Amelia Edwards traveled up the Nile to sketch and measure ancient ruins along its shore. This shifted the focus of exploration from finding the Nile's source to learning about its history. When Edwards returned from her trip, she wrote *A Thousand Miles Up the Nile* (1877), which provided the fullest account of Egyptian ruins up until that time. She also established the Egypt Exploration Fund (1883), dedicated to archaeological exploration, and helped create the science of Egyptology. In 1889 she toured America, lecturing about Egypt. She subsequently published her lecture notes as *Pharoahs, Fellahs, and Explorers* (1891). All of her activities furthered women's interest in Egypt, and the country continues to be an attractive destination for women tourists even in periods of political unrest and terrorism.

In the late nineteenth century, big game hunting also attracted women to Africa, although today there is less interest in this activity among women. In particular, the publication in 1892 of May French-Sheldon's *Sultan to Sultan: Adventures among the Masai and other Tribes of East Africa* increased upper-class women's desire to go on African safaris. A best-seller in both England and the United States, the book describes the Englishwoman's experiences leading a safari into the interior of East Africa from the coastal city of Mombasa, Kenya, in 1891, accompanied by over 100 porters and servants. French-Sheldon was one of the first women to lead a safari. Another Englishwoman, Etta Close, claimed to have been the first woman ever to lead an African safari, although she offered no proof. Nonetheless, she repeated her claim in a popular book about her experiences, *A Woman Alone in Kenya, Uganda, and the Belgian Congo* (1924).

But perhaps the best known female big game hunter is American Delia Denning Akeley, who traveled throughout Africa during the years 1905 to 1929. On her first trip, which lasted over a year, she shot two elephants, along with 17 other large animals. On a second expedition, which lasted two years, Akeley began to study animal behavior as a naturalist rather than a hunter, and in 1924 she began studying indigenous tribes as well, venturing through the African interior to study Pigmy tribes. By the end of her expedition, she had gone from the Indian Ocean to the Atlantic Ocean, making her the first Western woman to cross Africa latitudinally.

The first woman to journey longitudinally through Africa was Mary Hall from Great Britain, who began her 1906 adventure in South Africa and went north to Cairo, Egypt, by railway, steamer, rickshaw, foot travel, and being carried by porters. Except for her servants, she was unaccompanied on her seven-month trip. This meant that she was responsible for all major decisions during her journey—a responsibility that was perhaps more precious to her than the solitude she might have experienced had she been a male explorer who would have traveled with far fewer—or perhaps even no—servants.

Hall wrote about her experiences in *A Woman's Trek from the Cape to Cairo* (1907). After her book was published she continued to travel throughout the world, alone except for her retinue of servants.

Still another notable traveler to Africa was American Osa Johnson, who ventured into the interior with her husband, fellow photographer Martin Johnson, to photograph wild animals. Their 1923 expedition, which was funded by the Kodak company and the American Museum of Natural History in New York, lasted four years, during which they lived near a heavily used watering hole and took pictures of big game. They also visited many other sites in Africa and created movies of their experiences. Several years later, Swiss filmmaker Vivienne De Watteville also visited Africa to film wildlife, writing about her experiences in *Speak to the Earth* (1935).

Serious scientists have been drawn to Africa because of its animals as well. Perhaps the most notable were Austrian-born Joy Adamson, American Dian Fossey, and Englishwoman Jane Goodall. Joy Adamson lived on a wildlife preserve in Kenya in the 1940s and became famous because of three popular books based on her experiences with wild lions there, *Born Free* (1960), *Living Free* (1961), and *Forever Free* (1963). Dian Fossey established the Karisoke Research Center in Rwanda in 1967 to study rare mountain gorillas, and her book about her experiences, *Gorillas in the Mist* (1983), became famous, particularly after gorilla poachers murdered her in 1985. Jane Goodall began studying chimpanzees in 1960 in the Gombe Stream area beside Lake Tanganyika in Tanzania and wrote *In the Shadow of Man* (1971) to document the first 10 years of her research; in 1977 she founded the Jane Goodall Institute, which is dedicated to the preservation of wild chimpanzees, and she remains active in African wildlife conservation today.

Whereas hunters, photographers, and scientists are typically attracted to Africa because of its wildlife, some women adventurers have gone there because of a challenging region known as the Sahara, the largest tropical desert in the world. Home to many nomadic tribes, the Sahara is located over much of northern Africa, bordered on the north by the Mediterranean Sea and on the south by a semidesert region called the Sahel. Its border on the west is the Atlantic Ocean, on the east, the Atlas Mountains. Eleven countries include parts of the Sahara within their boundaries: Morocco, Libya, Algeria, Tunisia, Egypt, Mauritania, Mali, Niger, Chad, Sudan, and Western Sahara. Women travelers who have visited this region include Dutch explorer "Alexine" Tinne (killed by a hostile tribe in the Sahara), British writer Rosita Forbes, Mildred Bruce, and American author Eleanor Clark. In 1920 Forbes became the first foreign woman to enter the sacred city of Kufara (in present-day Libya), writing about her experiences in *The Secret of the Sahara: Kufara* (1921). During the 1930s, motorist Bruce drove a car into regions of the Sahara where no one else had ever driven. In the 1980s, Clark spent 13 days riding camels across the Sahara, then climbed an Algerian mountain before returning home; she wrote about her trip in *Tamrart: Thirteen Days in the Sahara* (1984). All of these women sought the kind of adventure that only a difficult landscape could provide.

Meanwhile, politics drew many important women to African towns and cities. Some of the first arrived to report on Britain's policies regarding colonization. During the last two decades of the nineteenth century, European countries—particularly the Netherlands, England, and France—began to establish permanent settlements in Africa; by 1920 most of the continent was under some form of colonial rule. This resulted in conflicts between those who had taken over various regions and the settlers and indigenous peoples who had lived there previously. The largest of these conflicts was the Boer War (1899–1902), which was fought between British colonialists and Dutch settlers, known as Boers, in South Africa. British journalist Florence Dixie went to Africa specifically to cover this conflict for London's *Morning Post* and subsequently became involved in other aspects of South African politics.

English author Margery Freda Perham also addressed African political issues, not only in South Africa but also elsewhere on the continent. She made several trips to Africa between 1922 and 1932 and wrote books such as *African Apprenticeship: An Autobiographical Journey in Southern Africa* (1929), *East African Journey: Kenya and Tanganyika, 1929–30* (1976), and *West African Passage: A Journey through Nigeria, Chad, and the Cameroons, 1931–32* (1983). In addition, a collection of her correspondence with fellow Englishwoman and African traveler Elspeth Huxley, *Race and Politics in Kenya: A Correspondence between Elspeth Huxley and Margery Perham*, was published in 1944. Huxley wrote primarily about East Africa, where she grew up on her parents' 500-acre plantation. Danish author Isak Dinesen also lived on a plantation, and her experiences there became the basis of her classic book, *Out of Africa* (1937). American author Emily Hahn is representative of women who traveled to Africa to offer medical help. She went to the Belgian Congo in Africa in 1930 as a Red Cross worker, offering aid to Pygmy tribes. She wrote many articles about her experiences. Other aid workers to visit Africa were Christian missionaries, who established outposts with churches to bring their religion to indigenous tribes. The missionary presence in Africa had an enormous influence on the development of the continent, and women were a major part of this presence. Many of them traveled to the region under the auspices of groups like the Missionary Sisters of Our Lady of Africa, established in 1869 as a companion to a male organization founded the previous year as the Society of the Missionaries of Africa.

Tourism with no agenda other than sightseeing has also attracted many women to Africa, particularly experienced world travelers. Several notable books have resulted from these experiences, including Mary Eliza Bakewell Gaunt's *Alone in West Africa* (1912), which describes two trips to Africa in 1908 and 1910, and Christina Dodwell's *Travels with Fortune: An African Adventure* (1979), which reports on three years spent traveling through Africa. *See also* ADAMSON, JOY; AKELEY, DELIA DENNING; BAKER, FLORENCE; BRUCE, MILDRED MARY; CLARK, ELEANOR; DE WATTEVILLE, VIVIENNE; DINESEN, ISAK; DIXIE, FLORENCE; DODWELL, CHRISTINA; EDWARDS, AMELIA; FORBES, ROSITA; FOSSEY, DIAN; GOODALL, JANE; HAHN, EMILY; HALL, MARY; HUXLEY, ELSPETH JOSCELINE; JOHNSON, OSA; PERHAM, MARGERY FREDA; TINNE, "ALEXINE" (ALEXANDRINE).

Further Reading: Edwards, Amelia Ann Blandford. *Pharoahs, Fellahs, and Explorers.* http://www.cs.cmu.edu/~mmbt/women/edwards/pharaohs/pharaohs.html; ———. *A Thousand Miles Up the Nile.* Los Angeles: J.P. Tarcher; Boston: Houghton Mifflin, 1983; Huxley, Elspeth Josceline. *The Challenge of Africa.* London: Aldus, 1971; ———. *Four Guineas: A Journey Through West Africa.* Westport, CT: Greenwood Press, 1974; ———. *The Sorcerer's Apprentice: A Journey Through East Africa.* Westport, CT: Greenwood Press, 1975; Kamm, Josephine. *Explorers into Africa.* London: Gollancz, 1970; Perham, Margery Freda. *African Apprenticeship: An Autobiographical Journey in Southern Africa, 1929.* Reprint. New York: Africana Publishing, 1974; ———. *East African Journey: Kenya and Tanganyika, 1929–30.* London: Faber and Faber, 1976; ———. *Pacific Prelude: A Journey to Samoa and Australasia, 1929.* Edited and with an introduction by A. H. M. Kirk-Greene. Chester Springs, PA, 1988; ———. *West African Passage: A Journey Through Nigeria, Chad, and the Cameroons, 1931–1932.* Edited and with an introduction by A.H.M. Kirk-Greene. Boston: Peter Owen, 1983; Rotberg, Robert I., ed. *Africa and Its Explorers; Motives, Methods, and Impact.* Cambridge: Harvard University Press, 1970; Severin, Timothy. *The African Adventure: A History of Africa's Explorers.* London: Hamilton, 1973.

airplanes. *See* AIR TRAVEL.

Air Travel

Women have traveled by air since 1784, when Elizabeth Thible of France became the first woman to ride in a hot-air balloon. The trip took place in France, and she flew with a French artist known as Fileurant. Thible was just a passenger, but in 1798 Frenchwoman Jeanne Labrosse became the first woman to pilot a balloon on a solo flight, also in France. In 1805, another Frenchwoman, Marie Blanchard, soloed in a gas-powered balloon; she subsequently made her living as a balloonist, supervising a fleet of hydrogen balloons as "Aeronaut of the Empire" for Emperor Napoleon Bonaparte of France. She was also the first person to die in an air accident. During a ballooning exhibition at a French festival in 1819, she accidentally ignited her balloon with her own fireworks, whereupon it plummeted to the ground.

The first American woman balloonist was Madame Johnson, who flew in a hot-air balloon in 1825 from New York to New Jersey and landed unceremoniously in a swamp. Another American balloonist to experience a rough landing was Lizzie Ihling, niece of famous balloonist John Wise. She made a solo flight in 1875 in the eastern United States, but when the skin of her balloon ripped she plummeted to the earth. Unlike Blanchard, however, she survived the fall. But perhaps the most famous woman balloonist of the late 1800s was American Mary Hawley Myers, who set a world altitude record in a balloon in 1885 by soaring four miles above the town of Franklin, Pennsylvania. Myers made her first ascent in 1880 in Little Falls, New York, and in the next decade she made more balloon ascents than any other person, male or female, of her time.

In 1906, ballooning made the transition from a novelty and an exhibition event to a serious sport when American publisher James Gordon Bennett organized the first long-distance competition. From the outset, the sport was dominated by males, but women have continued to participate in ballooning for recreation, sightseeing, publicity, and record setting. One of the best known publicity-seeking balloonists was Muriel Matters, a suffragette who flew over the British Parliament in 1908 to drop fliers that urged people to support women's right to vote.

Two years later, in 1910, Americans Blanche Scott and Bessica Raiche independently became the first women to pilot airplanes; the following year, Harriet Quimby became the first woman in the United States to become a licensed pilot. Since then many women have become aviators. Some of them have done so as a practical way to get from place to place or as a way to earn a living. For example, Fay Wells was the first airplane saleswoman, flying across the United States to demonstrate aircraft during the 1920s and 1930s. Louise Sacchi started her own air transport company and made the most ocean crossings of any woman pilot between 1955 and 1980. Willa Brown (1906–1992) and Bessie Coleman (1893–1926) were the first African American aviators to get a commercial pilot's license and an international pilot's license, respectively, so they could fly passengers. Brown obtained her license in 1937; Coleman got hers in 1921. During the second decade of the twentieth century, Americans Katherine and Marjorie Stinson flew as the first female airmail carriers. During the early 1930s, German medical student Hanna Reitsch was one of the first people to cross the Alps in a glider, and she set several gliding distance and altitude records. She was also the first female test pilot in the world; while working in this capacity for the Third Reich she set even more flying records. She also flew one of the first helicopters.

In 1930 American Ellen Church became the first flight attendant, once known as a stewardess. A pilot and a nurse, she sat with passengers on a Boeing Air Transport flight from Oakland, California, to Cheyenne, Wyoming, and while en route she diagnosed a case of acute appendicitis in one of the passengers. She radioed ahead to have doctors standing by when the plane landed, thereby saving a man's life. After Boeing publicized this story, other airlines added stewardesses to their planes. For example, National Airlines hired its first stewardess, Charlotte "Georgie" Robbins, in 1937, for flights to Miami from other cities in Florida. Robbins was not a nurse as Church had been; instead her function was to calm the fears of passengers unused to flying. On her first flight, Robbins provided her eight passengers with gum, magazines, and cigarettes, and pointed out interesting sights through the windows. But more important, she reassured them about every bump and turn the plane made. On that first flight's return trip, the plane developed engine trouble and had to land in a field.

Robbins was successful at her job, but National fired her after just one year, when it bought a new plane and could no longer afford her $75 monthly salary. National Airlines did not hire stewardesses again until 1940. During the 1940s and 1950s, the number of stewardesses grew steadily, and today every major airline employs flight attendants, most of them women. Women also dominate in other service aspects of the airline industry, working in large numbers as ticketing agents and customer service representatives. However, they remain in the minority in the upper levels of management; most airline executives are male. Moreover, relatively few women pilot commercial aircraft. In 1997, women represented just 5.9 percent of the 622,261 licensed airplane pilots in the United States.

Setting Records

Many of the first female pilots took to the air not as part of their job but simply for the adventure and/or to set records in competition with men. Sometimes they traveled great distances to do so. For example, American "Pancho" Barnes, the first woman stuntpilot, competed in transcontinental air races. American Ruth Nichols set a world distance record in 1931 by flying nonstop from Oakland, California, to New York City. In 1934 American aviator Jacqueline Cochran became the first woman to win the Bendix Trophy Transcontinental Race. The first solo cross-country air race specifically for women was the Women's Air Derby, a race from Santa Monica, California, to Cleveland, Ohio, which was initially run in 1929 with 19 competitors, including Americans Amelia Earhart, "Pancho" Barnes, and Ruth Nichols. Before the event began, newspaper editorialists called for it to be cancelled, objecting to the idea of allowing women to race planes. Afterward, some of the women who participated in the race formed an organization called the Ninety-Nines, whose aim was to support women aviators. The group still exists today.

In addition to participating in long-distance races, women aviators have also traveled great distances in attempts to set various records for solo flight. For example, in 1928 Mildred Bruce set a record for the longest solo airplane flight by a woman, traveling from England to Japan. In 1930 Englishwoman Amy Johnson set out to be the first woman to fly solo from England to Australia after hearing about a male aviator's success at accomplishing the same feat; Johnson also set speed records for flights from England to Japan and from England to Capetown, South Africa. British aviator Beryl Markham, one of Africa's first pilots, wrote one of the most popular and well-received travel books of her time, *West With the Night* (1942), about her experiences making a solo flight across the Atlantic Ocean in 1936. Born in 1927, British aviator Sheila Scott was the first person to fly solo from the equator in the eastern hemisphere to the equator in the western hemisphere by way of the North Pole, a feat she accomplished in 1971. She also circumnavigated the globe alone in 1966.

Not all record-setting attempts ended in success. In 1937, for example, Amelia Earhart set out to be the first woman to fly solo around the world. However, she disappeared during her attempt. Although various theories have been developed regarding what might have happened to her plane, most people believe she crashed into the sea. Amy Johnson was also killed in a flying accident; in 1941 she was delivering a plane for the British military when she became disoriented during her flight and crashed into the River Thames in London.

Flying is much safer today because of improvements in planes and instrumentation. Consequently there was little concern that American aviator and airshow pilot Linda Finch would crash when she set out in 1997 to duplicate but complete Amelia Earhart's flight. Finch's restored 1935 Lockheed Electra was the same type of plane that Earhart used, but it was equipped with a satellite terminal so Finch could send and receive e-mails, faxes, digital images, and weather information while in flight. Finch completed the flight successfully. Even more sophisticated was the plane of American aviator Jeana Yeager, who in 1986 became the first woman to circumnavigate the globe nonstop in an airplane without refueling. She was accompanied by Dick Rutan, the brother of the designer of the aircraft, *Voyager,* which was a new design that could fly over 28,000 miles nonstop. Hot-air balloons have also improved in design, making it possible to fly over areas of extreme cold. Consequently American Eleanor Conn was able to become the first woman to fly a hot-air balloon over the North Pole in 1980. *See also* BARNES, "PANCHO"; BRUCE, MILDRED MARY; EARHART, AMELIA; FINCH, LINDA; JOHNSON, AMY; MARKHAM, BERYL; NICHOLS, RUTH; QUIMBY, HARRIET; SACCHI, LOUISE; SCOTT, BLANCHE; STINSON, KATHERINE AND MARJORIE; WELLS, FAY GILLIS; YEAGER, JEANA.

Further Reading: Bell, Elizabeth S. *Sisters of the Wind: Voices of Early Women Aviators*. Pasadena, CA : Trilogy Books, 1994; Cochran, Jacqueline, and Maryann Bucknum Brinley. *Jackie Cochran: An Autobiography*. New York: Bantam Books, 1987; Cochran, Jacqueline, with Floyd Odlum. *The Stars at Noon*. New York: Arno Press, 1980; Dwiggins, Don. *The Air Devils: The Story of Balloonists, Barnstormers, and Stunt Pilots.* Philadelphia: Lippincott, 1966; Fisher, Marquita O. *Jacqueline Cochran: First Lady of Flight*. Champaign, IL: Garrard Publishing, 1973; Freydberg, Elizabeth Hadley. *Bessie Coleman, The Brownskin Lady Bird*. New York: Garland Publishing, 1994; Haynsworth, Leslie, and David Toomey. *Amelia Earhart's Daughters: The Wild and Glorious Story of American Women Aviators from World War II to the Dawn of the Space Age*. New York: William Morrow, 1998; Reitsch, Hanna. *Flying is My Life*. Translated by Lawrence Wilson. New York: Putnam, 1954.

Akeley, Delia Denning (1875–1970)

Born in 1875, American hunter Delia Denning Akeley, nicknamed "Micky," was the first Western woman to cross Africa. Akeley had always been resourceful and independent;

she had lived on her own since the age of 13. In 1924 she led an expedition across Africa from the Indian Ocean to the Atlantic Ocean, venturing through the interior to study Pygmy tribes.

Akeley made her first trip to Africa in 1905. Three years earlier she had married a sculptor and taxidermist for museums, Carl F. Akeley, who collected his own specimens. He took her along on his next hunting trip (a year and a half expedition), intending for her to collect birds and butterflies while he shot big game. However, Akeley wanted to hunt large animals too, so she taught herself to shoot and soon brought down two elephants, along with 17 other large animals.

In 1909 she and her husband went on a two-year expedition during which Carl fell ill for most of the trip and was also gored by an elephant. While he recovered, Akeley ran the expedition. She also studied the behavior of the animals she was hunting.

Akeley and her husband returned to the United States in 1911 to prepare their specimens for exhibition. However, the marriage grew quarrelsome and in 1923 they divorced. The following year Akeley got her own museum assignment, and during this trip to gather specimens she began studying the Pygmies. She lived among them for several months, the first white woman to do so, and revisited them on another expedition in 1929, participating in their lives and teaching them to jump rope with a tree vine. She also continued to collect animal specimens; in addition, upon returning to civilization, she provided museums with native artifacts.

Akeley wrote several articles about her African experiences, as well as two books: *J.T, Jr.: The Biography of an African Monkey* (1928), about a monkey she made into a pet during her first expedition, and *Jungle Portraits* (1930), which describes her adventures as an African hunter. A member of the Society of Women Geographers, she also gave lectures about her adventures. In 1939, Akeley remarried and settled in Florida, where she died in 1970. *See also* AFRICA.

Further Reading: Akeley, Delia J. *J.T., Jr.: The Biography of an African Monkey*. New York: Macmillan, 1928; ———. *Jungle Portraits*. New York: Macmillan, 1930; Olds, Elizabeth Fagg. *Woman of the Four Winds*. Boston: Houghton Mifflin, 1985.

Allison, Stacy (1958–)

Born in 1958, Stacy Allison was the first American woman to reach the summit of Mount Everest, a feat she accomplished in 1988. She climbed several other peaks before attempting Everest, including Mount Washington in Oregon and Mount Huntington and Mount McKinley in Alaska. She was also a member of a 1982 all-woman expedition that climbed Ama Dablam in the Himalayas. All members of this expedition reached the summit.

In a 1993 autobiography, *Beyond the Limits: A Woman's Triumph on Everest*, Allison writes about opposition to all-woman expeditions, saying: "In the early seventies the majority of male climbers were up-front about their sexism. Women, they said, just didn't have the muscle power to take on the big mountains. That attitude had dominated mountaineering since the sport's beginnings."

Beyond the Limits recounts Allison's experiences during her Mount Everest climb. She had previously attempted to climb Everest in 1987, but the expedition was abandoned because of bad weather. Shortly thereafter she signed up for the second, successful expedition, in which it took her 29 days to reach the summit. In retrospect, Allison realized that she embarked upon her first attempt to climb Everest for the wrong reasons. In *Beyond the Limits* she says:

> When I let the First American Woman [to climb Everest] race capture my spirit, when I let myself believe that one title would change my life, I lost sight of my true reasons for climbing. . . . Now I had to come to terms with the truth: that a personal triumph does not come from winning a race. No, it comes only after you can take a good, honest look at yourself. The real triumph comes when you can accept yourself in any weather and in any state, and still be able to say: That's me, and I'm okay. (p. 191)

Beyond the Limits offers many insights into the reasons why women enjoy mountaineering. Calling mountain climbing "the key that unlocks my spirit, the clearest representation of who I am," Allison is very open about her feelings about her travel adventures. For example, she explains her motive for climbing in part by saying:

> When life gets tangled there's something so reassuring about climbing a mountain. The challenge is unambiguous. Ice and snow and rock. Self-discipline. Concentration. Focus. As you push higher you work yourself into a trance. Can I reach that ledge? Are my fingers strong enough to hold on to this crack? Will this ice screw hold? Eventually the weight of the world—the stalled career, the broken marriage, the shattered confidence—slides away. For those moments when it's just you and the rock and the ice and the snow, life always makes sense. (p. 6)

Allison also talks about her experiences of being part of an all-woman climbing expedition and about all-woman teams in general. She says:

> [One mountaineering expert] knew mixed climbs usually worked better, since the varied physical abilities and character traits of men and women can make a more complete team. But an all-woman team had its benefits too. . . . The sexual politics of the climb held little interest for me, but once we got to the mountain it was fun to be in a group of women. When men are on the team, it's easy to defer to their strength. If the going gets really tough, a woman knows she can hang back and let the more muscular men take the lead. A group of women doesn't have this luxury, which is both sobering and freeing. We had to do everything for ourselves. (p. 84)

Shortly after climbing Everest, Allison led an expedition to K2, a mountain in Central Asia. She currently tours the United States giving inspirational speeches. *See also* MOUNTAINEERING.

Further Reading: Allison, Stacy, with Peter Carlin. *Beyond the Limits: A Woman's Triumph on Everest*. New York: Delta (Dell), 1996.

Americas. *See* LATIN AMERICA, NORTH AMERICA.

Anthropologists and Archaeologists

Anthropologists study human cultures, archaeologists study the remains of past human cultures. To conduct such studies, both types of scientists typically travel to remote regions of the world. In the case of anthropologists, the reason for this is the need to see humans within a context; or, as American anthropologist Margaret Mead explains in her book *A Way of Seeing*:

> Anthropologists, whatever problem they may be working on, necessarily deal with a whole way of life. There is no facet of a people's existence that need not be taken into account. What their homes are like, how they classify their relatives, how they make a living, what they eat and who does the cooking, how they sing their babies to sleep, what they teach their children, how they think about illness and accidents, how they bury the dead, how they choose their leaders, how they see the past and the future, and how they think about the world and the nature of the universe—all these things enter into an interpretation of a people's way of life. However, it is not the details, as such, that matter but the relationship among them. This is why, I think, the life of a small primitive village, studied intensively, so that one comes to know every individual from the newest baby to the oldest grandparents, provides a perfect background for thinking about the problems of our own complicated world. (p. xi)

Some anthropologists, like Mead, study just one general location throughout their careers, perhaps living in the study area for long periods, but usually going back and forth between the study area and their home. For example, Mead traveled frequently to the South Pacific to study tribal cultures beginning in 1925, but she lived in New York. Similarly, German anthropologist Birute Galdikas has traveled back and forth between Canada and Borneo for most of her life. More rare are situations like that of Irish anthropologist Daisy May Bates, who became so taken by Australia's Aborigines during the late nineteenth and early twentieth centuries that eventually she decided to live among them permanently.

The majority of anthropologists, however, choose to study a variety of cultures throughout the world, perhaps while examining a broader issue. For example, during the early 1930s American Ruth Benedict traveled to North America, Europe, and Asia to study indigenous cultures in order to develop her theory that culture rather than biology is responsible for differences among races and nations, a premise she expressed in her book *Patterns of Culture* (1934). Her first research was into Native American tribes, which earned her a doctorate from Columbia University in New York; in 1930 Benedict became an assistant professor at Columbia and, in 1948, a full professor but continued to travel intermittently.

Similarly, from the 1950s until the late 1970s, American sociologist and anthropologist Vera Mae Green traveled throughout the Caribbean and South and Central America studying African American and indigenous cultures. She also taught anthropology at several universities and promoted racial equality and human rights.

Many women anthropologists have chosen to travel with their husbands or a male companion because of cultural differences; in some parts of the world it is unacceptable for a women to travel alone. As an example, Lilian Brown, a wealthy Englishwoman, exchanged her financial support for the right to accompany anthropologist Frederick Mitchell-Hedges on his expeditions to Central America. From 1921 to 1931, the two made several trips to study indigenous tribes and collect material for the British Museum; afterward Brown helped found the Pacific Geographical Society. During the 1970s, Englishwoman Marika Hanbury-Tenison traveled with her anthropologist husband to South America, Indonesia, and Malaysia to write about the ways in which tribal cultures are threatened by the spread of civilization.

Archaeologists

Archaeologists have experienced similar problems related to their gender. Moreover, whereas some of the most prominent female anthropologists, such as German Birute Galdikas, Englishwoman Jane Goodall, and American Dian Fossey, were encouraged in their profession by a male mentor, women archaeologists have received little support from their male counterparts. In fact, archaeology as a profession was initially restricted to males; therefore some early female archaeologists had trouble even entering the field. For example, American archaeologist Harriet Boyd Hawes was denied entrance into an archaeological field study program at the American School of Classical Studies in Athens, Greece, again because of her gender. Undeterred, she took a leave of absence in 1901 and financed her own expedition, traveling to the Greek island of Crete to find the site of a Minoan town. During the 1930s British archaeologist Margaret Alice Murray was one of the foremost experts in Egyptian artifacts, yet when she first became interested in archaeology she was not allowed to study it in college because of her gender. She studied linguistics instead, specializing in Egyptian hieroglyphics.

Such pioneers made it increasingly acceptable for woman to embark on archaeological expeditions and head major archaeological studies. British archaeologist Gertrude Caton-Thompson worked in Egypt from 1927 to 1928 as a field director for the Royal Anthropological Institution, and American archaeologist Hetty Goldman led excavations in Greece, Asia Minor, Turkey, and Yugoslavia between 1911 and 1939. Another pioneer archaeologist was Englishwoman Dorothy Garrod, the first woman to study prehistoric people from the Paleolithic era, or Old Stone Age. She began this research in Gibraltar in 1925, but her excavations of Old Stone Age sites also took her to France, Palestine, Lebanon, Kurdistan, and Bulgaria.

Amateurs were notable in archaeology, particularly during the period when the profession was closed to women. Perhaps the best known are Englishwomen Amelia Edwards and Gertrude Bell, because they wrote books about their archaeological discoveries. Edwards's *A Thousand Miles Up the Nile* (1877), which describes a trip to Egypt in 1873, provided the fullest account of Egyptian ruins up until that time. Bell's works, which include *The Desert and the Sown* (1907), *Palace and Mosque at Ukhaidir* (1914), and *The Thousand and One Churches* (1909, with archaeologist William Ramsay), describe ruins in the Middle East.

Modern women archaeologists continue to work on field projects throughout the world, but they enjoy greater acceptance in their careers than their predecessors. Both they and anthropologists are also able to dress as comfortably as men while in the field, whereas their nineteenth-century counterparts were expected to wear the long skirts and undergarments typical of their era and could therefore not move and work as freely. *See also* AFRICA; ASIA; BATES, DAISY MAY; BELL, GERTRUDE; CATON-THOMPSON, GERTRUDE; EDWARDS, AMELIA; FOSSEY, DIAN; GALDIKAS, BIRUTE; GARROD, DOROTHY; GOLDMAN, HETTY; GOODALL, JANE; HANBURY-TENISON, MARIKA; HAWES, HARRIET BOYD; MEAD, MARGARET; MURRAY, MARGARET ALICE.

Further Reading: Benedict, Ruth. *An Anthropologist at Work: Writings of Ruth Benedict*. Margaret Mead, ed. Boston: Houghton Mifflin, 1959; ———. *Patterns of Culture*. Boston and New York: Houghton Mifflin, 1934; Golde, Peggy, ed. *Women in the Field: Anthropological Experiences*. Berkeley: University of California Press, 1986; Mead, Margaret. *A Way of Seeing*. New York: McCall Publishing, 1970.

Arctic. *See* POLES, NORTH AND SOUTH.

Artists. *See* PHOTOGRAPHERS AND ARTISTS.

Asia

The world's largest continent, extending for approximately 6,000 miles from east to west and 4,000 miles from north to south, Asia's boundaries are the Arctic Ocean on the north, the Indian Ocean on the south, Europe on the west, and the Pacific Ocean on the east, although several islands, including those of Japan, are also considered part of Asia. The region encompasses many countries, but, in general, the diverse Asian cultures have not encouraged women to travel. One notable exception was the large-scale transportation of Chinese women to California during the mid-1800s, when they were sold into slavery to work as prostitutes, laundresses, or cooks in Gold Rush towns, before a later treaty (1880) and the Chinese Exclusion Act (1882) limited Chinese immigration to the United States.

In addition to encouraging women to stay home, Asian countries also have a history of restricting their access to foreigners. Until the late nineteenth century, for example, Japan was closed to foreigners. After Matthew Perry forced Japan to sign a treaty that opened the country to commerce (1854), Englishwoman Isabella Bird Bishop became one of the first European women to visit the country. She wrote *Unbeaten Tracks in Japan* (1880) to share her experiences and encourage other women to travel to Asia. Similarly, Anna Leonowens's 1870 book *The English Governess at the Siamese Court*, which describes her experiences as a governess for the wives and children of King P'hra Paarmendr Maha Mongkut of Siam (present-day Thailand), encouraged European interest in that country.

Another country that attracted a significant number of European women travelers was Tibet—not because it lifted its restrictions against foreigners but precisely because it so firmly enforced those restrictions. Concerned about the growing power of England and Russia, in the late nineteenth century Tibet decided to close its borders, placing guards in the mountain passes to keep non-Tibetans from entering the country. Many outsiders took this as a challenge, and within a short time people from nine other countries were vying to be the first foreigner to reach Tibet's capital city of Lhasa.

In 1904, this "race" was won by Sir Francis Younghusband, who led a small army of British soldiers there under the excuse that his government needed to be sure the Tibetans were not under Russian control. After his arrival, many Europeans expected the Tibetans to open their borders; when this did not happen, average travelers became frustrated by their continued exclusion from Lhasa. In particular, many missionaries wanted to go there to preach Christianity, and among them were several women who decided to try to become the first woman to enter Lhasa.

The first to try for Lhasa was Englishwoman Annie Royle Taylor, a 36-year-old missionary who set out for the capital in 1892. Disguised as a Tibetan nun, she was discovered and forced to leave the country just days before reaching Lhasa. Three years later a second woman explorer, Mrs. St. George Littledale, got within 49 miles of Lhasa before being expelled from Tibet. In 1898, Canadian missionary Susie Rijnhart and her husband were attacked by bandits en route to Lhasa and became lost; he died during their failed attempt. No woman tried again to reach Lhasa until 1923. In that year, Frenchwoman Alexandra David-Neel successfully reached the Tibetan capital disguised as a beggar. Meanwhile, other women were attempting to preach Christianity in China, supported by the China Inland Mission, a British organization. In 1902 the mission sent Briton Mildred Cable and Algerians Evangeline and Francesca French to Asia to establish missions and Christian schools. Once arrived, however, they decided their time was better spent traveling widely to preach the gospel, and they set out for the Gobi Desert.

But tourism and religion were not the only reasons that European women traveled to Asia in the late nineteenth and early twentieth centuries. Many mountaineers were also attracted to the region, largely because of the Himalayan mountain range; Englishwoman Elizabeth Mazuchelli was the first European woman to explore the eastern Himalaya, traveling there in 1869 and again in 1871 accompanied by her husband and numerous servants.

Other women were drawn to Asia because of a personal quest. For example, from 1848 to 1853 Englishwoman Lucy Atkinson traveled among the nomadic tribes of Russia's mountain regions because her husband was determined to sketch the region's scenery; in 1890 English nurse Kate Marsden went to Siberia to look for a cure for leprosy. Both women encountered difficult travel conditions, such as bad roads and bandits, and were in many cases the first European women ever seen by indigenous people.

By the 1920s traveling to Asia for European as well as American women was much more common; they usually went as tourists but sometimes as explorers and journalists. For example, Briton Ethel Tweedie traveled through Russia and China from April 1925 to January 1926, and Swiss explorer Ella Maillart traveled extensively through Asia in the 1930s. Another Asian traveler was American journalist Anna Louise Strong, who in the 1920s, 1930s, and 1940s traveled to places that were experiencing political unrest. She spent perhaps the greatest amount of time in Russia and China and sometimes wrote Communist propaganda for those countries' governments. Marguerite Harrison also traveled to Russia for political reasons, but in her case it was to act as a spy for the United States after World War I. Both women wrote books about their experiences abroad, as did travel writer American Emily Hahn. Hahn lived in China from 1935 to 1944 and wrote such books as *China to Me* (1944) and *Times and Places* (1970).

Today Asia remains an important travel destination for women, and because of the global economy women are increasingly traveling there to conduct business. However, cultural differences often cause difficulties for Western women in Asia. As an example, in many parts of Asia it is not considered proper for men to socialize with unmarried women, which means that businesswomen cannot conduct business over restaurant lunches. Accordingly, some corporations prefer to send male rather than female executives to the region. This will most likely change as Western influences become stronger in the region, but for now Asia retains many aspects of its traditional gender biases. *See also* Atkinson, Lucy; Bishop, Isabella Bird; Cable, Mildred; David-Neel, Alexandra; French, Evangeline and Francesca; Hahn, Emily; Harrison, Marguerite; Leonowens, Anna Harriette; Maillart, Ella; Marsden, Kate; Mazuchelli, Elizabeth; mountaineering; Rijnhart, Susie Carson; Strong, Anna Louise; Taylor, Annie Royle; Tweedie, Ethel Brilliana; West, Rebecca.

Further Reading: Adams, W. H. Davenport. *Celebrated Women Travellers of the Nineteenth Cen-*

tury. London: W.S. Sonnenschein, 1883; Alec-Tweedie, Mrs. *An Adventurous Journey (Russia-Siberia-China)*. London: Hutchinson, 1926; Bird, Isabella L. *Unbeaten Tracks in Japan*. 1880. Reprint, with an introduction by Pat Barr. London: Virago Press, 1984; David-Neel, Alexandra. *My Journey to Lhasa*. 1927. Reprint. Boston: Beacon Press, 1986; Foster, Shirley. *Across New Worlds: Nineteenth-Century Women Explorers and Their Writings*. New York: Harvester Wheatsheaf, 1990; Leonowens, Anna. *The English Governess at the Siamese Court*. New York: Roy Publishers, 1870; Maillart, Ella. *The Land of the Sherpas*. London: Hodder and Stoughton, 1955; McLoone, Margo. *Women Explorers in Asia: Lucy Atkinson, Alexandra David-Neel, Dervla Murphy, Susie Carson Rijnhart, Freya Stark*. Mankato, MN: Capstone Press, 1997; Murphy, Dervla. *The Waiting Land: A Spell in Nepal*. London: John Murray, 1967.

Atkinson, Lucy (1820–1863)

Born in 1820, Lucy Atkinson was an Englishwoman who traveled through Russia from 1848 to 1853 and wrote about some of her experiences in *Recollections of Tartar Steppes and Their Inhabitants*. This 1863 book was one of the first to describe the nomadic tribes of Russia's mountain regions.

Little is known about Atkinson's early life, except that she was forced to leave home and earn her own living while still fairly young. At some point she began working as a governess in Russia, where she met and married an English architect who was planning to travel east from Moscow through Siberia sketching the sights. The two set off together in February 1848, and along the way Atkinson learned to shoot a gun and to ride a horse astride instead of sidesaddle. She also canoed around the Altin Kool, or Golden Lake, and down the Biya and Katun rivers. After traveling through the Altai Mountains, Atkinson gave birth to a son, whom she named Alatau after a peak in the mountain range. She rested for six months in a primitive village before continuing her journey. Accompanied by their son, she and her husband traveled thousands of miles in search of subjects to sketch; they also collected botanical and mineralogical specimens.

After the couple returned to England in 1854, Atkinson's husband wrote two books about their adventures, illustrated with his sketches. However, his works never mentioned his wife or son; it was as though he had traveled alone. Atkinson corrected this omission in her own book, which was published two years after her husband died. In *Recollections*, she says a great deal about her family and its adjustment to life in a foreign country. For example, she writes:

> In Kopal [a Russian village] they considered me very silly for washing [my baby] so often, saying once in two days was quite enough to change [clothing]: but the maxims of a mother are not easily forgotten; and mine had so instilled into my mind the necessity of cleanliness in my youth, that I determined to follow her injunctions. (Robinson pp. 69–70)

Shortly after *Recollections* was published, Atkinson fell into obscurity. She died in 1863. *See also* ASIA; SPOUSES.

Further Reading: Atkinson, Lucy. *Recollections of Tartar Steppes and Their Inhabitants*. 1863. Reprint. London: Cass, 1972; McLoone, Margo. *Women Explorers in Asia: Lucy Atkinson, Alexandra David-Neel, Dervla Murphy, Susie Carson Rijnhart, Freya Stark*. Mankato, MN: Capstone Press, 1997; Robinson, Jane. *Unsuitable for Ladies: An Anthology of Women Travelers*. Oxford: Oxford University Press, 1995.

Australia and New Zealand

The smallest continent but one of the largest countries in the world, Australia has long been a prime travel destination. Indigenous peoples called Aborigines first inhabited it around 40,000 to 60,000 years ago, but these people were gradually subjugated by Dutch and the English colonialists in the seventeenth and eighteenth centuries. In 1770, Britain claimed Australia as its own, and in 1787 it began transporting criminals to a colony there.

Accordingly, most of the earliest European women to travel to Australia were convicts. Of these, the most notable is Englishwoman Mary Bryant, who became famous in England after she escaped from a penal colony at Botany Bay, in 1792. Bryant spent 69 days with some fellow convicts in an open rowboat, heading north along the coastlines of

Australia and New Guinea. Because she and her comrades frequently went ashore to rest or replenish their provisions, they were the first nonindigenous people ever to visit some parts of Australia.

Convict transportation to Australia continued well into the nineteenth century, but by this time many voluntary settlers had arrived in the country as well. Women were encouraged to come to Australia by groups such as the Female Middle Class Emigration Society, which was established in London in 1862 to help well-educated British women, particularly governesses, immigrate to British colonies. The society provided interest-free loans to women who could not afford their passage. However, many of the immigrants who became indebted to the society were later unhappy with their decision. Australia did not offer enough jobs for governesses, and these women usually did not want to take the more menial jobs offered to them instead.

Meanwhile, married women traveled to Australia and the nearby islands of New Zealand with husbands who sought to settle and farm the new land. One of the most popular travel books of the period, *Station Life in New Zealand* (1870), came out of this type of experience. Written by Lady Mary Anne Barker, an Englishwoman who traveled to New Zealand in 1865 with her husband, it describes life on a New Zealand sheep farm. After its publication, tourism to the region increased, particularly after Thomas Cook added Australia as a stop on his Grand Tours in the late nineteenth century.

The tourism industry remains strong in Australia today; the country is a popular destination for women travelers throughout the world. Most tourist attractions are concentrated along Australia's coast, as is most of the country's population. Most of western and central Australia is a dry, relatively barren region known as the Outback. Most tourists are advised to visit the Outback as part of a well-organized group tour, because of the dangers associated with traveling through an area with little water or shade. However, some people have undertaken a journey across the Outback alone. For example, in her book *Tracks* (1980), Australian Robyn Davidson describes her journey across the Australian desert with only four camels and a dog for company. Irish anthropologist Daisy May Bates also went solo into the Outback to study Aborigines.

In addition to traveling within their own continent, many Australian women have ventured out into the world as sailors, aviators, and other types of travelers seeking adventure. For example, Australian sailor Kay Cottee is one of only seven women to have circumnavigated the globe alone, and the first woman to circumnavigate Australia nonstop, beginning and ending in Sydney. She accomplished this feat in 1988 after traveling for 189 days and more than 22,000 nautical miles. *See also* BARKER, MARY ANNE; BATES, DAISY MAY; BRYANT, MARY; CIRCUMNAVIGATORS AND ROUND-THE-WORLD TRAVELERS; DAVIDSON, ROBYN; GRAND TOURS.

Further Reading: Davidson, Robyn. *Tracks.* New York: Pantheon Books, 1980; Erickson, Joan, ed. *Australia Travel Guide.* Menlo Park, CA: Lane Publishing, 1987; Grattan, C. Hartley. *The Southwest Pacific to 1900: A Modern History: Australia, New Zealand, the Islands, Antarctica.* Ann Arbor: University of Michigan Press, 1963; Inglis, Kenneth Stanley. *The Australian Colonists: An Exploration of Social History, 1788–1870.* Carlton, Australia: Melbourne University Press, 1974; Klein, Renate, and Susan Hawthorne, eds. *Australia for Women: Travel and Culture.* New York: Feminist Press at the City University of New York, 1994.

Automobiles. *See* TRANSPORTATION, GROUND.

B

Baker, Florence (1841–1916)

Born in 1841, Florence Barbara Maria Finnian Von Sass Baker discovered Lake Albert, the source of the Nile River in Africa, with her explorer husband Sir Samuel White Baker in 1864. The Bakers' expedition, which included 45 men, was very difficult. Many people and pack animals died from illness or attacks by wild animals and indigenous peoples, and Baker herself was once near death from sunstroke.

Florence Baker was accustomed to a hard life; when Samuel first met her, she was a 17-year-old slave of the Turks. According to rumors, her parents had been members of European nobility killed during one of the many revolutions in the region in 1848. However, some historians believe that Baker grew up a slave in a noble household. Either background would explain why she was well-educated, and since neither Florence nor Samuel Baker would ever confirm the rumors the truth remains a mystery. What is known is that Samuel Baker bought Florence at a slave auction in Hungary, freed her, and later married her. She remained his constant companion while he worked in the area supervising construction of a railroad, and in 1861 she traveled with him to Abyssinia (present-day Ethiopia) to search for the source of the Nile. During their travels, she added Arabic and English to her German and Hungarian.

In 1865, after completing their African expedition, the Bakers went to Paris and then to London, where they were married. However, because of her background, Florence Baker was never accepted by British society. In 1870 the couple decided to leave London for another African expedition, this time to fight slave trading along the Nile. The Egyptian government had asked Baker's husband to lead troops into the remote regions along the river where slave traders were hiding; along the way the Bakers freed slaves being sent to market.

During this trip Florence Baker kept a diary, which was published in 1972 as *Morning Star: Florence Baker's Diary of the Expedition to Put Down the Slave Trade on the Nile, 1870–1873*. The work includes letters that Baker sent to her husband's grown children, whose mother had died when they were young, and provides insights into the difficulties that Baker encountered. For example, in one series of diary entries she tells of the hostility that the people indigenous to the Lake Albert region had for her expedition, saying: "As usual the natives ran away after . . . [we arrived] here, and there is not a man to be found belonging to the village, and no food. . . . Sam sent the soldiers out to look for potatoes, but there was none near here. The natives of this village took them all away, and cut all the plantains [banana plants] down when they heard we were on the route." (Robinson, p. 204)

The Bakers' expedition ended in 1873, whereupon they returned to England. There Samuel Baker received wide acclaim for his exploits, while Florence was ignored. He died in 1893, and, until her own death in 1916, she lived quietly.

Further Reading: Baker, Florence. *Morning Star: Florence Baker's Diary of the Expedition to Put Down the Slave Trade on the Nile, 1870–1873*. Edited by Anne Baker. London: William Kimber, 1972; Hall, Richard Seymour. *Lovers on the Nile: The Incredible African Journeys of Sam and Florence Baker*. New York: Random House, 1980; Robinson, Jane. *Wayward Women: A Guide to Women Travellers*. Oxford: Oxford University Press, 1990.

Barker, Mary Anne (1831–1911)

Born in 1831, Lady Mary Anne Barker was an Englishwoman who traveled to New Zealand in 1865 to establish a sheep farm there with her husband. Her writings about her experiences increased interest in New Zealand among her peers in England, thereby encouraging travel to the region. Barker's books include *Station Life in New Zealand* (1870) and *A Christmas Cake in Four Quarters* (1872).

In *Station Life*, Barker offers commentary not only on daily activities such as shearing sheep but also on the differences between her and her male companions. For example, in discussing preparations for an outing, she says:

> Of course the gentlemen were very fussy about their equipments, and hung themselves all over with cartridges and bags of bullets and powder-flasks; then they had to take care that their tobacco-pouches and match-boxes were filled; and lastly, each carried a little flask of brandy or sherry . . . I dispensed with all little feminine adornments . . . , tucked my hair away as if I was going to put on a bathing-cap, and covered it with a Scottish bonnet . . . [None of the men] looked at me, they were all too much absorbed in preparations for a great slaughter, and I only came at all upon sufferance. . . . (Morris, pp. 79–80)

Barker left New Zealand in 1868 when her husband received a government post on a nearby island; eventually he became governor of Trinidad, where Barker died in 1911.

Further Reading: Barker, Mary Anne. *Station Life in New Zealand*. 1870. Reprint, with a new introduction by Fiona Kidman. Boston: Beacon Press, 1987; Morris, Mary, ed. *Maiden Voyages: Writings of Women Travelers*. New York: Vintage, 1993.

Barkley, Frances (1769–1845)

Born in 1769, Englishwoman Frances Hornby Barkley was among the first European women to visit western Canada. Her husband was a fur trader and ship's captain. Four weeks after their marriage in 1786 they embarked on a trading expedition from Belgium to the Pacific Northwest to acquire otter skins. The following year they reached Nootka Sound in Canada, then continued south to Vancouver Island, making Barkley perhaps the first European woman to see Clayoquot Sound. They also discovered a new sound, which they named Barkley Sound, an island that they named Frances Island, and a peak that they named Hornby Peak. In addition, they rediscovered the Strait of San Juan de Fuca. They then landed to replenish their supplies, and indigenous peoples there killed four of their crew.

From the Pacific Northwest, the ship sailed to the coast of China, where Barkley's husband traded 800 otter skins for goods that could be traded on the island of Mauritius in the Indian Ocean. However, once the Barkleys reached the island, their ship was taken over by their business partners, who left them stranded on Mauritius. While there, Frances Barkley gave birth to her first child. Eventually the family found an American ship willing to take them to England, but along the way their ship ran aground on the coast of France. They reached England in November 1789.

Seven months later, the Barkleys set sail on another trading expedition, traveling from Denmark to India. Frances gave birth to her second child en route. After India they went to Micronesia, Japan, Russia, Alaska, Hawaii, Vietnam, and back to Mauritius. Along the way one of their children died. In December

1794, the family returned to England, and Frances never traveled again. In all, she had spent more than eight years at sea, circumnavigating the globe twice. She died in 1845. *See also* NORTH AMERICA; SEA TRAVEL.

Further Reading: Harding, Les. *The Journeys of Remarkable Women: Their Travels on the Canadian Frontier*. Waterloo, Ontario: Escart Press, 1994; Hill, Beth, ed. *The Remarkable World of Frances Barkley, 1769–1845*. Sidney, B.C.: Gray's Publishing, 1978.

Barnes, "Pancho" (1901–1975)

Born Florence Leontine Lowe in Pasadena, California, in 1901, "Pancho" Barnes was the first woman to fly to Mexico City. In 1924, after receiving an inheritance of $500,000 (the approximate equivalent of over $4 million today), she left her fiancé—an Episcopalian minister—and embarked on a cruise to South America. After a few months she returned, built a mansion, and began entertaining Hollywood celebrities. In 1927 she again went to South America, this time on a banana packet, wanting to experience life as person without means. She traveled across Mexico dressed in old trousers, thereby earning the nickname "Pancho" after a similarly impoverished character in the Spanish novel *Don Quixote*. But even this was not enough adventure for Barnes, so in 1928 she decided to learn to fly and bought a plane. Within a few weeks she was good enough to solo. Before the year was out she had formed her own stuntshow, "Mystery Circus of the Air," and performed in it every Sunday afternoon. In 1929 she also began competing in air races, winning some and performing well in all. Her plane, the Travel Air Model R, was one of the fastest in the world, and she was the only woman ever to fly one.

Barnes was also the first woman stunt pilot to perform in movies. Her film work includes *Hell's Angels*, a 1930 movie about World War I aviators directed by Howard Hughes. She not only performed on screen in this movie but also provided audio material, flying around a balloon while recording her plane's sounds. During the 1930s Barnes helped stunt pilots experiencing financial hardship, allowing them to stay at her home. She also worked to establish the Motion Picture Stunt Pilots Association. In 1934 she bought an 80-acre ranch near what is now the Edwards Air Force Base in California and established a motel, saloon, and restaurant. She hosted various activities for pilots at her ranch, and she profited from the Air Force's decision to establish a pilot training center in the area. However, despite a court battle, the Air Force eventually took over her land as well. Barnes lived in poverty for a time but in later years worked as a flight instructor and popular lecturer. She died in 1975. *See also* AIR TRAVEL.

Further Reading: Keslem, Lauren. *The Happy Bottom Riding Club: The Life and Times of Pancho Barnes*. New York: Random House, 2000; Schultz, Barbara Hunter. *Pancho: The Biography of Florence Lowe Barnes*. Lancaster, CA: Little Buttes Publishing, 1996.

Barton, Clara (1821–1912)

Born in 1821, American Clara Barton traveled to help people throughout the world. She established a permanent Red Cross Society in the United States and provided disaster relief on battlefields in many countries. Her first experience with combat came during the American Civil War, when she was working as a clerk in the Patent Office in Washington, D.C. Realizing that soldiers were having a difficult time obtaining supplies, Barton single-handedly organized a series of supply depots, an activity that earned her much recognition. She then began serving as a battlefield nurse, having had experience nursing an invalid brother when she was only 11 years old. In 1864 she was appointed a superintendent of nurses. After the Civil War she helped locate soldiers who had been reported missing in action.

Barton's interest in the Red Cross was sparked during a trip to Europe at the time of the Franco–Prussian War (1870–1871). After observing the works of the International Red Cross, she returned home in 1872 determined to set up a division of the Red Cross in the United States. After this organization was established in 1881, Barton directed its

Clara Barton, perhaps in 1910. *Library of Congress.*

relief efforts for over 23 years. She retired in 1904 and died in 1912.

Further Reading: Barton, Clara. *The Story of My Childhood*. 1907. Reprint. New York: Arno Press, 1980; Boylston, Helen. *Clara Barton: Founder of the American Red Cross*. 1955. Reprint. New York, 1986; Oates, Stephen B. *A Woman of Valor: Clara Barton and the Civil War.* Reprint. New York: Free Press, 1995; Pryor, Elizabeth. *Clara Barton: Professional Angel.* Philadelphia: University of Philadelphia Press, 1987.

Bates, Daisy May (1863–1951)

Born in 1863, Irishwoman Daisy May Bates traveled throughout Australia and promoted the anthropological study of that country's Aborigines. She first went to Australia in 1884, but after marrying an Australian and having a son she grew unhappy there. In 1894 she moved to London, but five years later she decided to return to her husband and child. On her way back to Australia, Bates met a priest who encouraged her to visit his monastery, which ministered to Aborigines. She spent six months at the monastery and among the indigenous peoples, studying their culture and language, before joining her family on a cattle station.

Bates soon discovered that she disliked ranch life, and her marriage remained problematic. She again separated from her husband and moved to the city of Perth, where she lectured and wrote about her experiences with the Aborigines. In 1904 the government hired her to compile information about Aboriginal behavior and language. Bates lived among the Aborigines for approximately two years doing this work.

Because of her efforts, other anthropologists—all of them male—grew interested in studying Aborigines, and Bates soon lost her job to them. Unwilling to give up her connection to the Aborigines, she continued, with the support of friends, to live among them; in 1912 the government named her their Honorary Protector.

Bates compiled a great deal of information about Aborigines, but the only work she had published during her lifetime was *The Passing of the Aborigines: A Lifetime Spent among the Natives of Australia* (1938). By the time it appeared in print, another Australian author, Ernestine Hill, had written some articles about Bates and made her a popular public figure. Her identity was forever linked with the Aborigines, but eventually she grew too infirm to live among them. She died in 1951. *See also* ANTHROPOLOGISTS AND ARCHAEOLOGISTS; AUSTRALIA.

Further Reading: Bates, Daisy. *The Long Shadow of Little Rock: A Memoir*. With a new preface by Willard B. Gatewood, Jr. Fayetteville: University of Arkansas Press, 1987; ———. *The Passing of the Aborigines: A Lifetime Spent Among the Natives of Australia*. 1938. Reprint, Melbourne: Heinemann, 1966; Blackburn, Julia. *Daisy Bates in the Desert*. New York: Pantheon Books, 1994; Hill, Ernestine. *Kabbarli: A Personal Memoir of Daisy Bates*. Sydney: Angus & Robertson, 1973; Salter, Elizabeth. *Daisy Bates*. New York: Coward, McCann & Geoghegan, 1972.

Bates, Katharine Lee (1859–1929)

Born in 1859, American Katharine Lee Bates traveled extensively in the United States, Europe, and the Middle East, using her experiences as inspiration for stories, poems, and travel books. Her best known work is the lyrics of "America the Beautiful," which she composed while standing atop the summit of Pikes Peak in Colorado. Bates had climbed the mountain as part of a 1893 expedition from Colorado College, where she was teaching. In 1904 the work was published in *The Boston Evening Transcript*, and shortly thereafter people began singing the lyrics to various tunes. Eventually the music of *Materna* by Samuel Ward became the popular choice of accompaniment. In 1926, several groups lobbied to make "America the Beautiful" the national anthem of the United States, but in 1931 the official selection was "Star-Spangled Banner." Bates died in 1929, four years after her retirement as a professor of English literature at Wellesley College.

Further Reading: Bates, Katherine Lee. *America the Dream*. New York: Thomas Y. Crowell, 1930.

Bell, Gertrude (1868–1926)

Born in 1868, British explorer and archaeologist Gertrude Margaret Lowthian Bell became famous for her travels throughout the Middle East. However, she also went around the world twice and climbed the Alps on several occasions. A Fellow of the Royal Geographical Society, Bell wrote several scholarly articles and books about her adventures. Her books include *The Desert and the Sown* (1907), which concerns Syria; *Amurath to Amurath* (1911), which describes Asia Minor; *Palace and Mosque at Ukhaidir* (1914), which describes Asia Minor's ancient ruins; and, *The Thousand and One Churches* (1909, with archaeologist William Ramsay), which is about the region that later became Iraq. Bell also translated Persian poetry in *Poems from the Divan of Hafiz* (1897).

Bell came from a wealthy family and was well educated. She enrolled in a woman's college when she was only 16 years old, and in 1888 she became the first woman to earn a top academic degree at Oxford University, where she studied history. Bell spoke several languages, including Arabic, and in 1892 she decided to visit an uncle who had been named British ambassador to Persia. There she fell in love with a British diplomat, but because the young man had many debts, Bell's father refused to give her permission to marry and ordered her to come home. Back in England, she wrote a book on Persia entitled *Safar Nameh: Persian Pictures, A Book of Travel*, which was published anonymously in 1894. She then accompanied her brother on two world tours.

In 1899, Bell again traveled to the Middle East, where she explored various cities and archaeological sites. She also traveled through the Syrian desert to Asia Minor, a journey that was the basis for *The Desert and the Sown*. Over the next three years, she climbed several peaks in the Alps and took a world tour with her brothers. Eventually she became known as one of Britain's most famous female mountaineers.

In 1904 Bell's father died, leaving her with more than enough money to pursue her archaeological studies. She embarked on an extensive exploration of the Middle East, working closely with archaeologist Sir Will-

Gertrude Bell, probably 1910. © *Hulton-Deutsch Collection/CORBIS.*

iam Ramsey, and in 1914 went deeper into the Arabian Desert than any previous female British explorer had ever traveled. She also studied and photographed ancient ruins and wrote extensively about her experiences.

Bell's writings focus on the people as much as on the sights. In the preface of *The Desert and the Sown*, she writes: "I desired to write not so much a book of travel as an account of the people whom I met or who accompanied me on my way, and to show what the world is like in which they live and how it appears to them." (Birkett, p. 171)

Bell also noted the difference between her own life and that of the women in the Arab countries she visited. In *The Desert and the Sown* she tells of a conversation with an Arab woman who said to her: "You go forth to travel through the whole world, and we have never been to [a nearby city]." (Birkett, p. 168)

But Bell had another interest besides travel; for several years she had an affair with a married British military officer she met in Arabia. When he was killed in battle during World War I, she was devastated. She enlisted in the military herself as a nurse, but soon became an officer with the Arab Intelligence Bureau in Cairo, Egypt. From 1916 on, her permanent home was in Baghdad. In her later years, she became a political advisor to the leader of Iraq and established the Iraq Museum. She was honorary director of antiquities for Iraq at the time of her death. Bell also helped found the Anti-Suffrage League in England. She remained despondent over her lover's death, and in 1926 at the age of 58 she died of a drug overdose. *See also* Asia.

Further Reading: Bell, Gertrude. *Amurath to Amurath*. London: William Heinemann, 1911; ———. *The Desert and the Sown*. London: William Heinemann, 1907; Birkett, Dea. *Spinsters Abroad: Victorian Lady Explorers*. Oxford: Basil Blackwell, 1989; Kann, Joseph. *Daughter of the Desert: The Story of Gertrude Bell*. London: Bodley Head, 1956; Tibble, Anne. *Gertrude Bell*. London: A. & C. Black, 1958; Wallach, Janet. *Desert Queen*. New York: Doubleday, 1996.

Bicycles

Prior to 1893, few women rode bicycles, but today bicycling is the preferred method of travel for many women throughout the world, particularly in countries where automobiles are costly and/or roads are poor. The 1890s brought changes that made bicycling appealing to women. First, advances in technology and design made bicycles more efficient and comfortable to ride. Second, clothing styles were modified to allow women the ease of movement necessary for bicycling.

The pioneers of women's bicycling endured criticism because of their style of dress. In 1893, for example, 16-year-old Tessie Reynolds shocked British society by riding a bicycle 120 miles from Brighton to London and back wearing a type of pants known as knickers, although they were concealed, in part, by a long jacket. Two years later, Mrs. Frank Sittig introduced the pantskirt specifically for lady bicyclists, and in 1896 American feminist Susan B. Anthony declared that the bicycle had done more for women's rights than any other modern development. That same year, Miss F. J. Erskine—herself an avid cyclist—published the first practical guide for women cyclists, *Bicycling for Ladies*.

By this time, women's interest in cycling for recreation had increased because of travel writers William and Fanny Workman, who embarked on a succession of bicycle tours throughout the world from 1890 to 1900. The two went to such places as Algeria in 1895 and Asia and India in 1897–1898, writing several books about their travels. As the twentieth century approached, some women began participating in bicycling as a sport. For example, mountaineer Elizabeth Le Blond became an accomplished alpine bicyclist, cycling up mountain paths to challenge herself physically.

As the sport of bicycling grew in popularity, competitions among cyclists began to be staged, some of them involving distance travel. In 1896, New York City hosted the first bicycle race for women; it lasted for six days at Madison Square Garden. A more demanding bicycle competition was held in 1898, when three New York women tested

Riders at Mountain Biking Championships, 1995. © *Ales Fevzer/CORBIS.*

their endurance by riding as far and as long as they could. Irene Bush went 400 miles in 48 hours; Jane Yatman went 500 miles in 58 hours, and Jane Lindsay rode 600 miles in 72 hours. In 1899, Yatman set a cycling endurance record by riding 700 miles in 81 hours, 5 minutes, over a three-day period with only two hours of rest. Shortly thereafter, Lindsay beat Yatman's record by cycling 900 miles in 91 hours, 48 minutes.

Women continue to compete in cycling events today, either against women or against both men and women. One of the most important distance-related cycling competitions is the Race Across America, which has taken place annually since 1982. In 1993 *Outside* magazine voted the race the most difficult of all major distance competitions, which include an around-the-world sailing race and the Iditerod dog sled race. Women have competed in the Race Across America since 1983, but no woman was able to finish the race until 1989, when cyclist Susan Notorangelo placed seventh overall. The first Tour de France, a major European cycling competition, to include women was held in 1984, when American cyclist Marianne Martin won the women's division after covering the course's 616 miles in 29 hours, 39 minutes, 2 seconds.

A particularly demanding cycling activity began on January 1, 2000: Odyssey 2000, a 250-person, yearlong, around-the-world trip on customized bicycles. The planned route passed through 45 countries on six continents, necessitating 21 airplane flights and 11 ferry rides. A scientific study was planned to be conducted on participants to measure physiological and psychological changes over the course of a year's daily cycling. Participants' ages range from 19 to 79 years old, and included 50-year-old American Ann Sutter, who not only helped organize the event and but also attracted riders by advertising it.

Other women have embarked on more leisurely bicycle tours much like Workman did. For example, in the 1960s Irishwoman Dervla Murphy went on a cycling trip across Europe to India, staying in inns and hostels along the way. She dressed in men's clothing and carried a gun to protect herself on the road.

When the terrain was difficult, she rode the train or hitchhiked a ride on a truck, stowing her bicycle in the back. Murphy wrote about her experiences in *Full Tilt: Ireland to India with a Bicycle* (1965). Similarly, in the late 1970s, when she was nearly 60 years old, Christian Miller decided to cross the United States on a bicycle. Her journey took her 4,500 miles, from Washington, D.C., to Eugene, Oregon, and she wrote a book about her experience, *Daisy, Daisy: A Journey Across America on a Bicycle* (1981). During the 1980s and early 1990s, Englishwoman Bettina Selby went on several bicycle tours and wrote about her travels in such books as *Riding to Jerusalem* (1986), locale—Israel; *Riding the Desert Trail* (1988), locale—the Nile River Valley; *The Fragile Islands: A Journey Through the Outer Hebrides* (1989), locale—Scotland; *Frail Dream of Timbuktu* (1991), locale—the Niger River Valley; and *Riding the Mountains Down* (1984), locales—Pakistan, India, and Nepal. Selby also retraced a popular route for medieval pilgrims, writing about her experiences in *Pilgrim's Road: A Journey to Santiago de Compostela* (1994).

Many modern guidebooks are devoted to helping women design their own bicycle tours. They include *Bicycle Touring in the '90s* by the editors of *Bicycling* magazine (1993); *Bicycle Touring: How to Prepare for Long Rides* by Steve Butterman (1994); and *Bicycle Across America* by the editors of *Bicycle USA* magazine (1996). *See also* Le Blond, Elizabeth; Murphy, Dervla; Workman, Fanny Bullock.

Further Reading: *Bicycling Magazine's Bicycle Touring in the '90's*. Emmaus, PA: Rodale Press, 1993; Butterman, Steve. *Bicycle Touring: How to Prepare for Long Rides*. Berkeley, CA: Wilderness Press, 1994; Calif, Ruth. *The World on Wheels: An Illustrated History of the Bicycle and Its Relatives*. New York: Cornwall Books, 1983; Miller, Christian. *Daisy, Daisy: A Journey Across America on a Bicycle.* Garden City, NY: Doubleday, 1981; Selby, Bettina. *Beyond Ararat: A Journey Through Eastern Turkey*. London: J. Murray, 1993; ———. *The Fragile Islands: A Journey Through the Outer Hebrides*. Glasgow: R. Drew, 1989; ———. *Frail Dream of Timbuktu*. London: J. Murray, 1991; ———. *Pilgrim's Road: A Journey to Santiago de Compostela*. Boston: Little, Brown, 1994; ———. *Riding the Desert Trail*. London: Abacus, 1988; ———. *Riding the Mountains Down*. London: V. Gollancz, 1984; ———. *Riding to Jerusalem*. NY: Peter Bedrick Books, 1986; Smith, Robert A. *Merry Wheels and Spokes of Steel: A Social History of the Bicycle*. San Bernardino, CA: Borgo Press, 1995; ———. *A Social History of the Bicycle, Its Early Life and Times in America*. New York: American Heritage Press, 1972.

Bishop, Isabella Lucy Bird (1831–1904)

Born in 1831, English explorer Isabella Lucy Bird Bishop was one of the most popular travel writers of the nineteenth century. She wrote a total of nine books, most of them written under the name Isabella Bird (although they have sometimes been reissued under the name Isabella Bishop). Two of her best known works are *A Lady's Life in the Rocky Mountains* (1879) and *Unbeaten Tracks in Japan* (1880).

Bishop traveled not only to North America and Japan but also to many other parts of the world, including the South Pacific, Hawaii, Asia, and the Middle East. Her first trip was to Canada and the United States in 1854, a journey she undertook to lift her spirits after a serious illness. After several months abroad she returned to England to write her first book, *The Englishwoman in America* (1856). Shortly after its publication she resumed her travels, and other books soon followed. They recounted adventures that few other women of her time had experienced. In addition, Bishop took her own photographs during her travels—not an easy task, given the size and weight of camera equipment at that time.

In discussing Bishop's accomplishments in an introduction to her book, *A Lady's Life in the Rocky Mountains*, historian Daniel J. Boorstin says:

> It is hard to recall another woman in any age or country who traveled as widely, who saw so much, and who left so perceptive a record of what she saw. There was nothing faddish or snobbish, and very little that was romantic in her travel interests. She did not try to retrace the steps of famous adventurers before her. Unlike Richard Halliburton, she was not interested in fol-

> lowing the track of [famous earlier travelers like] Cortés, of Hannibal, or of Alexander the Great. Unlike John Lawson Stoddard or Burton Holmes, she did not go gather raw material for entertaining lectures or travel books. Her writing was more often the by-product than the purpose of her trip. Strictly speaking, she was not an explorer or geographical discoverer. Yet the eminent British geographers of her day considered her one of themselves. (p. xviii)

Published in 1879, *A Lady's Life in the Rocky Mountains* describes Bishop's tour through the Rocky Mountains of the United States in the autumn and early winter of 1873, at a time when the region was inhospitable to visitors. Written in epistolary form, *A Lady's Life* was first published in 1878 serially in an English weekly magazine called *Leisure Hour* as "Letters from the Rocky Mountains."

The book has 17 letters in all, in which Bishop writes about her experiences more as a resident would. As Boorstin explains:

> [Bishop] never suffered from the tourist's nearsightedness. She avoided the romantic jargon of the professional world traveler. Nor was she confined by the specialized ways of observation that would later distinguish the professional anthropologist. She had no list of items to check off when she arrived in a new country. She was not seeking curiosities to sate the provincial complacency of people back home. . . . Yet she always seemed most interested in discovering what it felt like to live in other places. She had an amazing capacity quickly to become a resident. (pp. xix–xx)

For example, her fourth letter—dated September 10 and written in Fort Collins, Colorado—gives an idea of the hardships of life in a settlement called the Greeley Temperance Colony, where she stayed in a boarding house. She writes:

> I got a small upstairs room at first, but gave it up to a married couple with a child, and then had one downstairs no bigger than a cabin, with only a canvas partition. It was very hot, and every place was thick with black flies. . . . The kitchen was the only sitting room, so I shortly went to bed, to be awoke very soon by crawling creatures apparently in myriads. I struck a light, and found such swarms of bugs that I gathered myself up on the wooden chairs and dozed uneasily till sunrise. Bugs are a great pest in Colorado. They come out of the earth, infest the wooden walls, and cannot be got rid of by any amount of cleanliness. Many careful housewives take their beds to pieces every week and put carbolic acid on them [to kill the bugs]. (p. 31–32)

Bishop also offers descriptions of the scenery and of her relationship with her guide Mountain Jim and her horse Birdie. Her fondness for the latter is plainly evident. For example, she says: "Birdie amuses every one with her funny ways. She always follows me closely, and to-day got quite into a house and pushed the parlor door open. She walks after me with her head laid on my shoulder, licking my face and teasing me for sugar. . . . Her face is cunning and pretty, and she makes a funny, blarneying noise when I go up to her." (p. 155–156)

She also has good things to say about Mountain Jim, although she acknowledges his faults as well. For example, she reports:

> This man . . . is one of the famous scouts of the Plains, and is the original of some daring portraits in fiction concerning Indian Frontier warfare. So far as I have at present heard, he is a man for whom there is now no room, for the time for blows and blood in this part of Colorado is past, and the fame of many daring exploits is sullied by crimes which are not easily forgiven here. . . . His besetting sin is indicated in the verdict pronounced on him by my host [in Estes Park, Colorado]: "When he's sober Jim's a perfect gentleman; but when he's had liquor he's the most awful ruffian in Colorado." (p. 80)

Nonetheless, Bishop felt comfortable alone in Jim's presence. She explains that a friend "asked me if I should not be afraid of being murdered [by Jim], but one could not be safer than with him I have often been told." (p. 228) Consequently she is distressed to inform her readers that after her trip she learned Jim was shot and killed in a quarrel.

In her next work, the 1880 book *Unbeaten Tracks in Japan,* Bishop also focuses more on interpersonal relationships than on scenery. *Unbeaten Tracks* is a series of 44 letters that Bishop wrote to her sister during her travels in Japan for several months in 1878. It was the time of the Meiji Era, a period when Japan had opened itself to foreigners for the first time in more than 200 years. But the book is not a treatise on the Meiji Era. Instead it describes the people and places Bishop encountered on her journey. As she explains in her preface:

> This is not a "Book on Japan," but a narrative of travels in Japan, and an attempt to contribute something to the sum of knowledge of the present condition of the country, and it was not till I had travelled for some months in the interior of the main island and in Yezo that I decided that my materials were novel enough to render the contribution worth making. From Nikko northwards my route was altogether off the beaten track, and had never been traversed in its entirety by any European. I lived among the Japanese and saw their mode of living, in regions unaffected by European contact. As a lady travelling alone, and the first European lady who had been seen in several districts through which my route lay, my experiences differed more or less widely from those of preceding travellers. . . . These are my chief reasons for offering this volume to the public. (pp. 1–2)

Given such motivations, Bishop took great care to present accurate observations of Japanese life, saying: "I write the truth as I see it." Consequently, although she finds much to praise in Japan, she does not gloss over its failings. For example, in writing about one town she says:

> Shinjô is a wretched place. It . . . has an air of decay, partly owing to the fact that [its] . . . castle is either pulled down, or has been allowed to fall into decay. . . .The mosquitoes were in thousands. . . . There was a hot rain all night, my wretched room was dirty and stifling, and rats gnawed my boots and ran away with my cucumbers. . . . I find it impossible in this damp climate, and in my present poor health, to travel with any comfort for more than two or three days at a time, and it is difficult to find pretty, quiet, and wholesome places for a halt of two nights. Freedom from fleas and mosquitoes one can never hope for. . . . In some places the hornets are in hundreds, and make the horses wild. I am also suffering from inflammation produced by the bites of "horse ants," which attack one in walking. . . .These are some of the drawbacks of Japanese travelling in summer, but worse than these is the lack of such food as one can eat when one finishes a hard day's journey without appetite, in an exhausting atmosphere. (pp. 144–145)

This type of honesty was uncommon in travel books of the period, as were Bishop's details regarding how much preparation was needed for her trip. But travel was not the only concern in Bishop's life. She remained close to her sister Henrietta, whom she called Hennie, and when Hennie died after an illness, Bishop was devastated. In her grief she turned to Hennie's physician, John Bishop, who was already a close friend. In 1881 the two were married, and Isabella Bishop remained in England with her husband until his sudden death in 1886. Two years later she embarked on a series of trips to Asia, touring such places as China, India, Tibet, Persia, Syria, and other parts of Asia and the Middle East. While in the region she established two medical facilities in memory of her husband and sister. Bishop's last trip was to Morocco in 1897; shortly thereafter she became seriously ill. She died in Scotland in 1904. *See also* ASIA.

Further Reading: Bird, Isabella L. *A Lady's Life in the Rocky Mountains.* 1879. Reprint, with an introduction by Daniel J. Boorstin. Norman: University of Oklahoma Press, 1960; ———. *Unbeaten Tracks in Japan.* 1880. Reprint, with a new introduction by Pat Barr. London: Virago Press, 1984; Kaye, Evelyn. *Amazing Traveler Isabella Bird: The Biography of a Victorian Adventurer.* Boulder, CO: Blue Penguin Publications, 1994.

Bisland, Elizabeth (1861–1929)

Born in 1861, American journalist Elizabeth Bisland was the editor of *Cosmopolitan* magazine when she decided to travel around the world as quickly as possible and write about it; in doing so she was trying to beat Nellie Bly, another journalist attempting the same feat. Bisland left in late 1889; the trip lasted for 76 days, four days more than Bly's, running into 1890, with the first leg of the journey taking her from the United States to Asia. She published a book about her experiences, *A Flying Trip Around the World* (1891). In later years she wrote fiction as well as nonfiction. Bisland died in 1929. *See also* BLY, NELLIE; CIRCUMNAVIGATORS AND ROUND-THE-WORLD TRAVELERS; VICTORIAN ERA.

Further Reading: Marks, Jason. *Around the World in 72 Days: The Race Between Pulitzer's Nellie Bly and Cosmopolitan's Elizabeth Bisland.* New York: Gemittarius Press, 1993.

Blessington, Marguerite (1789–1849)

Born in 1789 in Ireland, Marguerite Blessington encouraged other women to make Grand Tours of Europe by writing about her own Grand Tour experiences in *The Idler in Italy* (1839–1840, three volumes) and *The Idler in France* (1841, two volumes). She also wrote *Journal of a Tour through the Netherlands in 1821* (1822), to recount her adventures during a short trip to the Netherlands. An Irish countess, her travel books were, according to historian Jane Robinson, "all *langeur* and sardonicism and with nothing of the distasteful practicalities of getting from A to B," adding that because of Blessington's writings "Continent travel became a coveted accomplishment, suave, a little risqué, and utterly fashionable." (p. 83) *See also* GRAND TOURS.

Further Reading: Blessington, Marguerite, Countess of. *The Idler in Italy*. Vol. 2, *Lady Blessington at Naples.* 1839–1840. Reprint, edited by Edith Clay with an introduction by Harold Acton. London: H. Hamilton, 1979; ———. *The Works of Lady Blessington.* New York: AMS Press, 1975; Leslie, Doris. *Notorious Lady: The Life and Times of the Countess of Blessington.* London: Heinemann, 1976; Robinson, Jane. *Wayward Women: A Guide to Women Travellers.* Oxford: Oxford University Press, 1990.

The Bloomer Girls

The Bloomer Girls was an American baseball organization that offered many young women who would not otherwise have traveled a unique opportunity to do so. In operation from 1890 to 1934, the organization comprised hundreds of baseball teams that sprang up informally during the 1890s and traveled throughout the United States competing in exhibition games against local or professional men's teams. Although most of the Bloomer Girls were women, each team usually had at least one male player.

The name of the organization came from the first women's pants, "bloomers," which feminist Amelia Bloomer began promoting in 1850. Female Bloomer Girls wore these loose-fitting trousers while playing, although in the organization's final years most of the women wore fairly standard baseball uniforms. Given the chance to leave their home-

Bloomer Girls in 1925. © *Bettmann/CORBIS.*

towns and travel throughout the United States at a time when many young women did not venture out into the world, many of the Bloomer Girls were deeply affected by their experiences in the organization.

In 1943, Chicago Cubs owner Philip K. Wrigley established the All-American Girls Professional Baseball League, which also sent women ballplayers touring the country. This group, whose members were also sometimes referred to as Bloomer Girls, but did not, in fact, wear bloomers, was disbanded in 1954. The 1992 movie *A League of Their Own* offers some insights on the adventures shared by these women. *See also* CLOTHING.

Further Reading: Berlage, Gai Ingham. *Women in Baseball: The Forgotten History*. Westport, CT: Praeger Publishing, 1994.

Blum, Arlene

In 1978, American mountaineer Arlene Blum, a biochemist specializing in studying the impact of chemicals on the environment, led the first all-woman expedition to climb Annapurna I in the Himalayas. She also led the first all-woman climb of Mount McKinley in 1970 and participated in an American expedition to climb Mount Everest in 1976. Blum has participated in more than 15 expeditions in Africa, Asia, and North and South America.

Blum wrote *Annapurna: A Woman's Place* (1980), which describes the Annapurna I expedition, a 13-woman team. Known as the 1978 American Woman's Himalayan Expedition, it was also the first all-American team, men or women, to attempt to reach the summit of Annapurna I, one of only 15 peaks in the world higher than 8,000 meters.

During the Annapurna I expedition only two women actually succeeded in their goal: Vera Komarkova and Irene Miller, who became both the first women and the first Americans to stand atop the peak. Two other women, Alison Chadwick-Onyszkiewicz and Vera Watson, fell and died during the ascent. In reporting their deaths, Blum emphasizes the hazards of mountaineering:

> Between them, Alison and Vera had climbed hundreds of mountains, and they knew that a fall on a steep slope can happen to any climber at any time. One foot placed insecurely just a fraction of an inch from safety, a shift of weight, a slip, a fall—the infinite difference between life and death. But why did it have to happen to them? Why could they not have put that foot just a little to the left or the right in a more secure place? (Blum, p. 231)

Blum also offers a glimpse of the reasons that people climb mountains, saying: "Recently a Swedish team wrote to me for advice about their . . . [own] attempt on . . . Annapurna. In my reply I warned that the mountain was extremely avalanche prone but did not try to discourage them. I could not tell them the risk was too great; they would have to make that decision for themselves. In mountaineering, the truism holds: the greatest rewards come only from the greatest commitment." (p. 245) *See also* ASIA; MOUNTAINEERING.

Further Reading: Blum, Arlene. *Annapurna: A Woman's Place*. San Francisco: Sierra Club Books, 1980.

Blunt, Anne (1837–1922)

Born in 1837 in England, Lady Anne Blunt became the first Western woman to visit some parts of Arabia. Between 1877 and 1880 she traveled throughout the country with her husband Wilfrid Scawen Blunt, a retired diplomat, and became close to the Bedouins who lived there. In fact, at times the Blunts lived like Bedouins themselves, dressing and eating like them. However, Blunt did have some difficulty adjusting to the native foods. In writing about the slaying of a hyena, she says: "I confess the look of the carcass was not appetising . . . and on examination the stomach was found to be full of locusts and fresh gazelle meat. Wilfrid pronounced it eatable, but I, though I have just tasted a morsel, could not bring myself to make a meal off it." (Hamalian, pp. 140-141)

Blunt wrote two important books based on her experiences: *Bedouin Tribes of the Euphrates* (1879) and *A Pilgrimage to Nejd, The Cradle of the Arab Race* (1881). These works not only provide descriptions of the place and its people but also discuss the Arabs' attitudes towards Europeans in general and the Blunts in particular. During the time of the Blunts' travels, several desert tribes were warring with another, and most Europeans felt the area to be too dangerous for travel. However, the Blunts were generally well treated by their hosts; they visited several different tribes and penetrated deep into central Arabia, also called the Nejd.

The Blunts' travels through Arabia ended in 1879, whereupon they went by land to India so that Wilfred Blunt could offer advice on a railroad project. After 1881, they settled in England, but spent winters on a 40-acre estate in Cairo, Egypt. The two separated in 1906 but reconciled in 1915. During World War I, she provided information and advice regarding the Middle East to Great Britain's Arab Intelligence Bureau. She died in 1922. *See also* AFRICA; ASIA; GREAT BRITAIN; VICTORIAN ERA.

Further Reading: Assad, Thomas J. *Three Victorian Travellers: Burton, Blunt, Doughty*. London: Routledge & Kegan Paul, 1964; Blunt, Anne. *Bedouin Tribes of the Euphrates*. Edited, with a preface and some account of the Arabs and their horses, by W. S. Blunt. 1881. Reprint. London: Murray, 1968; ———. *A Pilgrimage to Nejd, The Cradle of the Arab Race*. London: Cass, 1968.

Nellie Bly, probably in 1890. *Library of Congress.*

Bly, Nellie (1867–1922)

Nellie Bly was the pseudonym of American reporter Elizabeth Cochrane, who decided to travel around the world in less than 80 days after reading about a fictional character who did the same in Jules Verne's 1872 book *Around the World in 80 Days*. Her 1889 adventure was funded by her employer, Joseph Pulitzer, who owned the *New York World* newspaper; Bly was one of his top reporters. Adding even more challenge to this adventure was the decision of another women reporter, Elizabeth Bisland, to race with Bly to see which of them could make the journey the fastest.

Bly's travels took her from New York to London, then through Europe, Asia, and North America by various modes of transport, including steamboat, rickshaw, horse, and railroad. Along the way she sent stories to the *New York World*, which offered a trip to Europe to whoever could guess how long her journey would take. In all, it lasted 72 days, 6 hours, 11 minutes, and 14 seconds, beating Elizabeth Bisland's journey by four days.

Bly later wrote a book about her adventures entitled *Nelly Bly's Book: Around the World in Seventy-Two Days* (1890). It was not her first major work. Three years earlier she had written a series of articles exposing inhuman conditions at a mental asylum, posing as a patient to do so. This exposé made her famous, as did her subsequent reports about the realities of slum life. *See also* BISLAND, ELIZABETH; CIRCUMNAVIGATORS AND ROUND-THE-WORLD TRAVELERS.

Further Reading: Marks, Jason. *Around the World in 72 Days: The Race Between Pulitzer's Nellie Bly and Cosmopolitan's Elizabeth Bisland*. New York: Gemittarius Press, 1993.

Boats. *See* SEA TRAVEL.

Bosanquet, Mary (1918–1969)

Born in 1918, Englishwoman Mary Bosanquet gained fame by riding her horse, Timothy, 3,000 miles across Canada when she was only 20 years old, going from Vancouver to New York in 1938–1939. She decided to embark on this journey solely because she wanted an adventure and had done little traveling in her life. Funded by her parents, she sailed from London to Halifax, Nova Scotia, and journeyed by train to Vancouver, British Columbia, where she bought Timothy. She then rode him east, camping or staying with friendly farmers along the way. She wintered at a farm for eight months, working to earn her keep until the weather cleared, and she could continue on her way. In all, her adventure took 18 months. She later wrote about her experiences in *Canada Ride: Across Canada on Horseback* (1944). Bosanquet subsequently traveled through Italy to study art and lecture about it; these experiences were recounted in her 1947 book *Journey into a Picture*. She died in 1969. *See also* EQUESTRIENNES.

Mary Bosanquet with her horse "Jonty," at the end of a 4,000-mile horseback ride that began in Vancouver, B.C., in 1939. © *Bettmann/CORBIS.*

Further Reading: Bosanquet, Mary. *Journey into a Picture.* London: Hadder and Stroughton, 1947; Robinson, Jane. *Wayward Women: A Guide to Women Travellers*. Oxford: Oxford University Press, 1990.

Box-Car Bertha (ca. 1900–?)

Box-Car Bertha is the pseudonym of the author of *Sister of the Road. Box-Car Bertha: An Autobiography, As Told to Ben L. Reitman* (1937), an account of an American vagabond's travels by rail during the 1930s. While many people accept the book as an accurate portrayal of Bertha's life, some aspects of the narrative suggest it might have been fiction created by Dr. Reitman, who claimed to have met Bertha and written down her story. For example, although Bertha purports to have been a hobo for 15 years, at times she narrates the story more like an outsider. For example, she writes:

> "Girls and women of every variety seemed to keep Chicago as their hobo center. . . . The bulk of these women, and most all women on the road, I should say, traveled in pairs, either with a man to whom by feeling or by chance they had attached themselves, or with another woman. . . . These women were out of every conceivable type of home. But even [during my] first summer [as a hobo] I could see what I know now after many years, that the women who take to the road are mainly those who come from broken homes, homes where the father and mother are divorced, where there are step-mothers or step-fathers, where both parents are dead, where they have had to live with aunts and uncles and grandparents." (Morris, pp. 160–161)

Based on dates provided by the author, historians estimate that "Bertha" was born in approximately 1900. Her death date is unknown. No other information about her is available.

Further Reading: Box-Car Bertha. *Sister of the Road. Boxcar Bertha: An Autobiography, As Told to Ben L. Reitman*. 1937. With an Introduction by Kathy Acker and an afterword by Roger Bruns. New York: AMOK Press, 1988; Morris, Mary. *Maiden Voyages: Writings of Women Travelers*. New York: Vintage Books, 1993.

Boyd, Louise Arner (1887–1972)

Born in 1887, American explorer Louise Arner Boyd was the first woman to fly over the North Pole, a feat she accomplished in 1955. She also made several land expeditions to the Arctic, collecting detailed information about regions that had never been studied.

Boyd came from a wealthy family and traveled through Europe as a young woman. She first went to the Arctic in 1926 as a tourist and hunter, shooting several polar bears during her journey. On a second trip there in 1928, she crossed paths with a polar expedition and decided to turn to scientific pursuits herself. Her third expedition, in 1931, included a botanist, and Boyd spent more of her time recording scientific information and photographing the sights than she did hunting. In a 1933 expedition, she set out to study glaciers in Greenland and brought along several scientists in various disciplines.

Greenland was also Boyd's destination on 1937 and 1938 scientific expeditions. By this time, Boyd was a popular lecturer at scientific meetings, and her work was being supported in part by the American Geographical Society. Boyd wrote three books, all published by the National Geographic Society: *The Fiord Region of East Greenland* (1935), *The Coast of Northeast Greenland* (1937), and *Polish Countrysides* (1937). This last was a departure from her typical work. Based on a 1934 visit to rural Poland, it included Boyd's photographs of the region.

During World War II, Boyd provided photographs of the Arctic to the United States military, along with other information related to the region, and acted as a consultant on Arctic matters. She received a certificate of appreciation from the Department of the Army in 1949 for her contributions to the war effort. She received many other awards during her life as well. Boyd died in 1972. *See also* AIR TRAVEL; POLES, NORTH AND SOUTH.

Further Reading: Olds, Elizabeth Fogg. *Women of the Four Winds*. Boston: Houghton Mifflin, 1985; Tinling, Marion. *Women into the Unknown: A Sourcebook on Women Explorers and Travelers*. Westport, CT: Greenwood Press, 1989.

Bremer, Frederika (1801–1865)

Born in 1801, Swedish nurse Frederika Bremer wrote several popular novels that earned her invitations to visit social and political luminaries in America and Europe. Once she began traveling she discovered that she had a talent for travel writing as well. She wrote several books about her trips, including *The Homes of the New World* (1853), which focuses on the United States; *Life in the Old World* (1860), which focuses on Switzerland and Italy; and, *Travels in the Holy Land* (1862; two volumes), which concerns the Middle East. Her works are heavy on descriptive detail, although they also relate her personal experiences. For example, in *Travels in the Holy Land* she writes:

> I was conducted up into a little room furnished in Oriental taste with divans, red cushions, and many-coloured coverlets, all evidencing affluence; and, that which is more rare in the East, order and cleanliness. Close outside one of the windows a pomegranate was waving its branches laden with its brilliant flowers, whilst tall cane-plants formed a thick verdant screen outside the other; and from the roof or the piazza I had a free view across the lake. It was a pleasure to me to see from thence the sun take leave of the eastern mountains amidst a combat with shadows and cloud, as well as to watch the life which was going forward in the streets, and on the roofs round about me, for my dwelling lay high. It astonished me to see that the people who wandered amongst or who were upon the ruinous grey houses were so handsome, and generally so well dressed; to see signs of prosperity, and sometimes even of affluence, in their dress and their demeanour; even the drivers of asses had an elegant costume. (Hamalian, p. 118)

In her later years, Bremer not only continued to travel but became a feminist and philanthropist. She died in 1865.

Further Reading: Hamalian, Leo, ed. *Ladies on the Loose: Women Travellers of the 18th and 19th Centuries*. New York: Dodd, Mead, 1981.

Brown, Clara (1803–1885)

Born in 1803, Clara Brown is representative of African American slaves who traveled to the American frontier after receiving their freedom. Brown was freed by her Kentucky owners in 1857, whereupon she went to St. Louis, Missouri, to work as a cook. She then journeyed to Kansas to join a wagon train headed for Colorado. Brown was the only African American in the group of westward-bound pioneers.

Their trip was plagued by bad weather, but after two months they reached Denver, Colorado, where Brown again worked as a cook. She helped establish the first Methodist church in the city, and, after moving to nearby Central City, Colorado, she helped establish a Methodist church there too.

In 1866, Brown visited Kentucky, Tennessee, and Virginia to search for a relative and decided to help other African Americans travel to Denver as she did. She financed a wagon train for 16 African Americans, escorting them to Colorado herself. This and other charitable work depleted her finances, and in later years she had to rely on charity from others, including the Society of Colorado Pioneers. In 1881 Brown became one of the first African American women named a member of that group. She died in 1885. *See also* North America.

Further Reading: Stefoff, Rebecca. *Women Pioneers*. New York: Facts On File, 1995.

Brown, Molly (1867–1932)

Born in 1867, American Margaret (known as Molly) Brown is one of the most famous woman travelers because of her experiences on board the ocean liner *Titanic,* which sank in April 1912. Brown was reputed to have been the only female survivor of the disaster to have captained her own rescue lifeboat and handled the oars herself, after the sole male on board proved incompetent. However, since many of Brown's autobiographical stories have proven to be exaggerations, some historians doubt her account of the *Titanic* disaster.

At the time of her voyage on the ocean liner, Brown was one of the wealthiest women in America. In 1893 her husband, miner James Joseph "J. J." Brown, had found one of the largest and purest veins of gold in the country. Molly Brown then set out to transform herself from a simple country woman to a cultured, well-educated lady of society. She studied languages, art, music, and manners, bought lavish home furnishings and clothes, and went on European and world tours. She also gave parties for the elite in her Denver home.

Despite such efforts, Brown was looked down upon by upper-class society, and in 1909 her husband left her. The terms of their legal separation provided her with enough money to travel through Europe and the United States, and she booked expensive accommodations on the *Titanic.* The ship was on its maiden voyage from Southhampton, England, to New York City when it struck an iceberg and went down in the northern Atlantic, despite the fact that its builders had deemed it unsinkable. The *Titanic*'s 20 lifeboats were not enough for the 2,227 passengers on board, and more than 1,500 people drowned in the disaster. Most of the victims were third-class passengers; money and social standing helped determine who would live. However, social conventions of the time also dictated that women and children should be saved first, so among the first- and second-class passengers, most of the adult males drowned as well.

As a first-class passenger, Molly Brown had no trouble finding a place in a lifeboat, and she quickly took command, overriding the decisions of the seaman on board. After her rescue she also was an outspoken critic of the fact that there had not been enough lifeboats for all passengers. The ship's owners, the White Star Line, pointed out that they had provided enough lifeboats based on current regulations, which dictated that the number of boats be determined by a ship's size rather than its passenger capacity. As a direct result of the *Titanic* disaster, this regulation was changed; today all ocean liners must have enough lifeboat space for every passenger on board.

Once this issue was resolved, Brown turned her attention to other causes. An international celebrity, she found that she enjoyed fame, and when it began to fade she sought to remain in the public eye by running for the U.S. Senate in 1914. Her bid for political office failed, however, so she looked for other ways to remain visible. She joined the National Woman's Party in 1914 and worked for passage of the Equal Rights Amendment to the United States Constitution. During World War I Brown became an actress and entertained military troops. She also worked for several charitable causes. She remained estranged from her husband, and, after he died in 1922, she fought with her children over their inheritance. Brown died alone in 1932. The movie (1964) and musical *The Unsinkable Molly Brown* were based on her life.

Further Reading: Lord, Walter. *A Night to Remember.* New York: Holt, 1955; Whitacre, Christine. *Molly Brown: Denver's Unsinkable Lady.* Denver, CO: Historic Denver, 1997; Women of the West. http://www.over-land.com/westpers2.html May, 2000.

Bruce, Mildred Mary (1895–?)

Born in 1895, Mildred Mary Bruce was the first person to cross the Yellow Sea (between Korea and China) in a motorboat, a feat she accomplished in 1929. The following year, she set a record for the longest solo airplane flight by a woman, traveling from England to Japan. After this she joined a flying circus, then established her own air courier service. Bruce set several automobile-related records as well. In 1927, after winning the Coupe des Dames, a women's auto race at the Monte Carlo Rally, she broke several track speed and distance records of the speedway in Montlhéry, France. She also drove a car into remote regions of Lapland and the Sahara where no one else had ever driven. Bruce wrote books based on her experiences, including *Nine Thousand Miles in Eight Weeks: Being An Account of an Epic Journey by Motor-Car Through Eleven Countries and Two Continents* (1927). *See also* AIR TRAVEL; TRANSPORTATION, GROUND.

Further Reading: Bruce, Mildred (Mrs. Victor Bruce). *Nine Lives Plus: Record-Breaking on Land, Sea, and in the Air: An Autobiographical Account.* London: Pellham, 1977; Robinson, Jane. *Wayward Women: A Guide to Women Travellers.* Oxford: Oxford University Press, 1990.

Bryant, Mary (ca. 1765–?)

Born in approximately 1765, Englishwoman Mary Bryant was one of the most famous woman travelers of her time because of a dramatic sea journey she made after escaping from a penal colony at Botany Bay, Australia, in 1792. Five years earlier she had been sentenced to serve seven years in prison for stealing a woman's coat and became part of the first shipment of prisoners to Botany Bay. In all, 1,350 people in 11 ships left London for Australia in 1787; during the eight-month voyage, 48 people died of various illnesses.

When Mary boarded her ship she was pregnant. Some historians believe she was raped by a prison guard, while others believe she had a relationship with another prisoner. In either case, she gave birth before she landed in Australia and almost immediately married a fellow convict, William Bryant, with whom she had a second child.

Bryant's husband had been a fisherman, so he was soon set to work providing fish for the colony. This gave him access to free men at the docks. They sold him navigational tools and enough provisions for a long voyage, and one night in March 1791 the Bryants and a group of fellow convicts stole a rowboat and headed north along the coasts of Australia and New Guinea. Their journey lasted 69 days. Along the way, they rowed to shore to rest or replenish their provisions and were, consequently, the first non-natives ever to visit some parts of Australia. Hostile natives, bad weather, and low provisions made their journey extremely perilous. Nonetheless, they eventually reached the island of Timor in the East Indies, then occupied by the Dutch, where they told the governor that they had just survived a shipwreck.

Two months later, the Dutch discovered the truth when someone overheard Mary Bryant arguing with her husband about whether they should go back to England. The convicts were then arrested and shipped back to England for imprisonment; Bryant's husband, son, and daughter all died before reaching London.

In 1792 Bryant was sentenced to serve the rest of her prison term in an English prison. However, the public was so taken with the story of her adventures that they called for her release, and she was freed in 1793. *See also* AUSTRALIA; SEA TRAVEL.

Further Reading: Saxby, Maurice, and Robert Ingpen. *The Great Deeds of Heroic Women*. New York: Peter Bedrick Books, 1990.

Bullard, Supy (1968–)

Born in 1968, Supy Bullard, an American from Montana, led the first all-woman mountaineering team to climb the Cho Oyu in Nepal, the sixth-tallest peak in the world, without the assistance of local guides or supplemental oxygen. She accomplished this feat in 1999, along with teammates Caroline Byrd, age 39, Liane Owen, age 40, Cara Liberatore, age 28, Georgie Stanley, age 31, and Kathryn Miller Hess, age 33. Bullard is currently raising funds to return to the Himalayas for another all-woman expedition. She has also ascended peaks in Peru, Canada, and the United States. In addition, Bullard works as a professional guide, leading mountaineering expeditions and other outdoor activities in Alaska, Idaho, Nevada, Washington, Wyoming, and Montana. She acts as a consultant for companies making outdoor equipment and provides photographs on a freelance basis for outdoor magazines. She is also an experienced skier and ice climber. *See also* MOUNTAINEERING.

Further Reading: Expedition News. http://www.mountainzone.com/news/expedition/index.html. July, 1999; Howey, Noelle. "Women on Top." *New Woman,* May 1999.

Burlend, Rebecca (1793–1872)

Born in 1793, Rebecca Burton Burlend is representative of mothers who traveled from England to America in an attempt to improve their children's lives. In 1831 she, her husband, and their five youngest children embarked on a difficult three-month voyage on a ship, *Home,* to New Orleans. From New Orleans the Burlends took a 12–day steamboat trip up the Mississippi River to St. Louis, Missouri, and then went up the Illinois River to a Pike County, Illinois settlement called Phillips' Ferry, where they stayed with a friend until they could buy an 80-acre parcel of land that included a log cabin.

Life as pioneers was extremely difficult for the Burlends. They arrived in winter, and they had few provisions. The cabin was very cold, and they had to carry water from a nearby river after first breaking its crust of ice. When they ran out of candles they had to burn lard; they also used lard to make soap. In spring they found life no easier because they had no plow and had to plant their crops by hand. In the summer they were tormented by insects, and while harvesting their crops John Burlend injured himself with a sickle and nearly died from his infected wound.

However, as the years passed their circumstances improved. Their farm made money and they were able to buy more land and build a new cabin. Their oldest son, Edward, convinced Rebecca Burlend to write about her experiences as a pioneer, and together they produced a book, *A True Picture of Emigration* (1848). Although it offered a realistic portrayal of the difficulties of pioneer life, this work inspired many other English families to move to Pike County.

Further Reading: Burlend, Rebecca, and Edward Burlend. *A True Picture of Emigration*. 1848. Reprint, Lincoln: University of Nebraska Press, 1936.

Burton, Isabel (1831–1896)

Born in 1831, Englishwoman Isabel Burton was the wife of British explorer and diplomat Sir Richard Burton and accompanied him on some of his travels to Africa, South America,

Syria, India, and Egypt. But long before meeting her husband, Isabel Burton was attracted to exotic lands and people. In her posthumously published autobiography, *The Romance of Isabel Burton* (1897), she writes that as a girl she was "enthusiastic about gypsies, Bedawin [sic] Arabs, and everything Eastern and mystic, and especially about a wild and lawless life." (Burton, p. 21)

Raised by wealthy parents, as a young woman she toured Europe, where she met Richard Burton. She was immediately attracted to him because he reminded her of a gypsy. She said that "it was not only his eyes which showed the gypsy peculiarity; he had the restlessness which could stay nowhere long, nor own any spot on earth, the same horror of a corpse, deathbed scenes, and graveyards, or anything which was in the slightest degree ghoulish, though caring little for his own life, the same aptitude for reading the hand [i.e., fortune telling] at a glance. . . . He spoke Romany like the gypsies themselves. Nor did we ever enter a gypsy camp without their claiming him. . . . 'Why don't you join us and be our king?' [they would say.]" Isabel Burton adds that her husband thought himself distantly related to the gypsies. (Burton, pp. 54-55)

After the couple's marriage in 1861, Isabel Burton encouraged her husband's explorations. She shared his adventurous nature; many people considered the two a perfect match. She was also her husband's champion. When his name was omitted from a Royal Geographical Society speech praising great explorers, she complained publicly and in a letter to the editor of *The Times* newspaper in London. She also wrote several books about her experiences traveling with her husband, including *The Inner Life of Syria, Palestine, and the Holy Land: From My Private Journal* (1875) and *A.E.I. Arabia Egypt India: A Narrative of Travel* (1879). She died in 1896, six years after her husband. *See also* EXPLORERS; GREAT BRITAIN; SPOUSES.

Further Reading: Burton, Isabel. *The Romance of Isabel, Lady Burton*. Volume I. Edited by W.H. Wilkins. New York: Dodd Mead & Company, 1897; Lovell, Mary S. *A Rage to Live: A Biography of Richard and Isabel Burton*. New York: W.W. Norton, 1998.

Business Travel. *See* CHALLENGES FOR THE MODERN TRAVELER.

Butcher, Susan

Born in 1954, American dogsledder Susan Howlet Butcher is one of the most famous dogsledders in the sport. She has won the Iditarod—an annual Alaskan dogsledding race across 1,152 miles of wilderness—four times, in 1986, 1987, 1988, and 1990. From 1980 to the present, she has placed in the Iditarod's top five finishers every year except 1985, when a moose attacked and killed two of her dogs. Butcher has been competing in the Iditarod since 1978, but her interest began when she was a teenager in Boston, Massachusetts, and got her first Siberian husky dog. The people who sold her the dog told her that its mother had been a sled dog, and this sparked Butcher's interest. Shortly thereafter she went to Colorado State University to prepare for a job as a veterinary technician, but she found herself drawn to Alaska, and at the age of 20 she moved there to breed huskies herself. Four years later she was racing as well. Today she continues to race in dogsledding competitions; in fact, she and her husband, Dave Monson, have jointly won almost every dogsledding race in the world. She has also dogsledded to the summit of Mount McKinley, at 20,230 feet the tallest mountain in North America. Butcher and Monson live in Eureka, Alaska, with their daughter and sled dogs. *See also* IDITAROD TRAIL DOG SLED RACE.

Further Reading: Dolan, Ellen M. *Susan Butcher and the Iditarod Trail*. New York: Walker, 1993; Freedman, Lew. *Iditarod Classics: Tales of the Trail from the Men and Women Who Race Across Alaska*. Fairbanks, AK: Epicenter Press, 1992.

C

Cable, Mildred (1878–1952)

Born in 1878, English missionary Mildred Cable is representative of Victorian missionaries who traveled because of their religious beliefs. Cable was employed by the China Inland Mission to go to China in 1902 and help establish a school. She worked closely with two other missionaries, Algerians Evangeline and Francesca French, and eventually the three women decided to go into the Gobi Desert to preach the gospel. They traveled throughout the region for over 15 years, spreading Christianity and learning about Buddhism in return. Periodically they would return to civilization for supplies, then set out again.

Their freedom to travel ended, however, in 1931 when a political-religious conflict in China led to military rule and oppression. The new government restricted the women to the city of Tunhwang, but after several months they managed to escape across the desert. They left China shortly thereafter and did not return until 1935, when they could only travel in the company of an official Chinese guide. In their later years the women made two more trips: in 1946 they traveled to Africa, Australia, New Zealand, and Europe, and in 1947 they visited South America. Cable also wrote a first-person account of their experiences in Asia with help from Francesca French. Entitled *The Gobi Desert* (1943), the book includes rich descriptions of the desert's landscape and people, as well as commentary related to spirituality. For example, Cable writes:

> Desert dwellers have keener sight than other men, for looking out over wide spaces has adjusted their eyes to vastness, and I also learnt to turn my eyes from the too constant study of the minute to the observation of the immense. I had read about planets, stars and constellations, but now, as I considered them, I realised how little the books had profited me. My caravan guide taught me how to set a course by looking at one constellation . . . and to observe the seasons by the phase of Orion in the heavens. The quiet, forceful, regular progress of these mighty spheres indicated control, order and discipline. To me they spoke of the control of an ordered life and the obedience of a rectified mind which enables man, even in a world of chaos, to follow a God-appointed path with a precision and dignity which nothing can destroy. (Morris, pp. 216–217)

Cable and the French sisters remained together until Cable's death in 1952. *See also* French, Evangeline and Francesca; missionaries.

Further Reading: Cable, Mildred, with Francesca French. *The Gobi Desert*. 1943. Reprint, with a new introduction by Marina Warner. Boston: Beacon Press, 1987; Morris, Mary. *Maiden Voyages: Writings of Women Travelers*. New York: Vintage Books, 1993.

Caddick, Helen (ca. 1842–1926)

Born in 1842, Englishwoman Helen Caddick traveled throughout the world during the Victorian era, keeping detailed diaries about her experiences; these diaries are still available today. She wrote a book about one of her trips, an expedition to central Africa in 1898. Her diaries mention encounters with such pests as biting rats and burrowing maggots, but she tells of them matter-of-factly and without complaint. She is also noted for refusing to take quinine to prevent jungle fever—an illness she was likely to but never did contract. Entitled *A White Woman in Central Africa* (1900), the book describes Caddick's journey up the Zambesi and Shiré rivers to a region rarely visited by any European, let alone a woman traveling alone. *See also* AFRICA; VICTORIAN ERA.

Further Reading: Caddick, Helen. *A White Woman in Central Africa*. London: T. Fisher Unwin, 1900; Robinson, Jane. *Wayward Women: A Guide to Women Travellers*. Oxford: Oxford University Press, 1990.

Calderón de la Barca, Frances Erskine (1804–1882)

Born in 1804, Frances Erskine Calderón de la Barca, born Frances Erskine Inglis in Edinburgh, Scotland, wrote one of the first books about Mexico, *Life in Mexico, during a Residence of Two Years in that Country* (1843). Prior to her 1838 marriage, she lived in the eastern United States, although she was raised in Scotland and France. She traveled throughout much of Mexico from 1839 to 1841 as the wife of a Spanish diplomat. She observed two revolutions while in Mexico and wrote about the political climate in her book, along with information about Mexican culture and geography. Her work was so thorough that in 1847 the U.S. Army adopted it as a guide to Mexico. In her later years she lived in Spain, where she became a governess for Spanish royalty. She died in 1882. *See also* LATIN AMERICA.

Further Reading: Calderón de la Barca, Frances Erskine. *Life in Mexico during a Residence of Two Years in that Country*. 1843. Reprint, New York: AMS Press, 1980.

California Gold Rush. *See* GOLD RUSH.

Cameron, Agnes Dean (1863–1912)

Born in 1863, Canadian Agnes Dean Cameron was one of the first white women to reach the Arctic. She was originally an educator rather than an explorer. She began her teaching career at age 18 and is noted for being the first teacher in Vancouver, British Columbia, and the first female high school teacher as well as the first female principal in the province. Cameron lost her job in 1906 for allowing students in her school to use a ruler while taking an art exam that was supposed to be done entirely in freehand. She then accepted a position working for the Canadian Immigration Society in Chicago, encouraging Americans to move to Canada. In 1908 she decided to travel to the Arctic in order to write a book about it. The result of that trip, *The New North: An Account of a Woman's Journey through Canada to the Arctic*, was published in 1909. For the next three years Cameron toured Canada, England, and the United States, giving lectures about the Arctic. She died of appendicitis in 1912. *See also* NORTH AMERICA; POLES, NORTH AND SOUTH.

Further Reading: Cameron, Agnes Dean. *The New North: An Account of a Woman's Journey through Canada to the Arctic*. 1909. Reprint, Lincoln: University of Nebraska Press, 1986.

Cameron, Charlotte (?–1946)

English explorer Charlotte Cameron went on expeditions throughout the world, visiting Africa, Polynesia, Borneo, Asia, India, and South America. She also traveled through the Yukon and Alaska, taking a steamer through the Inner Passage of British Columbia from Vancouver. According to historian Jane Robinson, Cameron should be called "the patron saint of cruise liners," because most of her journeys were complimentary of shipping companies that wanted the publicity she gave them. Others were courtesy of railroad companies. In all, Cameron traveled approximately 250,000 miles between the years 1910 and 1925. She wrote several travel narratives of her experiences, including *A Woman's Winter in Africa, A 26,000 Mile Journey* (1913), *Mexico in Revolution,* (1925), *A Woman's Winter in South America* (1911), and *Two Years in Southern Seas* (1923). She died in 1946. *See also* LATIN AMERICA.

Further Reading: Cameron, Charlotte. *A Woman's Winter in South America*. London: St. Paul, 1911; Robinson, Jane. *Wayward Women: A Guide to Women Travellers*. Oxford: Oxford University Press, 1990.

Campbell, Fanny

American sailor Fanny Campbell captained a war ship during the American Revolution, perhaps the only woman to do so. By that time she had already served on ships for several years disguised as a man. She initially undertook this deception when she heard that her lover, a pirate, had been imprisoned in Cuba. She signed on as an officer on a ship sailing to a nearby location, then she instigated a mutiny and took over command of the vessel, heading it toward her goal. Along the way her crew embraced piracy. After freeing Campbell's lover, they began attacking British ships, increasing the size of their fleet in the process. Shortly thereafter, war broke out between the American colonies and the British, and Campbell's ships allied themselves with England. However, Campbell's duties as a captain were shortlived; she became pregnant and retired from sailing while the war was still being fought. *See also* DISGUISES; PIRATES; SEA TRAVEL.

Further Reading: De Pauw, Linda Grant. *Seafaring Women*. Boston: Houghton Mifflin, 1982.

Carr, Emily (1871–1945)

Born in 1871, Canadian artist Emily Carr was one of the most significant landscape painters of the early twentieth century. She traveled throughout British Columbia in search of subjects, which included not only landscapes but also Native American villages, people, and sacred totem poles, with which she was particularly fascinated. When she became too ill to travel, she began writing books about her experiences. They include *Klee Wyck* (1941) and *Growing Pains* (1946). The former, whose title means "Laughing One," was a collection of anecdotes that won the Canadian Governor General's Award for Nonfiction in 1941. In it she describes some totem poles in a village called Kitwancool by saying:

> The sun enriched the old poles grandly. They were carved elaborately and with great sincerity. Several times the figure of a woman that held a child was represented. The babies had faces like wise little old men. The mothers expressed all womanhood—the big wooden hands holding the child were so full of tenderness they had to be distorted enormously in order to contain it all. Womanhood was strong in Kitwancool. (Morris, pp. 206–207)

Carr eventually became one of the most beloved figures in Canada. She died in 1945. *See also* NORTH AMERICA; PHOTOGRAPHERS AND ARTISTS.

Further Reading: Carr, Emily. *Klee Wyck*. New York: Oxford University Press, 1941; Hembroff-Schleicher, Edythe. *Emily Carr: The Untold Story*. Saanichton, BC: Hancock House, 1978; Morris, Mary. *Maiden Voyages: Writings of Women Travelers*. New York: Vintage Books, 1993; Tippett, Maria. *Emily Carr, A Biography*. New York: Oxford University Press, 1979.

Cars. *See* TRANSPORTATION, GROUND.

Cashman, Nellie (ca. 1850–1925)

Born in approximately 1850, American pioneer Nellie Cashman emigrated from Ireland in the 1860s and, after a short stay in Boston, traveled west across North America, first to San Francisco and then to a series of mining camps in Nevada and British Columbia in search of fortune and adventure. To support herself, she worked as a cook and established several boarding houses. On one occasion, she joined an expedition to transport medical supplies to sick miners in a remote region, traveling 77 days through heavy snowdrifts. In 1879, she went to the Arizona silverfields and followed the same pattern there, moving from one town to another opening restaurants. In Tombstone, Arizona, she became involved in many charitable causes, including the Salvation Army, the Red Cross, and the Miner's Hospital, earning herself the nickname "The Angel of Tombstone." Nonetheless, she eventually moved again, traveling this time to mining camps in Wyoming, Montana, New Mexico, Arizona, and Canada. In 1898 she established her own mining operation just 60 miles from the Arctic Circle. Finally in 1923 Cashman settled in British Columbia, where she died two years later. *See also* GOLD RUSH.

Further Reading: Cahput, Don. *Nellie Cashman and the North American Mining Frontier*. Tucson, AZ: Westernlore Press, 1995; Ledbetter, Suzann. *Nellie Cashman, Prospector and Trailblazer*. El Paso: Texas Western Press, University of Texas at El Paso, 1993; Women of the West. http://www.over-land.com/westpers2.html (May 2000).

Cather, Willa (1873–1947)

Born in 1873, American author Willa Cather often wrote about the pioneer experience; therefore although she herself was not an extensive traveler she is strongly associated with women's travel. The inspiration for much of her work were the European immigrants of Red Cloud, Nebraska. Cather moved there from Virginia when she was nearly 10 years old, after her parents decided to live with relatives who had already traveled to Nebraska. Red Cloud, a railroad town, was primarily inhabited by Scandinavian, Bohemian, French, Russian, and German immigrants who worked to turn what was then frontier wilderness into civilization.

Willa Sibert Cather. © *Bettmann/CORBIS.*

Cather became a writer after graduating from the University of Nebraska in 1895, producing stories and poems while working as a journalist and editor for various magazines. For a brief time she was also a schoolteacher. Her first poetry collection was published in 1903, her first story collection in 1905. Her first novel, *Alexander's Bridge*, appeared in 1912. It focused on cosmopolitan life, but thereafter she concentrated on writing about the pioneers of her youth in such novels as *O Pioneers!* (1913) and *My Antonia* (1918). Later she decided to write about other types of pioneers as well. In *Death Comes for the Archbishop* (1927) she focuses on French Catholic missionaries who traveled to the American Southwest, and in *Shadows on the Rock* (1931) on the French Canadians who settled Quebec. Cather also wrote a few travel articles based on a trip to Europe during the early 1900s. She died in 1947.

Further Reading: Bennett, Mildred R. *The World of Willa Cather*. Lincoln: University of Nebraska Press, 1961; Brown, E. K. *Willa Cather, A Critical Biography*. Completed by Leon Edel. New York, Knopf, 1953; Woodress, James Leslie. *Willa Cather: Her Life and Art*. New York: Pegasus, 1970.

Caton-Thompson, Gertrude (1888–1985)

Born in 1888, British archaeologist Gertrude Caton-Thompson traveled extensively in North Africa and the Middle East. She was responsible for important advances in the knowledge of ancient Egypt, establishing that there were two prehistoric cultures, one dating to 5000 B.C.E. and the other to 4500 B.C.E., in Upper Egypt. Caton-Thompson worked in Egypt from 1921 to 1926 as a student of the

British School of Archaeology and from 1927 to 1928 as a field director for the Royal Anthropological Institution. In 1928 she went to Africa to direct archaeological studies on ruins in Zimbabwe (formerly Rhodesia), and a year later she returned to Egypt to conduct an excavation of the Kharga oasis. From 1937 to 1938 she conducted archaeological research at several sites in southern Arabia. During all of her fieldwork, she wrote numerous books and articles about her findings; her works include *The Zimbabwe Culture* (1931), *The Desert Fayum* (1935), *Kharga Oasis in Prehistory* (1952), and an autobiography entitled *Mixed Memoirs* (1983). She also served on the boards of major archaeological organizations, lectured extensively, and received numerous awards. Caton-Thompson died in 1985. *See also* AFRICA; ANTHROPOLOGISTS AND ARCHAEOLOGISTS.

Further Reading: Caton-Thompson, Gertrude. *Mixed Memoirs*. Gateshead, Tyne & Ware: Paradigm Press, 1983.

Caufield, Catherine

An American journalist living in London, Catherine Caufield travels extensively in order to report on environmental threats throughout the world. She has written many newspaper and magazine articles, as well as several books. The books include *In the Rainforest* (1981), *Tropical Moist Forests* (1982), and *Multiple Exposures: Chronicles of the Radiation Age* (1990).

In the Rainforest is Caufield's best-known work, as well as the project that involved the most traveling. To research the condition of rain forests worldwide, Caufield visited jungles in Africa, Central and South America, India, the Philippines, and Indonesia. Her work reports on their destruction because of such practices as cattle ranching, logging, and intentional flooding to create power-plant dams. *In the Rainforest* was one of the first books to focus on the issue of rain forest conservation, combining hard facts about rain forest extinction with Caufield's first-person experiences investigating these regions. *See also* AFRICA; ECO-TOURISM; LATIN AMERICA.

Further Reading: Caufield, Catherine. *In the Rainforest*. New York: Knopf, 1985.

Challenges for the Modern Woman Traveler

Modern women travelers face many challenges related to their gender, particularly in regard to safety and health concerns, equipment needs, and cultural differences that subject them to sexual harassment. Women on the road, particularly in foreign countries, are often fearful of being accosted by men. For this reason, the earliest women travelers went abroad in groups or with a male relative, or disguised themselves as men in order to pass unmolested among unknown males.

Modern women travelers increasingly travel alone or in pairs, but while this gives them greater freedom to make decisions regarding what sights to see, it can also lead to unwanted sexual overtures in certain countries. Italy, Morocco, India, Pakistan, and some parts of Indonesia and South America are reputed to be the least safe in regard to sexual harassment and/or rape. In Italy women traveling alone may be pinched or fondled while riding on public transportation, and in Columbia women who do not travel with a male companion are frequently harassed by men, including those in the police and military. In Egypt women traveling alone must ride in a separate, women-only bus or railroad car.

Risk of rape varies by region throughout the world, although it appears to be less common in rural areas with strong cultures. In her collection of travel essays, *A Woman's World,* Marybeth Bond reports:

> Many of my . . . correspondents, living in societies so shattered that rape has become a hobby, give *that* as their reason for fearing to travel. Some register incredulity when told that in unshattered rural societies, bound by tradition, rape is not a hobby and solitary women are not fair game. They may be approached by hopeful men who fancy them, but that is no threat. Having politely to decline sexual advances in Baltistan, Coorg, or Ecuador is no more stressful than having politely to decline them in San Francisco, London, or Paris.

> In my experience the only exceptions to the above are Eastern Turkey and the adjacent northwestern corner of Iran. There you must exert yourself to avoid rape; bring a heavy stick and don't hesitate to wield it with vigour. (Bond, pp. xxii–xxiii)

In some places, drugging is commonly used as a prelude to a rape. Women travelers dining in public who accept food or drink from strangers or who leave a drink unattended while they visit the restroom are particularly vulnerable to such attacks. Rapists also often target women who sightsee alone in remote locations or who choose a hotel room on the ground floor with a balcony or near stairs, because such rooms are easier to break into.

Even where rape is not a concern, modern woman travelers usually dress in conservative, unrevealing clothes to avoid attracting unwanted attention from males. Many also learn about the prevailing culture in their destination country—studying guidebooks, books on culture shock, and Internet sources like Journeywoman Online (http://www.journeywoman.com) that address harassment issues—so they can modify their behavior. Otherwise cultural differences can lead to misunderstandings regarding the kind of relationship a woman wants to have with a stranger. For example, in some countries when a woman makes eye contact with a man it means that she wants his company; some women deal with this problem by wearing dark glasses while traveling in such countries. Similarly, in some places if a woman sits in a restaurant alone she is considered to be sexually promiscuous.

Aside from concerns regarding sexual harassment, women have long had to restrict their behavior while traveling. For example, Englishwomen in the Victorian era were expected to ride horses sidesaddle; when they abandoned this practice to ride astride like men, it was a matter of controversy. In many Middle Eastern countries, women's behavior was—and continues to be—even more restricted. Regardless of their country of origin, women appearing in public cannot talk freely to men, nor can they dress immodestly; clothing must be loose and completely cover arms and legs. In Iran, women must also cover their hair and face with a veil, and in Egypt women must swim in their clothes when on public beaches. In Saudi Arabia, it is illegal for any woman, even a foreigner, to be alone with a man unless he is her relative. Consequently, some women travelers to these countries disguise themselves as men in order to experience more freedom on the road.

Business Travel

Businesswomen, however, do not have this option. Known in their professional capacity, they cannot appear in disguise and therefore are subject to discrimination based on their gender. For example, in many Asian countries, a traveling businesswoman at the executive or managerial level might be able to associate with men at an office during the day but would not be invited to business dinners at night. If she is invited, she might be expected to sit at a separate table with businessmen's wives. In some cultures, it is considered acceptable for men to flirt with foreign businesswomen. In others, it is so rare to find a businesswoman that such women print their complete job title, in both English and the language of the host country, on their business cards, so businessmen will understand that they hold a position of responsibility. In some areas of the Middle East, both men and women are forbidden to hand out a business card using the left hand because it is considered unclean, and women cannot allow their skin to touch the skin of the man receiving the card. In some Muslim countries, single women are not even allowed to enter the country. For example, Saudi Arabia refuses to issue visas to single women.

Businesswomen traveling in the United States also face challenges, particularly when they are involved in sales. The first traveling saleswomen were hired by U.S. companies in the 1910s to sell women's clothing, undergarments, or other products used by women. By the 1920s they were selling typically male products as well, because companies believed that men would be more likely to buy something from a beautiful woman.

This concept set the stage for a great deal of sexual harassment at that time, some of which continues today.

Smart Thinking

Another challenge facing women travelers, either in the United States or abroad, is theft. Women traveling alone are more likely to be assaulted and robbed than men, because they are perceived as weaker, easier targets. For this reason, many women seek to give themselves an advantage over a potential attacker by increasing their mobility, packing lightly so as not to encumber themselves. Some women also use male expectations to their advantage. They might put a man's name on their luggage tags, for example, to make it appear as though they are traveling with a male companion. They might also place expensive items in an ordinary-looking diaper bag in order to trick a would-be robber into believing they are carrying nothing more valuable than baby paraphernalia. In addition, many women travelers choose not to carry a purse, because purses are one of the most common targets of theft. Instead women use a small bag, commonly called a "fanny pack," that is strapped around the waist. Other choices include cotton money belts and security half-slips. The latter is a special undergarment worn under a skirt or dress; it typically has three zippered compartments in its hem.

Different Body, Different Equipment

Specialized clothing for women has also been designed for the unique challenges that women face while engaging in adventure travel. Until modern times, women have had to use the same equipment related to travel and exploration as men. This made it more difficult for them to participate in such activities as mountain climbing, where physical differences between the genders put women at a disadvantage in carrying heavy gear. With the advent of lighter-weight equipment, however, this disadvantage was eliminated. Such advances were often the result of new technology developed for the U.S. space program to keep the weight of equipment on a spacecraft at a minimum.

Today manufacturers of camping, hiking, and other equipment not only use lightweight materials that benefit women but also have begun making gear specifically designed for women's bodies. As an example, in 1995 sleeping bag manufacturers began making bags contoured to suit a woman's shape. Previously women campers used sleeping bags designed for short men, but these were too broad in the shoulders and too narrow in the hips for most women's comfort. New bags for women are shaped appropriately, which minimizes heat loss during sleep. In addition, these bags have additional insulation to compensate for the fact that women lose body heat 15 to 20 percent faster than men in cold weather.

Women also are more apt to suffer from overheating in hot weather during exertion. Normally the body keeps itself cool by sweating, because as the fluids and salt of sweat evaporate on the skin's surface, they reduce its temperature. Without enough liquid or in too much heat, however, this system does not work properly. Women are particularly prone to dehydration, so many travel guidebooks for women recommend that while traveling they wear hats, sunscreen, and loose-fitting, light-colored clothing, particularly natural cottons, and drink at least 10 liters of water a day to stay properly hydrated.

Traveling While Pregnant

Pregnancy increases the body's need for fluids and raises the risk of overheating still further. It offers many other challenges to the modern traveler as well. While pregnant women have always traveled, until modern times a significant number of them died from travel-related pregnancy complications before ever reaching their destination. Sunstroke, dysentery, and premature labor caused by rough roads along with problems related to an unhealthy environment made pregnancy even more dangerous for travelers than it was for those who remained at home. Thanks to advances in medicine, pregnant women no longer face such substantial risk from travel, but this does not mean they can travel without difficulty wherever they like. In fact, physicians commonly advise pregnant women not to travel to high altitudes while pregnant, particularly in the first trimester, because the fetus might not get enough oxygen to develop properly. Most airlines also ban women from flying during the last six weeks of pregnancy because air travel during this period can trigger labor.

Moreover, pregnant women expose themselves to dangerous illnesses when visiting undeveloped countries. They cannot use iodine to purify water, because iodine can damaged the thyroid of a developing fetus, so they run the risk of contracting dysentery from contaminated drinking sources. Yet they cannot take the antibiotics most commonly used to treat dysentery because these drugs can also damage the developing fetus.

For the same reason, pregnant women cannot be vaccinated against many of the diseases they might encounter during their travels. Most, if not all, live viral vaccines pose too great a risk, so pregnant women cannot receive the mumps, measles, and rubella (MMR) vaccine, nor can they be vaccinated against yellow fever and polio during the first trimester of pregnancy. Live bacterial vaccines have not been studied adequately to determine whether they pose an equally serious risk, but many doctors recommend that pregnant women not receive the oral typhoid fever vaccine. Physicians will, however, immunize pregnant women against hepatitis A if they are traveling to a region where they are at risk for contracting the disease, because hepatitis A poses far more danger to the developing fetus than does the vaccination. Malaria is another disease that poses great risk to the fetus—it can cause miscarriage, stillbirth, or premature delivery—so doctors allow pregnant women to take chloraquine, a medicine that prevents the type of malaria found in the Middle East, China, and Central America. But pregnant women are not allowed to take doxycycline, a medicine that prevents the strain of malaria found in South America, Africa, and Asia, because it can cause serious birth defects.

Illness

Illness is also a concern for women travelers who are not pregnant. Here, too, modern medicine has brought improvements for women travelers. In the past, they often suffered from serious illnesses during their trips. For example, American Florence Baker suffered from severe sunstroke while exploring the Nile River in Africa with her husband in 1864. That same year, Dutchwoman Alexine Tinne returned from an expedition into the African jungle to report that her mother, Harriet, had died of fever. Similarly, in 1855, Englishwomen Emmeline Stuart Wortley and her daughter Victoria grew ill from heat, exhaustion, and bad food, and Emmeline died. Now death during travel is no longer common, but there is still a risk of illness associated with travel, particularly in parts of the world where sanitation is a problem. Travelers to India, for example, are often cautioned not to drink unboiled water or eat unpeeled fruit, lest they suffer from dysentery or contract infectious hepatitis; in China, travelers are advised to carry medicines for sinus and breathing problems caused by the country's severe industrial pollution.

Most travel guidebooks offer warnings about specific diseases and recommend preventative measures for international travelers. The most common warnings in recent years have been for dengue fever in tropical countries; diphtheria in Russia; trypanosomiasis (also called sleeping sickness) in east Africa; meningitis in sub-Saharan Africa, Nepal, Delhi in India, and Mecca in Saudi Arabia; and rabies in Asia, Africa, and Central and South America. In rural areas throughout the world, travelers must take precautions against hepatitis A, as well as against a variety of parasites. Medical problems specific to women travelers include vaginal yeast infections, which often occur during travel in warm, moist climates, and cystitis, an infection of the urinary tract and bladder that can be caused by not drinking enough water. Women who get their ears pierced while abroad can also contract hepatitis B or AIDS.

Sanitation

In addition to illness, sanitary conditions are of great concern to modern travelers. Toilet facilities vary throughout the world. In all of North America and much of Europe, modern flush toilets are the norm, but in other places such facilities are rare. In developing countries, outhouses offer privacy, while in undeveloped countries there is usually little or no privacy and toilets are mere holes in the ground. In places where the latter are common, many women travelers wear long, full skirts for modesty

Toilet paper is also scarce in some places, or so coarse as to be considered unusable by Western travelers. In much of Asia and Africa it is nonexistent; a jug of water is beside the toilet instead. Soap is also scarce in many countries, so travelers often carry antiseptic towelettes or hand gels. In undeveloped countries, obtaining sanitary napkins and tampons can be difficult. Even in developing countries, the latter can be difficult to find. Women traveling to wilderness areas have the added problem of how to dispose of such products. Hiking and camping magazines and guidebooks sometimes recommend that women carry plastic bags to pack out such trash, after placing a cotton ball soaked in ammonia or some other aromatic substance into the bag to control odor, because burying menstrual waste is harmful to the environment. Alternately, some women travelers take birth control pills or have an injection of the contraceptive Depro-Provera in order to alter their menstrual cycle so they will not have to deal with the issue while on a trip. Others prefer to use washable, reusable sanitary products such as the "Keeper," a rubber cup inserted much like a diaphragm, or homemade pads of cotton cloth that can be washed in cold water. The latter were the forerunner of commercially made sanitary pads and have been used by women travelers for hundreds of years.

Further Reading: Axtell, Roger. *Do's and Taboos Around the World for Women in Business*. New York: Wiley, 1997; Bond, Marybeth, ed. *A Woman's World*. San Francisco: Travelers' Tales, 1997; Borman, Laurie D. *The Smart Woman's Guide to Business Travel*. Franklin Lakes, NJ: Career Press, 1999; Fraser, Keath. *Bad Trips*. New York: Vintage Books, 1991; Hoekstra, Elizabeth M. *Keeping Your Family Close When Frequent Travel Pulls You Apart*. Wheaton, IL: Crossway Books, 1998; Howarth, Dr. Jane Wilson. *Healthy Travel: Bugs, Bites, and Bowels*. Old Saybrook, CT: Globe Pequot Press, 1995; Knitter, Harry. *101 Stupid Things Business Travelers Do to Sabotage Success*. Irvine, CA: Richard Chang Associates, 1998; Langhoff, June. *The Business Traveler's Survival Guide: How to Get Work Done While on the Road*. Middletown, RI: Aegis Publishing, 1997; Swan, Sheila, and Peter Laufer. *Safety and Security for Women Who Travel*. San Francisco, CA: Travelers Tales, 1998.

Chapman, Olive (?–1977)

During the 1930s, English photographer Olive Chapman traveled throughout the world photographing and filming various subjects under the auspices of the Royal Geographical Society. She also wrote a series of books about her adventures: *Across Iceland, the Land of Frost and Fire* (1930), *Across Lapland with Sledge and Reindeer* (1932), *Across Cyprus* (1937), and *Across Madagascar* (1943). According to historian Jane Robinson, these were "encouraging" accounts that made it sound like it would be easy for every traveler to follow in Chapman's footsteps—when in fact she went to fairly inaccessible places. (pp. 85-85) *See also* PHOTOGRAPHERS AND ARTISTS.

Further Reading: Robinson, Jane. *Wayward Women: A Guide to Women Travellers*. Oxford: Oxford University Press, 1990.

Cheesman, Lucy Evelyn (1881–1969)

Born in 1881, Englishwoman Lucy Evelyn Cheesman was an insect collector who traveled extensively to find specimens and wrote about her experiences. She originally wanted to be a veterinarian, but she could not gain admittance to a veterinary college because of her gender. Consequently she became a veterinary assistant. She then did some secretarial work before being hired to manage the Insect House of the London Zoo in Regent's Park. In this capacity she gave lectures, created exhibits, and studied entomology in depth. In 1923 she joined a South Pacific expedition to collect insects but left the group while in Tahiti so she could explore on her own. She traveled to several islands before returning to England with hundreds of insects.

Her success led to a job as insect collector for the British Museum of Natural History, and she subsequently made seven more expeditions to the South Pacific, traveling to islands in the New Hebrides, New Guinea, and New Caledonia between 1928 and 1955. Her work was interrupted by World War II, during which she worked in various civil service jobs. She also contributed information about South Pacific islands to the military.

Throughout her career, Cheesman lectured about her experiences and wrote books based on her travels. Her works include *Hunting Insects in the South Seas* (1932), *Backwaters of the Savage South Seas* (1933), *The Two Roads of Papau* (1935), *Camping Adventures in New Guinea* (1948), *Camping Adventures on Cannibal Islands* (1948), and two autobiographical works, *Things Worth While* (1957) and *Time Well Spent* (1960). Cheesman died in 1969.

Further Reading: Cheesman, Evelyn. *Backwaters of the Savage South Seas*. London: Jarrolds, 1933; ———. *Time Well Spent*. London: Hutchinson, 1960; ———. *The Two Roads of Papau*. London: Jarrolds, 1935.

Children, Traveling With

Since prehistoric times, women have traveled with children in tow. During the Victorian era, however, it became unfashionable for Western women to take their children with them on trips, unless a governess was along—and even then, the children were expected to be kept apart from adults for most of the journey. This attitude remains prevalent in many places today, and mothers continue to be poorly represented in travel literature. As Marybeth Bond and Pamela Michael report in their book *A Mother's World* (1998):

> Most women in travel literature "leave behind" motherhood, children, and family ties when they travel. Women travelers are generally portrayed as childless, solo adventurers—rootless, tough, fearless, competitive, strong, assertive, and brave. Do women really become men when they travel? Is it necessary for a woman to shed all her uniquely female and maternal attributes in order to venture beyond her doorstep? (p. xvi)

These authors do not think so, and in fact they encourage women to travel with children. They also believe that the tendency to leave children behind is diminishing. They say:

> Women of all ages are traveling in unprecedented numbers to places near and far, alone, in groups, with babies on their hip. . . . There are few difficulties, physical or emotional, real or imagined, that they have not encountered and, generally, overcome. Women have learned that being a mother is one of the greatest assets a traveler can have—it makes connecting with other women much easier and, in some situations, can provide a patina of respect that can ward off unwanted [sexual] advances. (p. xvi)

In response to mothers' growing desire to bring their children with them on trips, travel agents, tour companies, and wilderness groups have endeavored to create positive travel experiences for families with children. Information about such activities can be found at numerous Internet Web sites, including http://www.family.com and http://www.familytravelforum.com, which provide articles, links, and message boards related to traveling with kids and other parenting subjects. Http://travelwithkids.miningco. com provides links and travel planning information.

There are also many books related to family travel. These include Frommer's Family Travel Guide series, which devotes each volume to a particular region or state; *Traveling with Your Baby* by Vicki Lansky (1985); *Kids and Cars: A Parent's Survival Guide for Family Travel* by Ellyce Field and Susan Shlom (1988); *Traveling with Children and Enjoying It: A Complete Guide to Family Travel by Car, Plane, and Train* by Arlene Kay Butler (1991); *Travel with Children* by Maureen Wheeler (1995); several books by Carole Terwilliger Meyers, including *The Family*

Travel Guide (1995); and a book focusing on adventure travel, *Adventuring with Children: The Complete Manual for Family Adventure Travel* by Nan Jeffrey (1990). The Travelers' Tales series also offers a collection of essays by people traveling with children, *Family Travel: The Farther You Go, The Closer You Get* (1998), edited by Laura Manske. *See also* ADVENTURE TRAVEL.

Further Reading: Bond, Marybeth, and Pamela Michael, eds. *A Mother's World: Journeys of the Heart*. San Francisco: Travelers Tales, 1998; Butler, Arlene Kay. *Traveling with Children and Enjoying It: A Complete Guide to Family Travel by Car, Plane, and Train*. Chester, CT: Globe Pequot Press, 1991; Field, Ellyce, and Susan Shlom. *Kids and Cars: A Parent's Survival Guide for Family Travel*. Aberdeen, SD: Melius and Peterson Publishing, 1988; Jeffrey, Nan, with Kevin Jeffrey. *Adventuring with Children: The Complete Manual for Family Adventure Travel*. Marston Mills, MA: Avalon House, 1990; Lansky, Vicki. *Traveling with Your Baby*. New York: Bantam Books, 1985; Manske, Laura, ed. *Travel: The Farther You Go, the Closer You Get*. New York: Travelers', 1988; Meyers, Carole Terwilliger. *The Family Travel Guide*. Albany, CA: Carousel Press, 1995; Wheeler, Maureen, *Travel with Children*. Oakland, CA: Lonely Planet Publications, 1995.

China. *See* ASIA; MISSIONARIES.

Ching Yih, Hsi Kai

During the early nineteenth century, Hsi Kai Ching Yih, also known as Madame Ching or Madame Cheng, was a well-known pirate who sailed the seas in search of treasure. She began her life on board ship as the captive of a pirate captain. Taken prisoner during one of his raids on Chinese villages, Ching agreed to marry him on the condition that she become co-captain of his large fleet. He was impressed with her boldness and agreed; the two worked together until he died in 1807. From that point on, Ching commanded the fleet alone, and under her direction it became stronger than ever. In part this was because Ching made an agreement with certain Chinese villages, exchanging her protection of their coast for a portion of their crops and other goods.

Eventually the Chinese government convinced Ching to trade piracy for the command of some official war ships, paying her well for her efforts. She successfully led part of the emperor's fleet for several years before retiring to live in a palace given to her by the government. However, Ching continued to be an outlaw for the rest of her life, secretly directing smuggling activities from her palace. *See also* PIRATES.

Further Reading: De Pauw, Linda Grant. *Seafaring Women*. Boston: Houghton Mifflin, 1982.

Circumnavigators and Round-the-World Travelers

The term circumnavigator refers to people who have circled the globe either by sailing ship or airplane; it is generally used in reference to those who have attempted to do so using just one vessel. People who take multiple trips and/or multiple conveyances in order to circle the globe are called round-the-world travelers.

It is difficult to know exactly how many women fall into either category of traveler, particularly in regard to sailing. During certain periods in history women have not been allowed on sailing vessels. They consequently resorted to disguising themselves as men in order to travel, and sometimes these disguises were never revealed.

Among the earliest women to travel openly on ships were the wives of sea captains. During the eighteenth and nineteenth centuries ships' officers were the only people allowed to bring wives on board a vessel, and many chose to exercise this option. For example, English woman Frances Barkley was the wife of a fur trader and ship's captain who took her to sea four weeks after their marriage in 1786. Their first voyage was from Belgium to the Pacific Northwest to acquire otter skins, which they then traded for other goods in China before traveling on to the island of Mauritius in the Indian Ocean. As their travels continued, Barkley gave birth to two children at sea, one of whom died, and visited Denmark, India, Micronesia, Japan, Russia, Alaska, Hawaii, and Vietnam. Her last voyage ended in 1794; in all, she had spent more than eight years at sea and circumnavigated the globe twice.

Solo by Ship

However, Barkley did not circle the globe alone, nor did any other women in the nineteenth century. The first woman to accomplish a documented solo circumnavigation was Naomi James of Ireland, who traveled in a 53-foot yacht from September 1977 to June 1978 (252 days). Since then, six other women have done the same. Krystyna Chojowska-Liskiewicz of Poland and Brigitte Oudry of France each completed a solo circumnavigation via a route south of Cape Horn and the Cape of Good Hope in 1977–1978. Kay Cottee of Australia went from and returned to the city of Sydney, Australia, on a 189-day circumnavigation in 1988, traveling more than 22,000 nautical miles. Isabelle Autissier of France completed her solo circumnavigation in 1990–1991. Samantha Brewster of England spent 161 days at sea to complete her journey in 1995, while Karen Thorndike of the United States spent two years at sea for her 1996–1998 circumnavigation.

Thorndike was 54 years old when she left San Diego, California, in August 1996. She arrived back there in August 1998 after going around all five Great Capes of the world: Cape Horn (South America), the Cape of Good Hope (South Africa), Cape Leeuwin (south of Perth, Australia), the South East Cape (Tasmania), and the Southwest Cape (off New Zealand). She made the voyage in a 36-foot boat, the *Amelia*, and traveled 33,000 nautical miles. Along the way she encountered severe weather and developed a serious flu-like illness that almost interrupted her travels. One year prior to her successful circumnavigation, in 1995, Thorndike failed in an attempt to solo circumnavigate the globe. She has been sailing over 20 years and has participated in several major sailing races; in 1988 she was part of the first all-woman team to compete in the Victoria–Maui Yacht Race.

Interestingly, while Thorndike was the first American woman to successfully complete a solo circumnavigation, another American woman, Tania Aebi, believed that she had the right to that title. She also erroneously believed herself to be the youngest woman—at age 18—to solo circumnavigate the globe by sea. However, a year after completing her trip and declaring her success, she discovered that she could not claim the solo title because she had a friend on board for 80 miles in the South Pacific.

Solo by Plane

The most famous woman to attempt a solo circumnavigation by airplane was American Amelia Earhart. She disappeared in 1937 while trying to complete her journey; various theories have been propounded to explain what might have happened to her. In 1997, American aviator Linda Finch recreated Earhart's journey using an identical airplane and route; she successfully finished the circumnavigation. However, she was not the first to complete a solo circumnavigation by air; in 1960 Sue Snyder went from Chicago to Chicago in 62 hours. In 1986 American aviator Jeana Yeager became the first woman to circumnavigate the globe nonstop in an airplane without refueling, beginning in and returning to southern California in nine days. Yeager used a new airplane that can fly over 28,000 miles without refueling.

Other Feats

The term circumnavigation can also be used to refer to a trip around a small body of water. One of the most accomplished women in this regard is American Ann Linnea, who in 1992 became the first woman to circumnavigate Lake Superior in a sea kayak; she accomplished this feat in nine weeks. However, women who circle the globe via means other than flying or sailing cannot accurately be called circumnavigators.

Travel Writers

Most of the famous round-the-world travelers have been writers who documented their experiences in articles or books. The majority of women who embarked on such trips did so at a leisurely pace. For example, Austrian Ida Pfeiffer spent 19 months on a world tour that began in 1847 and wrote a book about the experience, *A Lady's Voyage Round the World* (1850). Her second world tour lasted four years, from 1851 to 1855, and resulted in *A Lady's Second Voyage Round the World* (1856). Similarly, when she was in her late thirties, British Mrs. F. D. Bridges took three years, from 1877 to 1880, for a world tour with her husband and subsequently published her diary of her adventures as *Journal of a Lady's Travels Round the World* (1883). Like Pfeiffer, she offered interesting commentary on foreign cultures that encouraged other women to embark on world tours themselves.

Time Challenges

Some of these women added a time limit to their trip as a way to increase the sense of adventure. The first women to put a time limit on their round-the-world travels were Americans Elizabeth Cochrane (Nellie Bly) and Elizabeth Bisland, rival journalists who sought to recreate the journey depicted in Jules Verne's book *Around the World in 80 Days* (1872). Cochrane was a reporter who published articles about her adventures under the name Nellie Bly; she used several modes of transportation to complete a 73-day trip beginning and ending in New York City, including steamboat, rickshaw, and railroad. Bisland was the editor of *Cosmopolitan* magazine when she heard about Bly's trip and decided to undertake a similar one herself, although she utilized air travel more extensively. Her trip lasted for 76 days between 1889 and 1890, with the first leg of the journey taking her from the United States to Asia; in 1891 she published a book about her experiences, *A Flying Trip Around the World*.

Round-the-world travel that is undertaken simply to set a time record or to be able to write about such a trip has fallen out of favor in modern times. However, people with the time and money to make a round-the-world tour still do so. One of the most popular means of going around the world is on cruise ships like the *Queen Elizabeth II*. Another popular way to experience round-the-world travel is through guided, packaged tours offered by travel companies. *See also* AEBI, TANIA; AIR TRAVEL; BARKLEY, FRANCES; BISLAND, ELIZABETH; BLY, NELLIE; EARHART, AMELIA; FINCH, LINDA; LINNEA, ANN; PFEIFFER, IDA REYER; SEA TRAVEL; YEAGER, JEANA.

Further Reading: Bridges, Mrs. F. D. *Journal of a Lady's Travels Round the World*. London: J. Murray, 1883; Marks, Jason. *Around the World in 72 Days: The Race Between Pulitzer's Nellie Bly and Cosmopolitan's Elizabeth Bisland*. New York: Gemittarius Press, 1993; Pfeiffer, Ida. *A Lady's Second Journey Round the World*. London: Longman, Brown, Green, and Longmans, 1855; ———. *A Lady's Voyage Round The World: A Selected Translation from the German of Ida Pfeiffer*. 1850. Reprint, with an introduction by Maria Aitken. London: Century, 1988; ———. *Visit to the Holy Land, Egypt, and Italy*. Translated by H. W. Dulcken. London: Ingram, Cooke, 1853; Yeager, Jeana, and Dick Rutan with Phil Patton. *Voyager*. Boston, MA: G. K. Hall, 1989.

Clappe, Louise (1819–1906)

Born in New Jersey in 1819, Louise Amelia Knapp Smith Clappe is one of the best known of the pioneers who traveled to California during the Gold Rush of 1849. Her fame resulted from a collection of letters detailing her experiences in California from 1851 to 1852. They were originally published in *The Pioneer: Or, California Monthly Magazine* in 1854–1855, then appeared in book form as *The Shirley Letters* in 1933. Today this book is considered a classic of Gold Rush literature.

Clappe was well educated, having attended the best schools in New England as a girl. In 1849 she, her sisters, her husband, and her husband's uncle took a ship around Cape Horn to San Francisco, where the two men planned to establish a medical practice. However, once in California they learned that doctors were badly needed in the Gold Rush mining camps in the Sierra Nevadas, so Clappe and her husband headed for the region, living first in Marysville, then in Rich Bar, and later in Indian Bar.

Clappe wrote letters about her new life to a third sister, Mary Jane ("Molly"), who had remained in New England. When these letters were subsequently published they were attributed to "Dame Shirley," a pseudonym Clappe had used while writing for a Marysville newspaper. Her work offered one of the most thorough accounts of life in a California mining town, covering events in the years 1851 and 1852.

In *The Shirley Letters,* Clappe finds much to criticize regarding the behavior of the miners among whom she lived. For example, in her letter, dated September 30, 1851, from Rich Bar, she writes:

> I think that I have never spoken to you of the mournful extent to which profanity prevails in California. You know that at home it is considered *vulgar* for a gentleman to swear; but I am told that here, it is absolutely the fashion, and that people who never uttered an oath in their lives while in the "States," now "clothe themselves with curses as with a garment." . . . Whether there is more profanity in the mines than elsewhere, I know not; but during the short time that I have been at Rich Bar, I have *heard* more of it than in all my life before. (p. 38)

Clappe also laments the fact that she is one of very few women in the area and tells of the hardships that women had to endure in Gold Rush towns. In another letter from Rich Bar, dated September 22, 1851, she writes:

> It seems indeed awful . . . to be compelled to announce to you the death of one of the four women forming the female population of this Bar. I have just returned from the funeral of poor Mrs. B—, who died of peritonitis (a common disease in this place), after an illness of four days only. Our hostess herself heard of her sickness but two days since. On her return from a visit which she had paid to the invalid, she told me that although Mrs. B—'s family did not seem alarmed about her, in her opinion she would survive but a few hours. Last night we were startled by the frightful news of her decease. I confess that without being very egotistical, the death of one out of a community of four women might well alarm the remainder. (p. 34)

Clappe's letters end with her decision to leave the mining region. By the beginning of 1853 the gold in the area had largely been exhausted, and most miners had abandoned their claims, so Clappe and her husband moved back to San Francisco. The following year Clappe's husband returned to the eastern United States; Clappe remained in San Francisco and became a schoolteacher. The couple divorced in 1857. In 1878 Clappe retired and moved back east herself. She died in 1906. *See also* GOLD RUSH.

Further Reading: Clappe, Louise Amelia Knapp Smith. *The Shirley Letters: From the California Mines, 1851–1852.* 1933. Reprint. Edited with an Introduction by Marlene Smith-Baranzini. Berkeley, CA: Heyday Books, 1998.

Clark, Eleanor (1913–)

Born in 1913, American author Eleanor Clark is representative of women who participate in adventure travel, a modern pursuit that involves challenging physical activities in remote locations. She and her husband spent 13 days riding across the Sahara on camels, then climbed an Algerian mountain before returning home. She wrote about their trip in *Tamrart: Thirteen Days in the Sahara* (1984). Clark also wrote about her travel experiences in Italy in *Rome and a Villa* (1975) and is the author of several other nonfiction books, as well as fiction such as *Dr. Heart: A Novella and Other Stories* (1974) and *Gloria Mundi: A Novel* (1979). *See also* ADVENTURE TRAVEL; AFRICA.

Further Reading: Clark, Eleanor. *Tamrart: Thirteen Days in the Sahara.* Winston-Salem, NC: S. Wright, 1984.

Class Distinctions

Historically, women have had dramatically different travel experiences depending on their social class. Working-class women have had neither the time nor the money to travel long distances for pleasure or educational purposes, but have traditionally embarked on long voyages to help a family member in trouble, to better themselves financially, or to meet a religious obligation. In other words, they are more often in the role of immigrant or pilgrim than vacationer. They also travel by the cheapest possible means and endure many hardships during their journey.

In contrast, although upper-class women have also embarked on journeys to help family members or meet religious obligations, they have often taken advantage of their position to go on long trips purely for enjoyment. When wealth was concentrated in the hands of monarchs and noblemen, it was their wives and daughters who traveled for pleasure, appropriately chaperoned by large retinues. As wealth trickled down into the middle classes, the wives of merchants began to enjoy pleasure travel as well.

As tourism became less expensive and consequently more available to the masses, some upper- and middle-class travelers sought to set themselves apart from those they considered inferior by spending lavishly on their travel. They also typically chose destinations too expensive for most working-class people to afford. For example, during the Victorian era, when middle-class Englishwomen began traveling to Europe on vacations known as Grand Tours, upper-class Englishwomen increasingly vacationed in Egypt. When this region, too, was "discovered" by the ordinary traveler, upper-class women began embarking on round-the-world tours. In addition, it became popular to cruise across the Atlantic Ocean on lavish oceanliners like the *Titanic*, where "lower-class" immigrant passengers were physically kept apart—in greatly inferior quarters—from upper-class ones

Today such forced social stratification no longer exists, but money still determines the type of travel experience a woman will have. First-class cruise and airline passengers enjoy comforts that are superior to their second-class and economy-class counterparts, for example, and the most exotic tours are affordable only by people with ample wealth and leisure time. *See also* Grand Tours; Victorian Era.

Further Reading: Withey, Lynne. *Grand Tours and Cook's Tours: A History of Leisure Travel, 1750–1915*. New York: W. Morrow, 1997.

Clothing

Modern women travelers have the freedom to choose clothing based on practical concerns, wearing garments similar to men's for such activities as mountain climbing, hiking, and riding. Such was not always the case. Prior to the 19th century, British, American, and European society dictated a strict dress code that included a long skirt, which was impractical for certain physical activities. Therefore, in Western countries, the only women who dressed like men were doing so as part of a disguise to conceal their gender.

When early feminists began to encourage women to participate in traditionally male activities openly, rather than in disguise, clothing styles began to change. In 1850 feminist Amelia Bloomer promoted the wearing of pants underneath a woman's skirt; such pants came to be known as "bloomers" even though Bloomer did not design them. Several prominent feminists adopted the new garment immediately, but other women were more reticent, even though pants were clearly beneficial for equestriennes, bicyclists, mountaineers, and other sportswomen. Social pressures were so intimidating during the Victorian era that women had difficulty not only dressing like men but also riding astride a horse like men. Sidesaddle was the only "proper" way for ladies to ride.

Even adventurous travelers like Englishwoman Isabella Bird Bishop, who did adopt more masculine styles of dressing and riding horses, bowed to society's dictates when in a town. In *A Lady's Life in the Rocky Mountains*, she writes that upon coming near a settlement, even a rugged one: "I got off [my horse], put on a long skirt [over my trousers], and rode sidewise, though the settlement scarcely looked like a place where any deference to prejudices was necessary." (Bird, p. 152) In fact, prejudices were so strong that

Bird became defensive about her clothing when a reporter for *The London Times* in 1879 criticized her choice of male attire. In the second edition of *A Lady's Life in the Rocky Mountains,* she pointed out that she had indeed worn a lady's riding dress, despite the pants beneath it.

Meanwhile, other women explorers during this time retained their traditional clothing despite the discomfort. In an introduction to Englishwoman Mary Kingsley's book *Travels in West Africa,* scholar Elizabeth Claridge reports that Kingsley almost always wore "a moleskin hat . . . , high-necked white blouse, cummerbund, long black skirt of stout material, and lace-up boots, with the occasional addition of a red silk tie. (Kingsley p. x) She adds that such garments were not unusual among Kingsley's peers, explaining:

> In the matter of dress suitable for a woman traveller, Mary Kingsley was of like mind with Miss [Marianne] North, Miss Gordon Cumming, and Miss [Isabella] Bird. Miss North made no particular issue of what she wore but it clearly never occurred to her to dress other than as a lady. Describing a trek across a sodden part of Brazil in 1872, she observed in *Recollections of a Happy Life:* 'My dress was as good as any could be for such riding, namely a short linsey petticoat and a long waterproof cloak with sleeves.' According to Miss Bird, Miss Gordon Cumming maintained a formidable elegance at all times. (pp. ix–x)

But if the issue was important to travelers, it was critical for mountain climbers, whose inability to move fluidly could cause their death. Therefore long before it was fashionable, a few women climbers adopted men's clothing, although others stuck to social conventions. In her book *Grand Tours and Cook's Tours,* historian Lynne Withey explains:

> What to wear was a problem for . . . female mountaineers [in the 1800s]. [Henriette] D'Angeville, who continued to climb mountains until she was sixty-nine, wore flannel-lined tweed knickers and jacket, woollen stockings, heavy nailed boots, and a fur-lined cape. Together with her fur-lined bonnet, straw hat, velvet mask, veil, and green glasses—all for protection from the sun—she must have been quite a sight. By the 1860s and '70s, however, most women (including D'Angeville) wore skirts, in conformity to prevailing notions of feminine propriety. Mrs. H. W. Cole recommended a lightweight wool dress with rings sewn into the seams of the skirt and a cord strung through them; on difficult mountain paths, one could readily draw up the skirt several inches above the ground. A riding shirt made of waterproof cloth, to protect the dress from dirt and rain, a broad-brimmed hat, and strong boots with hobnailed soles completed her outfit. . . . Cole admitted that dresses were "inconvenient," but never hinted that ladies might properly adopt any other sort of costume. Some women wore riding breeches under their skirts, removing the skirts once they had passed the last villages on their routes. (p. 209)

While explorers had trouble with the idea of wearing pants, sportswomen adopted them fairly quickly. In 1858, Canadian Julia Archibald Holmes (1838-1887) climbed Pikes Peak in Colorado wearing bloomers, and in the 1890s a women's baseball organization was dubbed the Bloomer Girls because its female members all wore trousers. In 1893 16-year-old Tessie Reynolds became the first woman in England to ride her bicycle in public wearing a type of pants called knickerbockers, or knickers, and in 1895 American Annie Smith Peck wore knickers when she became the first woman to reach the peak of the Matterhorn. In 1909, Mrs. Adolph Ladenburg of New York introduced the first pantskirt specifically designed for equestriennes, which in turn encouraged more women to ride astride.

However, women who were not suffragettes or athletes stuck to fairly conventional dress well into the twentieth century. Pants did not become common as daily wear until actress Katherine Hepburn popularized them via her movies in the 1950s. Meanwhile, some women did wear specialized, somewhat unfeminine clothing for certain activities. For example, women motorists in the early 1900s adopted the practice of wearing long coats called dusters over their dresses, along with scarves, veils, hats, and gauntlet-style gloves. Before windshield wipers became standard equipment, they also wore goggles or headcoverings that looked somewhat like beekeepers hats, with a glass window to see through, or they carried tiny hand-held windshields.

The concept of specialized clothing still exists today; garments are manufactured expressly for certain sports or activities. In addition, clothing manufacturers take the differences between male and female bodies into account when designing clothes. For example, Patagonia Corporation of Ventura, California, offers its own line of shirts and jackets made specifically for women hikers and mountaineers. *See also* Bishop, Isabella Lucy Bird; Bloomer Girls; Cole, Mrs. H.W. (Henry Warwick); Cumming, Constance; equestriennes; Kingsley, Mary; mountaineering; Peck, Annie Smith; Victorian era.

Further Reading: Bird, Isabella L. *A Lady's Life in the Rocky Mountains* (with an introduction by Daniel J. Boorstin). Norman: University of Oklahoma Press, 1960; Kingsley, Mary H. *Travels in West Africa* (with an introduction by Elizabeth Claridge). London: Virago Press, 1982; Melinkoff, Ellen. *What We Wore: An Offbeat Social History of Women's Clothing, 1950–1980*. New York: W. Morrow, 1984; Withey, Lynne. *Grand Tours and Cooks' Tours: A History of Leisure Travel, 1750–1915*. New York: W. Morrow, 1997.

Coatsworth, Elizabeth (1893–1986)

Born in 1893, American author Elizabeth Jane Coatsworth spent much of her life traveling—often alone—and although she wrote many books that were not specifically about her travels, she incorporated her travel experiences into most of her works. She traveled to England, France, Greece, Spain, Italy, Egypt, Morocco, Japan, China, Mexico, the Philippines, and the Yucatan and wrote more than 90 books, many of them children's fiction, as well as poetry. Her 1976 book *Personal Geography: Almost an Autobiography*, which contains selections from her writings as well as reminiscences about her life, offers insights into the importance of travel in a woman's life. Coatsworth died in 1986.

Further Reading: Coatsworth, Elizabeth Jane. *Maine Memories*. Brattleboro, VT: S. Greene Press, 1968; ———. *Personal Geography: Almost An Autobiography*. 1976. Reprint. Woodstock, VT: Countryman Press, 1994.

Cobbold, Evelyn (1867–1963)

Born in 1867, Lady Evelyn Cobbold was the first Englishwoman to make a true religious pilgrimage to Mecca (in present-day Saudi Arabia). She converted to Islam, and, under its tenets, she had to travel to the holy city at least once during her lifetime. She wrote about this journey in a 1934 book, *Pilgrimage to Mecca*. Cobbold also traveled to the Middle East with her parents on vacations, an experience she describes in *Wayfarers in the Libyan Desert* (1912), and to East Africa as a tourist, which she writes about in *Kenya—The Land of Illusion* (1935). She died in 1963. *See also* Africa; Asia.

Further Reading: Cobbold, Lady Evelyn. *Pilgrimage to Mecca*. London: John Murray, 1934.

Cobham, Maria Lindsey

During the eighteenth century, English pirate Maria Lindsey Cobham sailed throughout the Atlantic Ocean with her husband Eric, attacking ships and stealing their cargo. They were a ruthless couple. As historian Linda Grant De Pauw says in her book *Seafaring Women*:

> The marriage of the Cobhams was truly a match made in hell. Maria set herself to mastering the pirate business and developed considerable flair. Once the pirates captured a ship that was carrying a young naval officer. She admired his uniform, so she had him stripped before she ran him through with her sword. Then she put his clothing on herself. She was so pleased with the costume that she wore it from then on. . . . Maria Cobham always took an active part in the fighting when the ship engaged a prize, but murder for her was not merely a business necessity in the pirate trade; it was a pleasure and a sport. Once

> she had the captain and two mates of a prize tied to the windlass and used them for target practice, firing eight pistols one after another from a distance of twenty feet. She never missed. (p. 45)

After Eric retired from piracy, Cobham took one trip without him, capturing a trading vessel and killing the entire crew by poisoning her prisoners' food. She then settled down with him at their estate near the coastal town of La Havre, France. But although the couple had a yacht and could sail for pleasure, Cobham grew depressed—some say over her husband's refusal to return to piracy, others say because of remorse for the murders she had committed. After a few years on land, she took poison and threw herself into the sea. *See also* PIRATES.

Further Reading: De Pauw, Linda Grant. *Seafaring Women*. Boston: Houghton Mifflin, 1982.

Colcord, Joanna Carver (1882–1960)

Born at sea in 1882, American Joanna Carver Colcord is representative of women who grew up on board ship. Her father was a sea captain, and in her parents' care Colcord visited ports throughout the world. While traveling she educated herself and passed a test that earned her a high school diploma. She also had chores to do, but like other women during her era, when she reached puberty she was no longer allowed to roam freely among the crew. Such a restricted life did not appeal to her, and when she became an adult she gave up life at sea entirely. However, she later became involved in American relief efforts abroad and wrote several books on the subject. She also wrote about social issues, such as broken homes, and used her childhood experiences to write *Roll and Go: Songs of American Sailormen* (1924) and *Sea Language Comes Ashore* (1945). Colcord died in 1960. *See also* SEA TRAVEL.

Further Reading: De Pauw, Linda Grant. *Seafaring Women*. Boston: Houghton Mifflin, 1982.

Cole, Mrs. H. W. (Henry Warwick)

Mountaineer Mrs. H. W. Cole wrote a book about her experiences climbing in the Alps that encouraged other women to explore the region. Published anonymously in 1859, the work is entitled *A Lady's Tour round the Monte Rosa: with Visits to the Italian Valleys . . . in a Series of Excursions in the Years 1850–56–58*. It offered Victorian women advice regarding the ascent of Monte Rosa, a 15,217-foot mountain in Switzerland, and included information on how to dress for mountain climbing. According to scholar Jane Robinson, who calls *A Lady's Tour* "a mountaineering classic," Cole's advice was "eminently sensible." (Robinson, pp. 8–9) As examples, Robinson says that Cole recommended that women travelers always wear wide-brimmed hats and leave their traditional but cumbersome sun-shading parasols at home. Cole also suggested that women travelers sew rings into their dresses and string cord through the rings in such a way that their skirts could be lifted up whenever they came to a stream or particularly rocky place. Such practical tips made Cole's book essential reading for women about to embark on a climb. *See also* MOUNTAINEERING.

Further Reading: Cole, Henry Warwick, Mrs. *A Lady's Tour Round Monte Rosa*. London: Longman, Brown, Green, Longmans and Roberts, 1859; Robinson, Jane. *Wayward Women: A Guide to Women Travellers*. Oxford: Oxford University Press, 1990.

Cook's Tours. *See* GRAND TOURS.

Costello, Louisa Stuart (1799–1870)

Born in 1799, Englishwoman Louisa Stuart Costello is considered by many historians to be the first woman to make a living by writing travel books. Her works, which include *Bearn and the Pyrenees: A Legendary Tour to the Country of Henri Quatre* (1844) and *A Tour to and from Venice, by the Vaudois and the Tyrol* (1846), feature her own artwork. After her father died in 1815, leaving them with no source of income, Costello intended to support her family with her paintings but found writing to be more lucrative. At first she wrote only about France, where she lived from 1815 to 1820, but as she became more successful she was able to tour and write about other European countries as well. Costello died in 1870. *See also* PHOTOGRAPHERS AND ARTISTS; TRAVEL WRITERS.

Further Reading: Robinson, Jane. *Wayward Women: A Guide to Women Travellers*. Oxford: Oxford University Press, 1990.

Crawford, Mabel Sharman (ca. 1830–1860)

Born in approximately 1830, Englishwoman Mabel Sharman Crawford wrote two works of travel literature popular during the nineteenth century, *Life in Tuscany* (1859) and *Through Algeria* (1863), which dealt with her experiences as a tourist in both areas. She is also notable in that her books did not just describe her travel adventures but also addressed the issue of women's travel in general. For example, in *Through Algeria* she says:

> In bygone days, the rule that no lady should travel without a gentleman by her side, was doubtless rational; but in a period of easy locomotion, and with abundant evidence to prove that ladies can travel by themselves in foreign countries with perfect safety, the maintenance of that rule certainly savours of injustice. For unquestionable as it is that woman's sphere, as wife and mother, lies at home, it is surely unreasonable to doom many hundred English ladies, of independent means and without domestic ties, to crush every natural aspiration to see nature in its grandest forms, art in its finest works, and human life in its most interesting phases; such being the practical result of a social law which refuses them the right of travel, save on conditions often wholly unattainable. (Morris, pp. 43–44)

Crawford herself was unmarried and usually traveled unaccompanied. She died in 1860.

Further Reading: Morris, Mary. *Maiden Voyages: Writings of Women Travelers*. New York: Vintage, 1993; Robinson, Jane. *Wayward Women: A Guide to Women Travellers*. Oxford: Oxford University Press, 1990.

Creesy, Eleanor ("Ellen") (1815–1900)

Born in 1815, Eleanor Creesy—commonly known as Ellen—was a navigator on all ships captained by her husband Josiah Creesy during the mid- to late 1800s. She served on several vessels, but is most famous for her work on the clipper ship *Flying Cloud.* In 1851, under her husband's command, it sailed from New York City, to San Francisco in a record 89 days, 21 hours, despite difficult storms and an unruly crew. The Creesys repeated their success in 1854, besting their record by sailing the same route in 80 days, 19 hours; this record stood until 1989, when a yacht named *Thursday's Child* completed the same journey in 80 days 19 hours. Creesy began working as a navigator in 1841, immediately after marrying her husband. According to historian David W. Shaw, she was not a navigator in name only. In his book *Flying Cloud* he says:

> She bore all of the responsibilities that came with the job. She was not out there dabbling with a sextant. Far from it. Her insights and talents as a navigator contributed much to the ultimate success *Flying Cloud,* and her husband, enjoyed. . . . Her accomplishments showed that even in the male-dominated arena of the merchant marine, a lady with wit and courage could leave a lasting mark on history and outdo all but the best men who picked up a sextant. That she owed her opportunity to her husband's willingness to count her as an equal detracts nothing from her accomplishments. (p. xxvii).

Further Reading: Shaw, David W. *Flying Cloud: The True Story of America's Most Famous Clipper Ship and the Woman Who Guided Her.* New York: William Morrow, 2000.

Cressy-Marcks, Violet Olivia (ca. 1890–1970)

The travel experiences of Violet Olivia Cressy-Marcks reflect the changing motivations of women traveling during the first half of the twentieth century. No one knows where Cressy-Marcks was born; that she was an Englishwoman is pure assumption. Other details about her background are equally sketchy; she was a divorced woman during her first years of travel (her first husband was Captain Cressy-Marcks). She traveled throughout the world as an amateur archaeologist, going from northern to southern Africa in 1925, from Scandinavia into Russia in 1928, and through Brazil in 1930. She married a second time in 1932, and this husband—Francis Fisher—accompanied her during a trip to China in 1938. This was the first time she had traveled with anyone but a guide or servant. In *Journey into China* she spoke about the sense of freedom she felt while traveling. This was so important to her that she was willing to take great risks while on a journey. According to historian Marion Tinling, when people asked her how she could be brave enough to risk her life for the sake of trips that included such difficulties as potential illness, poor food, and the dangers associated with political unrest, she answered that "there was nothing brave about risking it, that leading a normal life in Europe exposed one to thieves, lunatics, and accidents, and that no man knew the hour of his death." (Tinling, p. 91)

Cressy-Marcks became interested in political struggles throughout the world and traveled to China, Spain, India, Turkey, Tibet, and back to Africa to study various conflicts. She also worked as a war correspondent for the *Daily Express* newspaper of London from 1943 to 1945. She wrote two books related to her adventures: *Up the Amazon and Over the Andes* (1932) and *Journey into China* (1940). She died in 1970. *See also* ANTHROPOLOGISTS AND ARCHAEOLOGISTS.

Further Reading: Cressy-Marcks, Violet Olivia. *Journey into China*. London: Hodder and Stoughton, 1940; ———. *Up the Amazon and Over the Andes*. London: Hodder and Stoughton, 1932; Tinling, Marion. *Women Into the Unknown: A Sourcebook on Women Travelers*. Westport, CT: Greenwood Press, 1989.

Culture Shock

Culture shock is a psychological phenomenon experienced by some people when they come into contact with an unfamiliar culture. The symptoms most often associated with this phenomenon are anxiety; feelings of helplessness and frustration; fear of succumbing to an illness or of being cheated, robbed, or injured; excessive concern over the cleanliness of bedding and dishes and the sanitation of food and water; and, in extreme cases, aggression and hostility toward the people, languages, transportation systems, and other aspects of the host country.

According to Dr. Lalervo Oberg, an anthropologist with the United States Operations Mission to Brazil in 1999, more women experience culture shock than men, particularly after living abroad for long periods of time. He believes that this is attributable, in part, to the fact that women are more socially oriented and therefore feel the loss of their native culture more deeply. He and others believe that the best way to avoid culture shock is to learn about a country before traveling there. This educational process should be extremely thorough and include becoming familiar not only with the country's culture but also with its languages as well. Some guidebooks also recommend that women intending to travel to a particular country speak personally with women who have already gone there and/or who were born there.

In this regard, the Internet is particularly helpful in bringing together women who want to discuss different cultures. More than a dozen Internet newsgroups related to travel, as well as Web sites, bulletin boards, and mailing lists are available. Another helpful source is Journeywoman Online (http://www.journeywoman.com), an Internet magazine with travel articles, travel tips, and bulletin boards for women interested in exchanging information about their travel experiences.

Further Reading: Axtel, Roger. *The Do's and Taboos Around the World*. New York: Wiley, 1993; ———. *The Do's and Taboos Around the World for Women in Business*. New York: Wiley, 1997; *Culture Shock!* Portland, OR: Graphic Arts Center Publishing Company. This is a series with a volume on each of several countries throughout the world, published during the 1990s; Divine, Elizabeth, and Nancy Braganti. *Traveler's Guide to African Customs and Manners*. New York: St. Martin's Press, 1995; ———. *Traveler's Guide to Asian Customs and Manners*. New York: St.

Martin's Press, 1998; ———. *Traveler's Guide to Latin American Customs and Manners*. New York: St. Martin's Press, 1988; ———. *Traveler's Guide to Middle Eastern and North African Customs and Manners*. New York: St. Martin's Press, 1991.

Cumming, Constance (1837–1924)

Born in 1837, Scottish artist Constance Frederica Gordon Cumming is representative of women who travel in order to find subjects to draw and paint. Cummings was from a wealthy, noble Scottish family, who funded her first travels. She made her first overseas journey in 1868, when she toured India with her sister and brother-in-law; from that time on she continued to make trips abroad, often staying in a place for months or, in the case of Fiji, for a few years. During her later travels, she was often a guest of those who valued her work. In addition to Fiji, her travels took her to Ceylon, Tonga, Samoa, Tahiti, Japan, China, and the coasts of California and the Pacific Northwest. She wrote several books based on her experiences, including *Wanderings in China* (1881) and *At Home in Fiji* (1886) and became well known for the watercolors she produced during her travels. Cumming died in 1924.

Further Reading: Birkett, Dea. *Spinsters Abroad: Victorian Lady Explorers*. Oxford: Basil Blackwell, 1989; Robinson, Jane. *Wayward Women: A Guide to Women Travellers*. Oxford: Oxford University Press, 1990.

D

David-Neel, Alexandra (1868–1969)

Born in 1868 in France and raised both there and in England, Alexandra David-Neel was the first European woman to enter the Tibetan capital city of Lhasa, which had been closed to foreigners. Afterward she wrote about her successful 1923–1924 trek in *My Journey to Lhasa: The Personal Story of the Only White Woman Who Succeeded in Entering the Forbidden City*. First published in 1926 in five installments in the magazine *Asia* as "A Woman's Daring Journey into Tibet," this book appeared in 1927 and created great interest among both female and male travelers.

David-Neel first began to travel while a singer with an opera company that toured Europe and parts of the Middle East and Asia. While touring she met railroad engineer Philippe Francois Neel and left the performing company to marry him in 1904. Their union, however, was not a conventional one. As scholar Peter Hopkirk explains in his introduction to *My Journey to Lhasa*:

> Alexandra had felt instantly trapped by her marriage, and within days they had split up. Yet a mutual affection and respect remained for the rest of their lives, and they were to write regularly to one another until his death in 1941. But more extraordinarily, he continued to support Alexandra financially through all her years of travel, and to act as her literary agent. It was an arrangement difficult to explain, although when Philippe died she wrote of him: "I had lost the best of husbands and my only friend." A cynic, on the other hand, might accuse her of having used him as a lifelong meal ticket. (p. xii)

In 1910 David-Neel was asked by the French Ministry of Education to report on conditions in India and Burma. On this trip she met the Dalai Lama, the spiritual leader of Tibet, who had been exiled from his country after it was taken over by China. David-Neel was fascinated by the Dalai Lama, whom she met in Kalimpong, India, and two years later she visited him again. He was the inspiration for her trip to Lhasa.

In *My Journey to Lhasa* David-Neel explains that she made four attempts to reach Lhasa before succeeding on the fifth. In preparation for her first trip, she went to a Tibetan monastery in the Himalayas to study the language, then visited other monasteries in the region to study Buddhism and the Tibetan people. She also spent a winter in contemplative isolation in a Tibetan cave.

While at one monastery, David-Neel met a 15-year-old boy named Yongden, who became her guide. In 1915, the two ventured into the interior of Tibet, but David-Neel was quickly discovered by authorities and deported. Nonetheless, she made another visit to the Himalayas that same year and again the following year. She also went to Burma, Japan, and China. From 1922 to 1923 she traveled through the Gobi Desert in east Asia, and afterward she decided to go to Lhasa,

Alexandra David-Neel, in 1929, dressed as a nun of a Tibetan Cloister. © *Bettmann/CORBIS.*

which the British were by then calling "The Forbidden City" because of its restrictions against foreigners. This time David-Neel adopted the disguise of a Tibetan beggar and pretended to be Yongden's mother. Soon she found herself identifying so strongly with the role that she began to think like a beggar. In *My Journey to Lhasa* she reports:

> My concern about the alms we will or will not receive amuses me. I have developed the true beggar's mentality since I began to play the wanderer! But our begging is not, after all, wholly a sport. It has a serious side, for the offerings which fill our wallets relieve us of the necessity of buying food and showing that we possess money, and therefore help to preserve our incognito. (p. 102)

Because of her disguise, David-Neel was able to reach Lhasa in time for the city's New Year festivals of 1924. However, she had to struggle to remain hidden from authorities until the festivals began. She says:

> I was in Lhasa. No doubt I could be proud of my victory, but the struggle, with cunning and trickery as weapons, was not yet over. I was in Lhasa, and now the problem was to stay there. Although I had endeavoured to reach the Thibetan capital rather because I had been challenged than out of any real desire to visit it, now that I stood on the forbidden ground at the cost of so much hardship and danger, I meant to enjoy myself in all possible ways. (p. 258)

Although she was almost detected on at least one occasion, David-Neel finally succeeded in witnessing the festivals. She also visited several historical sites and religious shrines. She then left the city and made her way back through Tibet to a British outpost, where she announced her victory. David-Neel then decided to adopt Yongden as her son, and he went to live with her in France.

After *My Journey to Lhasa* was published, David-Neel spent a great deal of time lecturing on Tibet. She subsequently wrote other books related to her subject, including *Initiations and Initiates in Tibet* (1931), *With Mystics and Magicians in Tibet* (1931), and *Buddhism, Its Doctrines and Its Methods* (1939). However, she had no proof that she had ever reached Lhasa, and throughout her life her credentials were often questioned.

In 1936 David-Neel and Yongden, now named Arthur, returned to the border of Tibet, this time to make a home in the mountains rather than to travel. They remained there until 1942. David-Neel's husband had died in 1941, and she was having financial difficulties. She returned to France, where she received a medal from the French Geographical Society and lived simply until her death in 1969. *See also* ASIA; SPOUSES.

Further Reading: David-Neel, Alexandra. *My Journey to Lhasa*. 1927. Reprint. Boston: Beacon Press, 1986; Foster, Barbara M., and Michael Foster. *Forbidden Journey: The Life of Alexandra*

David-Neel. San Francisco: Harper & Row, 1987; Middleton, Ruth. *Alexandra David-Neel: Portrait of an Adventurer.* Boston: Shambhala, 1989.

Davidson, Robyn (1950–)

Born in 1950, Robyn Davidson embarked on a unique travel adventure in 1977, setting out alone across the Australian Outback with four camels to carry her supplies and a dog for companionship. It took her almost two years to prepare for the trip, which began inland at the Glen Helen Tourist Camp in the Northern Territory and ended at Hamelin Pool on the west coast of the continent. In all, Davidson traveled 1,700 miles in six months, much of it spent in the Gibson Desert. She describes her experience in a book entitled *Tracks*, portions of which were based on material that appeared in *National Geographic* magazine.

Published in 1980, *Tracks* relates details not only about Davidson's trip but also about her preparations. Before leaving on her journey, she spent over 18 months learning to train the camels and then training them—one of them pregnant. During this time, she tried to raise money to buy equipment and finance her adventure, but she had no success until she met a photographer who had worked for *National Geographic* magazine. He convinced her to apply for the magazine's sponsorship, and, once it agreed to fund the project, her adventure turned from something private done for personal satisfaction to something public done for acclaim and publicity. This made Davidson unhappy. In *Tracks* she describes her feelings:

> I . . . [was] in a lather of conflicting emotions. Was I being too precious about this thing? Why shouldn't I share it with people? Was I a selfish mealy-mouthed little child? A bourgeois individualist even? Suddenly it seemed as if this trip belonged to everybody but me. Never mind, I said, when you leave . . . it will all be over. No more loved ones to care about, no more ties, no more duties, no more people needing you to be one thing or another, no more conundrums, no more politics, just you and the desert, baby. And so I pushed it all down into the dim recesses of my mind, there to fester and grow like botulism. (p. 105)

Davidson insisted that the *National Geographic* photographer not accompany her during the entire journey, but instead meet her at certain points along her route. However, after she set off, she discovered that she could not escape the consequences of publicity about her trip. Whenever she was in areas that tourists visited, she would be approached by those who had heard about her plans. She says:

> By the time I got near Wallera Ranch . . . the tourists were beginning to drive me crazy. In overrigged vehicles they would come in droves to see Australia's natural wonders. They had two-way radios, winches, funny hats with corks on them, stubbies (beer bottles) . . . all this to travel down a perfectly safe road. And they had cameras. I sometimes think tourists take cameras with them because they feel guilty about being on holiday, and feel they should be doing something useful with their time. In any case, when otherwise perfectly nice people don their hats and become tourists, they change into bad-mannered, loud, insensitive, litter bugging oafs. . . . At first I treated one and all with pleasant politeness. There were 10 questions invariably asked me, and I unfailingly gave my pat reply. I posed for the inevitable snap snap of Nikons [still cameras] and the whirr of Super-eights [movie cameras]. It got so that I was stopped every half hour and by three in the afternoon . . . [I was disgusted with] these fools who would pile out, block my path, frighten the camels, hold me up, ask stupid boring questions, capture me on celluloid . . . [and] I would begin to get mean. (pp. 135–136)

Davidson's disgust deepened after the *National Geographic* photographer met up with her and took some pictures of a sacred Aboriginal ceremony. One of the reasons she had wanted to take her trip was in order to get closer to the Aborigine people. After this incident, however, they viewed her as a tourist who wanted to exploit them and would not accept her as a friend. Later she did improve her relationship with some of the Aborigines, but she never achieved the level of closeness she desired. In addition, she experienced a mild bout of mental illness (she heard voices and became disoriented and depressed) while in the desert. Once she recovered, she was more content with herself and had a new awareness of the land. But as she left the Outback and approached the town of Hamelin Pool, she found the press waiting for her with both still cameras and television cameras. As they published sensationalized reports about her, Davidson felt that the importance of her journey for other women travelers was being lost. She explains:

> It would seem that the combination of elements—woman, desert, camels, alone–ness—hit some soft spot in this era's passionless, heartless, aching psyche. It fired the imaginations of people who see themselves as alienated, powerless, unable to do anything about a world gone mad. . . . I was now a kind of symbol. I was now an object of ridicule for small-minded sexists, and I was a crazy irresponsible adventurer. . . . But worse than all that, I was now a mythical being who had done something courageous and outside the possibilities that ordinary people could hope for. And that was the antithesis of what I wanted to share. That anyone could do anything. If I could bumble my way across a desert, then anyone could do anything. And that was true especially for women, who have used cowardice for so long to protect themselves that it has become a habit. . . . And now a myth was being created where I would appear different, exceptional. Because society needed it to be so. . . . Had I been a man, I'd be lucky to get a mention in [a local newspaper], let alone international press coverage. (pp. 237–238)

Once her journey was over, she remained disappointed in many aspects of her experience, although she was proud of herself for setting a goal and reaching it. Nonetheless, she remained interested in travel and subsequently set out to investigate a worldwide decrease in nomadic behavior by documenting the effects of modern civilization on the migratory cycle of one type of nomad, the Rabari people of India and Pakistan. She wrote about this experience in *Desert Places* (1996). *See also* ADVENTURE TRAVEL; AUSTRALIA AND NEW ZEALAND.

Further Reading: Davidson, Robyn. *Desert Places.* New York: Viking, 1996; ———. *Tracks.* New York: Pantheon Books, 1980.

Davison, (Margaret) Ann

In 1953 Ann Davison became the first person to sail solo across the Atlantic, traveling from England to the West Indies in a 23-foot sloop. However, she did not do so nonstop; she put into port in Spain and the Canary Islands along the way. (The first woman to sail solo across the Atlantic nonstop was Nicolette Milnes Walker.)

Davison's book about her experience, *My Ship Is So Small*, was published in 1956. It was not her first account of a sailing adventure. In 1951 she wrote *Last Voyage*, the story of a 1949 attempt to cross the Atlantic with her husband. That time she intended to go from England to Cuba, but the journey was aborted when her ship struck some rocks and her husband was killed. Davison wrote two other accounts about subsequent adventures: *By Gemini: A Coastwise Cruise from Miami to Miami* (1962) and *Florida Junket: The Story of Shoestring Cruise* (1964). Davison settled permanently in Florida after her successful Atlantic journey. *See also* SEA TRAVEL.

Further Reading: Davison, Ann. *My Ship Is So Small.* London: Davies, 1956; ———. *By Gemini: A Coastwise Cruise from Miami to Miami.* London: Davies, 1962; ———. *Florida Junket: The Story of a Shoestring Cruise.* London: Davies, 1964.

De Erauso, Catalina (1592–1650)

Born in 1592, Spanish noblewoman Doña Catalina de Eraso was determined to live an adventurous life. At the age of 15 she disguised herself as a man and became a bandit and mercenary, traveling throughout Europe and the New World to steal and fight. She was skilled with a sword and received commendations from the Spanish government and the Roman Catholic church after battles with foreign invaders. De Erauso died in 1650.

Further Reading: Salmonson, Jessica Amanda. *The Encyclopedia of Amazons: Women Warriors from Antiquity to the Modern Era.* Anchor Books, 1992.

De Watteville, Vivienne (1900–1957)

Born in 1900, Swiss filmmaker Vivienne De Watteville wrote about her travels in Africa in two books, *Out in the Blue* (1927) and *Speak to the Earth* (1935). The former tells of a 1930s expedition to Mount Kilimanjaro that the author made with her father, a hunter and zoologist, who was killed by a lion during the trip. The second book describes De Watteville's solo trip to Africa to film Kenyan wildlife.

Subtitled *Wanderings and Reflections Among Elephants and Mountains, Speak to the Earth* includes some of the author's black-and-white photographs. However, it begins with a note by De Watteville explaining that most of her photographs could not be adequately reproduced in the book because they were shot with 16 mm movie film. In fact, the travels described in *Speak to the Earth* were undertaken because De Watteville wanted to film wild elephants. She spent five months in the Southern Masai Game Reserve in Kenya near the Tanganyika border, and while there she learned that it was unsatisfying to photograph large game. She explains:

> The closer you come to wild animals the greater the feeling of frustration. The stalk is successful, the prize so near; but whether as a hunter you kill it or as a photographer you take your picture of it, it has eluded you to the end. The trophy is dead, the picture is nothing. I believe that every one who has held the spoor through long hours of alternate hope and despair, be he ever so keen a collector, feels before all else as I felt when I lay watching the wildebeest, that the real reward would be to go among them without their minding. (pp. 43–44)

In addition to describing her contact with animals, De Watteville talks about her encounters with indigenous peoples and the way in which Africans relate to their environment. She also discusses elephants in depth, talking not only about her encounters with them but also about their natures in general. In fact, half of her book, Part I, is entitled "Among the Elephants." Part II is called "The Mountain (Mount Kenya)" because it describes a two-month climbing expedition De Watteville made after completing her elephant filming expedition. As she ascended the mountain she collected plant specimens, sketched and painted flowers, and took photographs. She spent much of the climb alone, remaining distant from her porters, and found the solitude rewarding. She concludes that "solitude is an ally, there is nothing to fear, for truly 'Nature never did betray the heart that loved her.'" (p. 328) De Watteville continued to enjoy wilderness experiences until her death in 1957. *See also* AFRICA; ECO-TOURISM; PHOTOGRAPHERS AND ARTISTS.

Further Reading: De Watteville, Vivienne. *Speak to the Earth: Wanderings and Reflections Among Elephants and Mountains.* 1935. Reprint, with an introduction by Alexander Maitland. New York: Penguin Books, 1988.

Dibble, Lucy Grace (1902–)

Born in 1902, British Lucy Grace Dibble was a popular 1980s travel writer whose works encouraged women to make their own journeys abroad. A school principal in Canada, India, and Africa, Dibble embarked on a series of vacations throughout the world after her retirement in 1965. Along the way she decided to share her experiences with other travelers. She wrote a series of books whose titles include *Return Tickets* (1968), *Return Tickets to Southern Europe* (1980), *Return Tickets to Scandinavia* (1982), *Return Tickets to Yugoslavia* (1984), and *Return Tickets Here and There* (1988). These works not only provided information about the places that Dibble visited but also—by virtue of Dibble's example—offered encouragement to older travelers.

Further Reading: Dibble, L. Grace. *Return Tickets.* London: Ilfracombe, Stockwell, 1968.

Dinesen, Isak (1885–1962)

Born in 1885, Isak Dinesen is the pseudonym of Karen Christence Dinesen, Baroness Blixen-Finecke, a Danish author who wrote a book, *Out of Africa,* that romanticized some aspects of Africa and drew other women to the continent. Dinesen first traveled to Kenya with her husband in 1914 to run their 6,000-acre coffee plantation. The couple divorced in 1921, but Dinesen remained there until 1931, when drought and financial difficulties forced her return to Denmark.

Although Dinesen's experiences had a bleak outcome, in 1937 she published *Out of Africa* to share her affection for Africa and its indigenous people. About her affinity for the latter she writes:

> From my first weeks in Africa, I had felt a great affection for the Natives. It was a strong feeling that embraced all ages and both sexes. The discovery of the dark races was to me a magnificent enlargement of all my world. If a person with an inborn sympathy for animals had grown up in a milieu where there were no animals, and had come into contact with animals late in life; or if a person with an instinctive taste for woods and forest had entered a forest for the first time at the age of twenty; or if some one with an ear for music had happened to hear music for the first time when he was already grown up; their cases might have been similar to mine. After I had met with the Natives, I set out the routine of my daily life to the Orchestra. (pp. 17–18)

Isak Dinesen. © *Bettmann/CORBIS.*

Dinesen also wrote about her relationship with her land and about the tourists who visited it. For example, she wrote:

> To the great wanderers amongst my friends, the farm owed its charm, I believe, to the fact that it was stationary and remained the same whenever they came to it. They had been over vast countries and had raised and broken their tents in many places, now they were pleased to round my drive that was steadfast as the orbit of a star. They liked to be met by familiar faces, and I had the same servants all the time that I was in Africa. I had been on the farm longing to get away, and they came back to it longing for books and linen sheets and the cool atmosphere in a big shuttered room; by their campfires they had been meditating upon the joys of farm life . . . (pp. 205–206)

Such writings did much to convince Dinesen's peers to visit Africa, and some of them were inspired to become colonists. Meanwhile, Dinesen continued to write. Her other works include *Seven Gothic Tales* (1934), *Winter's Tales* (1942), and *Letters from Africa, 1914–1931*, which was published posthumously in 1981. Dinesen died in 1962. *See also* AFRICA.

Further Reading: Dinesen, Isak. *Out of Africa*. 1937. Reprint, with an introduction by Bernardine Kielty. New York, Modern Library, 1952. Donelson, Linda. *Out of Isak Dinesen in Africa: Karen Blixen's Untold Story*. Iowa City, IA: Coulsong List, 1998; Thurman, Judith. *Isak Dinesen: The Life of a Storyteller*. 1982. Reprint. New York: Picador USA, 1995.

Diplomats. *See* PUBLIC SERVICE.

Disabilities

Women with disabilities are able to travel, and there are many books and organizations dedicated to helping them enjoy quality travel experiences. For example, since 1985 a travel company called Accessible Journeys (http://www.disabilitytravel.com) has been designing international holidays and guided tours for disabled individuals and their families. Flying Wheels Travel (http://www.flyingwheels.com) has also been arranging guided tours for the disabled for 27 years. Meanwhile, two important books, *Wheelchair Around the World* (1998) and *Wheelchair Down Under* (1999), have inspired many people in wheelchairs to embark on travel adventures. These books were written by Patrick Simpson to share his wife Anne's experiences traveling in a wheelchair during a nine-month world tour and a five-month trip to Australia and New Zealand.

Further Reading: Simpson, Patrick. *Wheelchair Around the World*. Raleigh, NC: Pentland Press, 1997; ———. *Wheelchair Down Under*. Raleigh, NC: Pentland Press, 1999.

Disguises

Until modern times, women in most cultures have not had the same opportunities for travel as men. Therefore women seeking adventure sometimes disguised their gender before taking a trip. For example, in 1607 Spanish noblewoman Doña Catalina de Erauso disguised herself as a man and became a bandit and mercenary, traveling throughout Europe and the New World to steal and fight.

De Eraso's experience was unusual for her time, but during the nineteenth century it became increasingly common for women to adopt male disguises, particularly if they wanted to travel by ship. Captains on some types of vessels, such as whalers, pirate ships, and war ships, had strict rules against women being on board, so there was no other way for a woman to gain access to these vessels unless she went as a man. Examples of women who adopted such a disguise include "Tom" Bowling, "William" Brown, and Ann Jane Thornton. Historical records related to a lawsuit show that "Tom," who was a woman, served for 20 years on a British war ship, and her true gender was not discovered until a dispute arose regarding her pension. "William" was an African American woman who served as a crewman on a British war ship from 1804 to 1815. No record has been left behind regarding her real identity, but officers noted that she was highly skilled in her work, and her true gender was not discovered until after she retired from the sea. Thornton served as a cabin boy and cook on board sailing ships from 1832 to 1835 before being discovered. When Thornton was 16 years old, her true gender was discovered and she was left on land. Her fate remains unknown.

Historian Linda Grant De Pauw, in her book *Seafaring Women*, reports that several more women like these are mentioned in historical records. She therefore concludes that there must have been even more, explaining: "That a woman could live in close quarters with men without revealing her sex is hard for many people to believe. Nevertheless, there is clear evidence that numerous women did so successfully. . . . [And] Because so many escaped detection for years, it is a safe assumption that many others were never discovered." (p. 84)

Women also disguised themselves for land adventures, for various reasons. For example, Englishwoman Hannah Snell enlisted in the British army in 1743, disguised as "James Gray," in order to travel throughout North America to find a runaway husband. She soon decided that life in the army was too difficult, but because she did not want to return home she simply ran away from her regiment and joined the marines, thinking they had easier duties. Wounded during a Revolutionary War battle, she was reassigned to a war ship after recovering from her injuries. She served on board until 1750, whereupon she revealed her deception and became an actress, primarily playing war heroes.

Snell's deception was grounded in romance, but Deborah Sampson's arose from a desire to become a soldier at a time when the profession was barred to women. In 1782, when she was 22 years old, Sampson disguised herself as a man so she could fight in the Revolutionary War as part of the 4th Massachusetts Regiment. She traveled throughout the eastern United States from 1782 to 1783, when her gender was discovered during an illness. Inspired by Sampson's story, "Lucy Brewer" (the pseudonym of a former prostitute) also decided to adopt a male disguise so she could serve in the military. Brewer was a marine during the War of 1812, sometimes serving on land and sometimes on a war ship; afterward she wrote a book about her experiences, *The Female Marine, Or, Adventures of Lucy Brewer, a Native of Plymouth County, Massachusetts . . . From the Time of Her Discharge to the Present Day: To Which Is Added Her Serious Address to the Youths of Boston* (1816). There is no proof that her story is true; nonetheless, she made money not only from the book but also by giving lectures about her adventures.

Other women disguised themselves to take on male occupations. For example, Charley Parkhurst was a stagecoach driver in California during the mid-1800s. Born in New Hampshire as Charlotte Parkhurst, she apparently escaped from an orphanage by disguising herself as a boy, a deception she maintained for the rest of her life. She traveled west during the Gold Rush and worked for a stagecoach line until her death in 1881, whereupon her acquaintances discovered the truth about her gender.

Another common reason for a woman to disguise her gender has been so she could travel without fear of being raped. For example, in 1890 Scotswoman Ménie Muriel Dowie dressed as a boy so she could tour Poland alone and unmolested. She wrote a book about her experiences, entitled *A Girl in the Karpathians* (1891), as well as a book about other women who had adopted male disguises in order to travel, *Women Adventurers* (1893). More recently, in 1970 British filmmaker Sarah Hobson explored the Middle East dressed as a man in order to travel alone without fear of assault. The disguise also made it possible for her to travel to areas where women were not normally allowed and to speak to men as a peer in a culture that considered women inferior.

Yet another woman who dressed like a man to travel through the Middle East was Englishwoman Hester Stanhope, who visited the region in 1810. However, her decision to adopt men's clothing was not a deception but a practical choice, because it allowed her to abandon the cumbersome skirts that women wore at the time. Stanhope made no effort to hide her true gender, and her behavior outraged Britains; however, it was accepted by the Arabs she met as being just another quirk of foreigners. *See also* De Erauso, Catalina; Dowie, Ménie Muriel; Hobson, Sarah; Parkhurst, Charley; pirates; Sampson, Deborah; Stanhope, Hester.

Further Reading: De Pauw, Linda Grant. *Seafaring Women*. Boston: Houghton Mifflin, 1982.

Dixie, Florence (1855–1905)

Born in 1855, English author Lady Florence Dixie was noted for her travels in South America and South Africa. She wrote articles and books based on her experiences, including *Across Patagonia* (1880), *A Defence of Zululand and Its King: Echoes from the Blue Books* (1882), and *In the Land of Misfortune* (1882).

Dixie grew accustomed to travel early in her life. When she was a girl, her mother took her from place to place in Europe, trying to hide from relatives who sought to gain custody of Dixie. Later she married a man who loved to travel, and in the late 1870s they went on a hunting trip to South America with friends. Dixie enjoyed the adventure but felt bad for the animals they killed; several years later she became involved with a group opposed to hunting.

Dixie's writings about her South American travels, published in 1880, earned her a job the following year as a newspaper reporter for the London *Morning Post*; her assignment was to cover the Boer War (1899–1902) in South Africa between the British and Dutch settlers that involved a local tribe called the Zulus. Shortly after Dixie and her husband traveled to the region, they met with the leader of a Zulu tribe who had been taken captive. He urged the couple to visit his people, and when they did so they felt sympathy for the Zulus' position. Once she was back in England, Dixie successfully argued for the Zulu leader's release from prison.

Dixie then became involved in a variety of other political causes, including women's rights and animal rights. She also wrote novels and poetry. In her later years she suffered from poor health and reduced funds, making it impossible for her to travel. She died in 1905. *See also* Africa; Latin America; travel writers.

Further Reading: Dixie, Florence, Lady. *Across Patagonia*. London: R. Bentley and Son, 1880; Robinson, Jane. *Wayward Women: A Guide to Women Travellers*. Oxford: Oxford University Press, 1990; Tinling, Marion. *Women Into the Unknown: A Sourcebook on Women Travelers*. Westport, CT: Greenwood Press, 1989.

Dodwell, Christina (1951–)

Born in 1951, Christina Dodwell is one of the best known British travel writers and has produced numerous works that encourage women to travel. Dodwell has journeyed throughout the world gathering experiences for her books. Her first effort, *Travels with Fortune: An African Adventure* (1979), was the result of three years spent crossing Africa. Her other books include *In Papau New Guinea* (1983); *A Traveller in China* (1985); *A Traveller on Horseback: In Eastern Turkey and Iran* (1987); *Travels with Pegasus: A Microlight Journey Across West Africa* (1990); *Beyond Siberia* (1993); and *Madagascar Travels* (1995). She has also written a guidebook for fellow travelers entitled *An Explorer's Handbook: An Unconventional Guide for Travelers to Remote Regions* (1986).

Of these, perhaps her best-known work is *A Traveller on Horseback: In Eastern Turkey and Iran*. As the title suggests, the book describes her journeys by horse through Turkey and Iran and also mentions some time the author spent in Afghanistan, Pakistan, and Greece. Dodwell was in Iran for approximately three months of 1986, after the Ayatollah Khomeni had taken over control of Iran's government. Under this oppressive regime, it was difficult for Dodwell to travel without being harassed by Revolutionary Guards, and on three occasions she was briefly arrested, suspected of being a spy. Nonetheless, she managed to do some sightseeing and to visit stables. *See also* AFRICA; TRAVEL WRITERS.

Further Reading: Dodwell, Christina. *An Explorer's Handbook: An Unconventional Guide for Travelers to Remote Regions*. New York: Facts On File, 1986; ———. *Madagascar Travels*. London: Hodder and Stoughton, 1995; ——. *A Traveller in China*. New York: Beaufort Books, 1985;———. *A Traveller on Horseback: In Eastern Turkey and Iran*. London: Hodder and Stoughton, 1987; ———. *Travels with Pegasus: A Microlight Journey Across West Africa*. New York: Walker, 1990.

Dogsledding. *See* IDITAROD TRAIL DOG SLED RACE.

Dowie, Ménie Muriel (ca. 1860–?)

In 1890, Scotswoman Ménie Muriel Dowie dressed as a boy so she could tour Poland alone and unmolested. She then wrote a book about her experiences, *A Girl in the Karpathians* (1891). Three years later she wrote about other women who had adopted male disguises in order to travel in *Women Adventurers* (1893). Dowie also traveled to Egypt and Europe, sometimes with her husband, Henry Norman, but more often alone. Eventually she and Norman divorced, and in 1903 Dowie married mountaineer Major Edward Arthur Fitzgerald. In her later years she continued to travel but primarily wrote fiction. *See also* DISGUISES; MOUNTAINEERING.

Further Reading: Birkett, Dea. *Spinsters Abroad: Victorian Lady Explorers*. Oxford: Basil Blackwell, 1989.

Dreier, Katherine (1877–1952)

Born in 1877, American artist Katherine Sophie Dreier is representative of women who travel to study and promote art. In 1902–1903 she went to Europe to study great paintings; in 1905 she was hired to produce a chapel painting for St. Paul's School in Garden City, New York. After its completion she returned to Europe to study in Paris and London. Her first major exhibition was in London, but she also toured Germany with her art and had an exhibition in New York in 1913. In 1916 she helped found the Society of Independent Artists, which involved both European and American artists; in 1920 she helped found New York's first center for modern art, the Société Anonyme, whose works eventually went to Yale University. Dreier died in 1952. *See also* PHOTOGRAPHERS AND ARTISTS.

Further Reading: Bohan, Ruth L. *The Société anonyme's Brooklyn Exhibition: Katherine Dreier* *and Modernism in America*. Ann Arbor, MI: UMI Research Press, 1982.

Du Bois, Cora (1903–1991)

Born in 1903, American anthropologist Cora Du Bois traveled extensively to study the relationship between culture and human psychology, administering standard psychological tests to various native peoples. Her first such trip was to the island of Alor in the East Indies in 1937–1939. Between 1954 and 1970, while an anthropology professor at Harvard University, she organized research expeditions to California, the Netherlands, India, and Indonesia. Du Bois was the first woman tenured in Harvard's Anthropology department. She also developed new ways to analyze research data. She was one of the first anthropologists to document fieldwork with photography and to emphasize interdisciplinary teamwork during research expeditions. Du Bois retired from Harvard in 1970, then taught at Cornell University until 1975. She died in 1991.

Further Reading: Du Bois, Cora Alice. *Social Forces in Southeast Asia*. Minneapolis: University of Minnesota Press, 1949.

Du Faur, Freda (1882–1925)

In 1910 Australian mountaineer Freda Du Faur became the first woman to climb Mount Cook on New Zealand's South Island. In 1913, she was a member of the first party to climb the island's three major peaks—Mount Cook, Mount Dampier, and Mount Sefton—in the same day, an experience that became known as the Grand Traverse of Mount Cook. *See also* MOUNTAINEERING.

Further Reading: Robinson, Jane. *Wayward Women: A Guide to Women Travellers*. Oxford: Oxford University Press, 1990.

Dunsheath, Joyce (1902–1976)

Born in 1902, mountaineer Joyce Dunsheath led several important all-woman climbing expeditions. In 1956 she and her team climbed several Himalayan peaks and were the first to reach the summit of Cathedral Peak; with the rest of her mountaineering team, Dunsheath co-authored a book about her experiences, *Mountains and Memsahibs* (1958). The following year she received permission from the Soviet government to become the first Englishwoman to climb Mount Elbrus in the Caucasus mountain range; she wrote a book about this adventure, *Guest of the Soviets: Moscow and the Causasus, 1957* (1959). In 1960 she led a climbing expedition in Afghanistan, writing about it in *Afghan Quest: The Story of the Abinger Afghanistan Expedition, 1960* (1961). In the early 1970s Dunsheath also climbed in the Peruvian Andes and ascended Africa's Mount Kilimanjaro and Mount Kenya. She died in 1976. *See also* MOUNTAINEERING.

Further Reading: Abinger Himalayan Expedition. *Mountains and Memsahibs*. London: Constable, 1958; Dunsheath, Joyce. *Afghan Quest: The Story of the Abinger Afghanistan Expedition, 1960*. London: Harrap, 1961; ———. *Guest of the Soviets: Moscow and the Caucasus, 1957*. London: Constable, 1959.

Durham, Mary Edith (1864–1944)

Born in 1864, English author Mary Edith Durham wrote about her many travels but is best known for her books about the Balkans, which she visited often during the early 1900s. Her first trip there was in 1900, after a doctor advised her to travel for her health. During the Balkan Wars she worked as a nurse in the region, and later she became involved in Balkan political causes. Most of her books include political commentary, although *High Albania* (1909) primarily describes the sights along a passage to the Balkans through Montenegro and northern Albania. Durham typically illustrated her books with her own paintings; as a young woman she attended the Royal Academy of Arts in London and did naturalist paintings. She made her last trip abroad in 1921 and died in 1944.

Further Reading: Durham, Edith. *High Albania*. 1909. Reprint, with a new introduction by John Hodgson. Boston: Beacon Press, 1987.

E

Earhart, Amelia (1897–1937)

Born in 1897, American aviator Amelia Earhart was the first woman to fly alone across the Atlantic Ocean, a feat she accomplished in 1932. Four years earlier, she was the first woman to cross the Atlantic as an airplane passenger. Both flights brought her notoriety, and she used her fame to promote women's aviation.

Prior to becoming an aviator, Earhart worked as a nurse during World War I and then as a social worker after the war. Her life changed once she met publicist George Putnam, who promoted Charles Lindbergh's solo flight across the Atlantic. It was Putnam who chose Earhart to make her historic flight as an airplane passenger in 1928, and in 1931 she married him. By that time, she had been flying solo for 10 years and had done some stunt flying as well. Nonetheless, some people believe that Putnam pushed Earhart too hard in his quest for publicity.

In 1935 she became the first person, male or female, to fly solo from Hawaii to California, from California to Mexico City, and from Mexico City to New Jersey. In 1937, with another flurry of publicity, she set off to fly around the world, accompanied by navigator Fred Noonan. Their plane disappeared somewhere over the ocean in the South Pacific. No one ever discovered what happened to them, although many theories have been proposed. *See also* AIR TRAVEL; CIRCUMNAVIGATORS AND ROUND-THE-WORLD TRAVELLERS.

Further Reading: Davis, Burke. *Amelia Earhart*. New York: Putnam, 1972; Earhart, Amelia. *The Fun of It: Random Records of My Own Flying and of Women in Aviation.* New York: Brewer, Warren, and Putnam, 1932; Loomis, Vincent V., with Jeffrey L. Ethell. *Amelia Earhart, The Final Story*. New York: Random House, 1985.

Amelia Earhart. *Library of Congress.*

Earle, Sylvia (1935–)

Born in 1935, American undersea explorer Sylvia Earle is preparing to launch a five-year series of expeditions designed to explore the deepest parts of 12 U.S. National Marine Sanctuaries. As leader of the National Geographic Society's Sustainable Seas Expeditions (SSE), she will pilot a one-person submarine into each sanctuary and broadcast information about her discoveries on the Internet. Her submarine is the first one-person research vessel capable of going 2,000 feet below the ocean's surface while retaining the same internal air pressure as on the surface of the ocean, eliminating the danger that the occupant will suffer from decompression sicknesses while resurfacing.

Earle has studied marine habitats throughout the world and holds the record for the deepest undersea walk ever accomplished without a tether. She earned a Ph.D. in botany from Duke University, and she specializes in aquatic plants in the Gulf of Mexico. In 1964, while still a student, she joined a scientific expedition in the Indian Ocean despite having two children ages two and four at home. She continued to embark on scientific expeditions throughout the 1960s, and in 1968 she undertook a deep-sea dive in the Bahamas although she was four months pregnant. In 1969 she spent several weeks in an enclosed underwater habitat, or biosphere, off the Virgin Islands as part of a government research project. In 1970 she returned there as part of a two-week all-woman expedition designed to study the physical and mental challenges of living at great depths. This experience made her famous.

In subsequent years Earle participated in scientific missions to the waters of the Galapagos, Panama, China, the Bahamas, and the Caroline Islands of the South Pacific. She also wrote articles for the *National Geographic* magazine. In 1977 she went on a series of expeditions to follow the migrations of sperm whales; her travels were documented in a film, *Gentle Giants of the Pacific* (1980). In 1979 she donned a pressurized suit that was then attached to a submersible that took her beneath the waters near Oahu, Hawaii, to a depth of 1,250 feet. There she was detached from the submersible so she could walk on the sea floor unaided, the only person to do so at such a depth. In the 1980s Earle co-founded a company to design and build undersea vehicles; in the 1990s she served for a brief time as chief scientist of the National Oceanographic and Atmospheric Administration, dealing with issues of water pollution. *See also* EXPLORERS; ORGANIZATIONS AND ASSOCIATIONS.

Further Reading: Friend, Tim, "Hearing the Seas' Siren Call," *USA Today*, 2 February 1999, D1; Rosenblatt, Roger. "Sylvia Earle: Her Deepness Welcomes Us into Her World of Wonders." *Time* (October 5, 1998): 58–60; Wexler, Mark. "Sylvia Earle's Excellent Adventure." *National Wildlife* 37 (April 1, 1999):

Eberhardt, Isabelle (1877–1904)

Born in 1877, Isabelle Eberhardt is representative of women who traveled to escape an unhappy life. In the late 1800s she left her home in Switzerland to get away from a brutal father. She went to North Africa, having twice visited it on pleasure trips, and eventually settled in Tunis, where she spent much of her time drinking in bars. She then decided to wander the desert as a nomad and wrote about her experiences in a diary that was later published as *The Passionate Nomad.* In 1900, Eberhardt disguised herself as a man, converted to Islam, and joined a religious brotherhood, the Qadriya. Just one year later she left the Qadriya to marry an Algerian soldier, but she spent most of her money on drugs and felt like an outcast, neither truly European nor truly Arab. She died on the streets in Ain Sefra, South Africa, during a flood in 1904. Near her body were the scattered pages from a manuscript, *In the Shadow of Islam.* Although incomplete, it was later published.

Further Reading: Eberhardt, Isabelle. *In the Shadow of Islam.* Sharon Bangert, trans. Chester, PA: P. Owen, 1993; ———. *The Passionate Nomad: The Diary of Isabelle Eberhardt.* Nina de Voogd, trans., Rana Kabbani, ed. London: Virago, 1987; Morris, Mary. *Maiden Voyages: Writings of Women Travelers.* New York: Vintage Books, 1993.

Eco-tourism

Travel experiences that stem from the traveler's commitment to environmentalism is known as eco-tourism. Women have long been at the forefront of environmentalism; in fact, much of the credit for the launch of the modern environmental movement is given to American biologist and conservationist Rachel Carson, whose 1962 book *Silent Spring* argued that pesticide use was destroying wildlife and habitats. Scientists like Carson continue to travel to study environmental problems firsthand, as do journalists who report on environmental crises and activists who fight anti-environmental activities. For example, American journalist Catherine Caufield travels throughout the world in order to investigate global environmental issues; her book *In the Rainforest* (1989) is a first-person account of her study of rainforest destruction. American activist Mardy Murie explored the Alaska wilderness and fought for the passage of laws to protect it. American Dian Fossey, Englishwoman Jane Goodall, and German Birute Galdikas all became involved in environmentalism as a result of their research into the behavior of primates in the wild. Many other famous environmentalists have also traveled far from home to study wildlife and natural ecosystems.

Eco-tourists, however, do not always travel as part of scientific or environmental work but as tourists on vacation. Nonetheless, they still want to use their travel experiences to express their environmental beliefs. To this end, they might sign up for eco-tours that provide money to indigenous people who might otherwise seek income by harvesting timber or hunting rare species. Alternatively, they might visit wilderness areas on their own to educate themselves and their children on the importance of maintaining these areas.

Many environmental groups promote eco-tourism and/or offer educational programs and recreational opportunities designed to encourage people to enjoy the wilderness. One of the foremost environmental groups involved in such activities is the Sierra Club, which has more than 550,000 members and publishes numerous books on outdoor activities and wilderness travel opportunities. The Girl Scouts also promotes wilderness activities among young women in the United States.

International eco-tourism is also popular among both women and men, and until political unrest in Africa decreased travel there in the late 1990s, African safaris and other wildlife-related activities drew particularly high numbers of eco-tourists. In Kenya, for example, more than 830,000 tourists visited wildlife areas in 1994, with 90 percent of them going on an African safari. According to the Ecosource Network (http://www.ecosourcenetwork.com), this generated over $400 million in American dollars in revenue. Eco-tourism dollars have benefitted many other third-world regions as well, and environmentalists hope that this money will either be used to protect habitat directly or make it unnecessary for habitat to be destroyed for monetary gain. *See also* CAUFIELD, CATHERINE; FOSSEY, DIAN; GALDIKAS, BIRUTE; GOODALL, JANE.

Further Reading: Baker, Christopher P, ed. *World Travel: A Guide to International Ecojourneys.* New York: Time Life Books, 1996; Bonta, Marcia. *Women in the Field: America's Pioneering Women Naturalists.* College Station: Texas A & M University Press, 1991; Elander, Magnus, and Staffan Widstrand. *Eco-Touring: The Ultimate Guide.* Somerville, MA: Firefly Books, 1997; Hubbard, Patricia, and Stan Wass. *The Outdoor Woman: A Handbook to Adventure.* New York: MasterMedia, 1992; Netzley, Patricia. *Environmental Literature: An Encyclopedia of Works, Authors, and Themes.* Santa Barbara, CA: ABC-Clio, 1999; Thomas, Lynn. *The Backpacking Woman.* Garden City, NY: Anchor Press/Doubleday, 1980.

Edwards, Amelia (1831–1892)

Born in 1831, English author Amelia Ann Blandford Edwards increased interest in and knowledge of Egypt among her peers because her book *A Thousand Miles Up the Nile* (1877) provided the fullest account of Egyptian ruins up until that time. Edwards was an independent traveler, touring Europe and the Middle East and supporting herself with the income she earned from her travel books and novels. Her first popular book was *Untrodden*

Peaks and Unfrequented Valleys: A Midsummer Ramble in the Dolomites (1873), an account of her journey through the Dolomite mountain range of northern Italy in 1872.

In 1873 Edwards made her first and only trip to Egypt, traveling up the Nile River in a type of boat known as a *dahabeeyah.* Along the way she sketched and measured the ruins along the Nile's shore; this experience was the basis for *A Thousand Miles Up the Nile.* A scholarly work, the book includes historical information as well as details about the ruins.

Upon her return to England, Edwards established the Egypt Exploration Fund and helped create the science of Egyptology, receiving honorary degrees from several universities in the United States. She also continued to write about Egypt, and in 1889 she toured America lecturing about the topic, subsequently publishing her lecture notes as *Pharoahs, Fellahs, and Explorers* (1891). Edwards died in 1892. *See also* AFRICA.

Further Reading: Edwards, Amelia. *A Thousand Miles Up the Nile.* London: Longmans, Green, 1877; ———. *Untrodden Peaks and Unfrequented Valleys: A Midsummer Ramble in the Dolomites.* London: Longmans, Green, 1873.

Emigration Societies (British)

Several emigration societies established by the British during the late nineteenth and early twentieth centuries were responsible for sending large numbers of women abroad. The earliest was the Female Middle Class Emigration Society, which was established in London in 1862 by a group of women led by Maria S. Rye. Rye wanted to help educated British women, particularly governesses, emigrate to British colonies; her organization provided interest-free loans to women who could not afford their passage. However, many of the women who became indebted were unhappy with the outcome of their decision to emigrate. In describing Society emigrants to Australia, historian Patricia Clarke says:

> There were some success stories, but there were at least as many who suffered great hardship and degradation. They had decided to leave Britain because there were so few opportunities for educated women to obtain employment and they'd been misled into believing that governesses were in great demand in the colonies. Instead they entered a buyers' market where there were always more governesses seeking employment than there were jobs. During times of economic depression so widespread in the 1860s, the market dried up almost completely, a situation that was to be exacerbated with the introduction of free State education. (p. 2)

Moreover, these governesses had trouble adjusting to the demands of colonial life, and once teaching jobs became scarce they often did not want to work at anything else. Clarke explains:

> These were women with deeply ingrained ideas of status. They almost all, for instance, objected to travelling second class—although they had to borrow money to pay even second-class fares—not mainly because of discomfort, but because they felt superior to the people with whom they had to mix. Comments such as that made by [one governess,] Miss Cary, who wrote, "no *Lady* should come out on those emigrant ships," were common. (p. 3)

Emigrants often included letters of progress with their loan payments to the Society, but they undoubtedly minimized complaints in their writings. The organization therefore continued to fund loans until 1908, at which time the British Women's Emigration Association took over.

Established in 1901, the British Women's Emigration Association was an organization dedicated to promoting the emigration of well-educated women to British colonies. The group was originally founded in 1884 as the United Englishwomen's Emigration Association, and within four years the group was including Scots and Irish women as well as British women in its efforts. At that time it was a loose organization of women who acted independently to promote emigration, offer emigrants protection and connections as they traveled, and connect emigrants with trustworthy families once they arrived at their destinations. To this end, in 1888 the group opened a waystation for emigrants in Liverpool, England, to shelter women waiting there for passage abroad, and in 1893 it opened another emigrant home in London. The group particularly encouraged emigration to South Africa, forming a special committee to handle emigrants to this region. In 1901, this committee formed its own emigration society, first known as the South African Expansion Scheme Committee and after 1902 as the South African Colonization Society.

Once the South African organization was created, the British Women's Emigration Association began focusing on having women emigrate to Canada, New Zealand, and Australia. In addition to taking over the duties of the Female Middle Class Emigration Society in 1908, it set up several hostels in British Columbia, Canada, and London in 1913. These hostels were closed during World War I, when emigration dropped dramatically. Also during World War I, the British Women's Emigration Association worked with the South African Colonial Society and another emigration society, the Colonial Intelligence League, to design a postwar emigration policy and present it to the British government. In 1919, these three societies united to become the Society for the Overseas Settlement of British Women.

Like the South African Colonial Society, the Colonial Intelligence League began as a committee of the British Women's Emigration Association: the Committee of Colonial Intelligence for Educated Women. Its purpose was to promote emigration among intelligent women in Great Britain. To this end, the group provided information and advice for would-be emigrants, particularly encouraging those who wanted to work as teachers or clerks rather than housekeepers. Although emigration to all British colonies was supported, once the committee became the Colonial Intelligence League in 1911, it primarily focused on travel to Canada: in 1913 the group established a Canadian settlement, The Princess Patricia Ranch, where new emigrants could reside while training for various jobs.

Once the Colonial Intelligence League merged with the British Women's Emigration Association and the South African Colonization Society to become the Society for the Overseas Settlement of British Women, its members worked to promote the emigration of educated British women to British colonies in the years following World War I. The Society for Overseas Settlement of British Women chose women deemed suitable for emigration and assumed full responsibility for their transportation overseas, their protection while in a British colony, and their efforts to find employment there. The group worked with other women's organizations to find appropriate candidates; these organizations included the Girls' Friendly Society, the World's YWCA, the Women's Catholic Emigration Society, the National Amalgamated Society of Women Workers, and the Ministry of Labour.

In 1928 and 1936, the group provided tours of Canada for schoolgirls and in 1934 it provided a similar tour of Australia, hoping to attract more emigrants in the future. However, the Great Depression and World War II decreased and finally ended emigration until the 1940s, when the Society for Overseas Settlement of British Women reorganized and began to work more closely with the British government, which provided financial support. At the same time, the Society worked with an even larger number of women's organizations, including the Girl Guides Association, the National Association of Girls' Clubs and Mixed Clubs, the National Council of Women, and the Women's Land Army. It also began to concentrate more on finding overseas employment for women rather than on emigration promotional activities, and overseas employers increasingly contacted the organization when searching for qualified female workers. To reflect this new emphasis on employment, in 1962 the Society changed its name to the Women's Migration and Overseas Appointments Society. That same year the British government reduced funding, and after experiencing serious financial difficulties it disbanded in 1964. *See also* AFRICA; AUSTRALIA.

Further Reading: Patricia Clarke. "Private Lives Revealed." http://www.nla.gov.au/events/ private/clarke.html (1999).

Equestriennes

An equestrienne is a woman who rides on horseback, one of the oldest methods of travel for both men and women. Although horseback riding has long been a common activity, some women have distinguished themselves through unique travel experiences as equestriennes. For example, in 1704 Sarah Kemble Knight traveled alone on horseback from Boston to New York, keeping a detailed diary of the Colonial settlements and people she encountered along the way. Her writings were published in 1825 as *The Journal of Madame Knight*, by which time other books had already advised women against horseback riding. The common argument was that the activity damaged women's reproductive organs, but most women ignored this warning.

In the mid- to late-nineteenth century, the number of women who chose to travel on horseback rose dramatically. At the same time, women increasingly abandoned the sidesaddle in favor of riding astride, particularly after actress Fanny Kemble introduced the idea of wearing pants under a skirt in the late 1840s. Many women travelers, including Isabella Bird Bishop, removed the skirt once far from a town for further ease of motion. In 1909, Mrs. Adolph Ladenburg of New York introduced the first pantskirt specifically designed for equestriennes, making riding astride even more comfortable.

The following year, Nan Jane Aspinall became famous for embarking on a solo cross-country ride from San Francisco; she arrived in New York in 1911. In 1938, Mary Bosanquet embarked on a ride across Canada, going 3,000 miles from Vancouver to New York. Two other famous cross-country riders were Bell Cook of California and Emma Jewett of Minnesota, who toured the country in 1881 competing against one another in a series of 20-mile horse races. Each one had to change mounts several times during the course of a race. *See also* BISHOP, ISABELLA LUCY BIRD; BOSANQUET, MARY; CLOTHING; KEMBLE, FANNY; KNIGHT, SARAH KEMBLE; TRANSPORTATION, GROUND.

Further Reading: Knight, Sarah Kemble. *The Journal of Madame Knight*. 1825. Reprint. Chester, CT: Applewood Books, 1992.

Essay Collections

One type of women's travel literature that is particularly common is the travel essay collection. Such books are anthologies of first-person essays, either by one woman traveler or by several, in which the author or authors describe not only their adventures but their emotional responses to those adventures. The Travelers' Tales book series, published by Travelers' Tales of San Francisco, offers a large number of essay collections by multiple authors, including *A Woman's World* (1997) edited by Marybeth Bond, *A Mother's World* (1998) edited by Marybeth Bond and Pamela Michael, and *Women in the Wild* (1998) edited by Lucy McCauley. (Another book that offers essays by multiple authors is *Without a Guide: Contemporary Women's Travel Adventures* [1996]). While these essays typically deal with exotic places, they place more emphasis on the authors' internal journeys of the spirit than on their external journeys of geography. The same is true for *Rivers Running Free* (1987). This book offers a selection of journal entries, stories, and essays about women's canoeing experiences from the early 1900s to the present. *Maiden Voyages: Writings of Women Travellers* (1993), offers essays on a wider variety of travel-related topics, while *An Inn Near Kyoto: Writings by American Women Abroad* (1998), *The House on Via Gombito: Travel Writing by American Women Abroad* (1997), *Tanzania on Tuesday: Writings by American Women Abroad* (1997), all focus on specific geographical regions. Each of these works includes essays by many authors, whereas *Femme D'Adventure: Travel Tales from Inner Montana to Outer Mongolia* (1997) is representative of travel essay collections that feature the writings of just one author. *See also* TRAVEL WRITERS.

Further Reading: Bond, Marybeth, ed. *A Woman's World*. San Francisco: Travelers' Tales, 1997; Bond, Marybeth, and Pamela Michael, eds. *A Mother's World*. San Francisco: Travelers' Tales, 1998; Govier, Katherine, ed. *Without a Guide: Contemporary Women's Travel Adventures*. St. Paul, MN: Hungry Mind Press, 1996; McCauley, Lucy, ed. *Women in the Wild*. San Francisco: Travelers' Tales, 1998; Maxwell, Jessica. *Femme D'Adventure: Travel Tales from Inner Montana to Outer Mongolia*. Seattle, WA: Seal Press, 1997; Morris, Mary, ed. *Maiden Voyages: Writings of Women Travellers*. New York: Vintage Books, 1993; Truesdale, C. W., and Kathleen Coskran, eds. *An Inn Near Kyoto: Writings by American Women Abroad*. Minneapolis, MN: New Rivers Press, 1998; ———. *Tanzania on Tuesday: Writings by American Women Abroad*. Minneapolis, MN: New Rivers Press, 1997; Truesdale, C.W., and Madelon Sprengnether, eds. *The House on Via Gombito: Travel Writing by American Women Abroad*. Minneapolis, MN: New Rivers Press, 1997; Wieser, Barbara, and Judith Niemi, eds. *Rivers Running Free: Stories of Adventurous Women*. Minneapolis, MN: Bergamot Books, 1987.

Etheria

According to an ancient manuscript, between 381 and 384 C.E. an unidentified woman went on a pilgrimage to the Holy Lands of the Middle East. Many historians believe her to be Etheria, an abbess known to have traveled outside her abbey in Western Europe. The manuscript included this woman's own description of her pilgrimage, offering many insights into the experiences of fourth-century pilgrims. In discussing the work, historian Jane Robinson says that this makes Etheria the first travel writer, as well as the honorary "patron saint" of women travelers (p. 159). *See also* PILGRIMS.

Further Reading: Robinson, Jane. *Wayward Women: A Guide to Women Travellers*. Oxford: Oxford University Press, 1990.

Europe, Continental

Continental Europe includes many countries that are among the most popular travel destinations for women vacationers throughout the world. As long ago as the Middle Ages, women went to the region to sightsee with male relatives or to visit shrines while on religious pilgrimages to the Middle East. Beginning in the late eighteenth century, women also traveled to Europe specifically to enjoy mountaineering, and in the nineteenth century Thomas Cook began organizing affordable tours to Europe that attracted many women travelers who sought the safety of traveling in numbers and the enjoyment of sightseeing in the company of other women. Such tours remain popular today, although a fairly equal number of men and women now participate. Women travel writers of the late nineteenth and early twentieth centuries were particularly drawn to the Iberian Peninsula.

Noted authors American Harriet Adams, Englishwoman Emmeline Stuart Wortley, Irishwoman Kate O'Brien, and Englishwoman Rose Macaulay all visited the region and wrote about their experiences. Women who have written first-person travel books about other parts of Europe include Americans Barbara Grizzuti Harrison and Margaret Fuller. Harrison is a modern author, and as such is not unusual in writing about Europe. Fuller, however, is one of very few nineteenth-century American women travelers to write about the Old World. Most felt that they had little new to add on the subject, given that many nineteenth-century British authors were producing books on their European travel experiences. *See also* ADAMS, HARRIET CHALMERS; FULLER, MARGARET; GRAND TOURS; HARRISON, BARBARA GRIZZUTI; MACAULAY, ROSE; O'BRIEN, KATE; PILGRIMS; WORTLEY, EMMELINE STUART AND VICTORIA.

Further Reading: Withey, Lynne. *Grand Tours and Cooks' Tours: A History of Leisure Travel, 1750–1915*. New York: W. Morrow, 1997.

Everest, Mount

At 29,035 feet, Mount Everest is the highest mountain in the world and also one of the top destinations for women adventure travelers seeking the ultimate challenge. The peak lies in the Great Himalayas range on the Nepalese-Tibet border and was named for the British surveyor general of India from 1830 to 1843, Sir George Everest. The first people to reach its peak were two men, New Zealand explorer Sir Edmund Hillary and his Nepalese Sherpa guide Tenzing Norkay, in 1953. Since then, both men and women have climbed the peak. The first woman to summit Everest was Junko Tabei in 1975; the first American woman to reach the peak was Stacy Allison in 1988. Tabei was part of a 15-woman Everest expedition. There have been several all-woman expeditions since hers, the most recent being the May 2000 Nepalese Women's Expedition. During this expedition, which was organized by a commercial company called Asian Trekking, four native Nepalese from an ethnic group known as the Sherpa reached the summit: Lhakpa Sherpa, age 26; Ang Pasang Sherpa, age 32; Ang Phurba Sherpa, age 41; and Ang Mingma Sherpa, age 32. Asian Trekking plans another all-woman expedition in spring 2001, the International Peace and Friendship Women Everest Expedition, which is for mountaineers of all skill levels.

In recent years there has been a sharp increase in the number of inexperienced women

Nepalese Mt. Everest climber and leader of the Neplease women Millenium-2000, Lhakpa Sherpa, 27 talks to the press after her safe return from the mountain base, in Kathmandu, 27 May 2000. © *AFP/CORBIS.*

mountaineers attempting to climb Everest. The only barrier to participation is money; a Mount Everest expedition is one of the most expensive vacations a person can choose. Nonetheless, in spring 1993 there were 15 expeditions, with a total of 294 people attempting to climb Everest from the Nepal side of the mountain, at a cost of $2,300 per team. The following year, Nepalese's government limited the number of expeditions to four at a time. Concurrently, Nepal raised its fee for a climbing permit from $10,000 for a nine-member team (with $1,000 additional for each extra climber) to $50,000 for a five-member team (with $10,000 for each additional climber). On the Tibet side of the mountain, the Chinese government set the fee at $15,000 for a team of any size.

Such amounts—added to the cost of traveling to Mount Everest and outfitting an expedition—have made it difficult for skilled women to climb the mountain unless they are wealthy or can find wealthy sponsors. As Jon Krakauer reports in his book *Into Thin Air,* this has caused many people to be "offended that the world's highest summit was being sold to rich parvenus—some of whom, if denied the service of guides, would probably have difficulty making it to the top of a peak as modest as Mount Ranier." (p. 23) The lower skill levels of recent climbers has also caused several deaths. On May 10, 1996, several people died in an unexpected storm while trying to descend the peak; they included Japanese mountaineer Yasuko Nanba, at 47 the oldest woman to have reached the summit of Mount Everest. Krakauer argues that many of the men and women on the mountain that day did not have the qualifications to be attempting the climb. *See also* ADVENTURE TRAVEL; ALLISON, STACEY; CHALLENGES FOR THE MODERN TRAVELER; MOUNTAINEERING; TABEI, JUNKO.

Further Reading: Da Silva, Rachel. *Leading Out: Women Climbers Reaching for the Top*. Seattle, WA: Seal Press, 1992; Krakauer, Jon. *Into Thin Air*. NY: Villard, 1997; Miller, Luree. *On Top of the World*. Seattle, WA: Mountaineers Books, 1985; Williams, Cicely. *Women on the Rope; The Feminine Share in Mountain Adventure*. London: Allen & Unwin, 1973.

Explorers

Explorers travel for discovery, generally to visit geographical regions never before seen by people of their own culture. Because of behavioral restrictions place upon women throughout history, the most famous explorers have been male. However, a few women have managed to be the first to reach certain regions of the world.

Africa

Of these, one of the best known is English explorer Mary Kingsley. On two expeditions in the 1890s, she traveled throughout sections of West Africa previously unknown to the Western world, studying indigenous cultures, collecting specimens of plants and fish (particularly freshwater ones) for the Royal Botanical Gardens, and making a pioneering trek through the Gabon Mountains to ascend Mount Cameroon. She later wrote about her experiences, *Travels in West Africa* (1897).

Explorer Florence Baker also visited Africa, discovering Lake Albert, the source of the Nile River in Africa, with her explorer husband Sir Samuel White Baker in 1864. This same year, Dutch explorer Alexine Tinne returned from an African expedition during which she had attempted to discover the source of the Nile River. Along the way she became the first white woman to see many parts of the African interior. In 1869 Tinne set out on another expedition, attempting to be the first European woman to cross the Sahara Desert, and was killed by a hostile desert tribe.

Another notable African explorer was American May (also known as Mary) French Sheldon, one of the first women to lead her own African safari. In 1891 she traveled with more than 100 porters and servants into the interior of East Africa from the coastal city of Mombasa, Kenya, studying indigenous tribes. As part of this trek, she became one of the first white people to visit Lake Chala, which rests inside a volcanic cone, and one of the few to travel partway up the slope of Mount Kilimanjaro. When she returned to England, she wrote a book about her travels, *Sultan to Sultan: Adventures among the Masai and Other Tribes of East Africa* (1892).

A later African explorer was American hunter Delia Denning Akeley, the first Western woman to cross Africa. In 1924 she led an expedition from the Indian Ocean to the Atlantic Ocean, venturing through the African interior to study Pygmy tribes and collect animal specimens. She lived among the Pygmies for several months, the first white woman to do so, and revisited them on another expedition in 1929.

American explorer and filmmaker Osa Johnson was the first woman to take a movie camera to Africa, traveling throughout the continent as well as in the South Pacific with her husband, fellow photographer Martin Johnson. They visited several South Pacific islands specifically looking for cannibals, making a movie from their footage, *Among the Cannibal Isles of the South Seas* (1917), as well as a book, *Cannibal-Land* (1922). The following year they embarked on a major African expedition at the urging of African explorer Carl Akeley, the husband of explorer Delia Akeley. That expedition kept Johnson and her husband in Africa for four years photographing big game.

Arabia

The Middle East also attracted women explorers. Of these, one of the most famous was British explorer and archaeologist Gertrude Bell. In 1899, Bell embarked on an expedition to study various cities and archaeological sites in the region. She wrote several books based on her experiences, including *The Desert and the Sown* (1907), which concerns Syria, *Amurath to Amurath* (1911), which describes Asia Minor, *Palace and Mosque at Ukhaidir* (1914), which describes Asia Minor's ancient ruins, and *The Thousand and One Churches* (1909, with archaeologist William Ramsay), which is about Persia (which was officially renamed Iran in 1935).

Yet another explorer in the Middle East was Lady Anne Blunt, the first Western woman and Englishwoman to visit some parts of Arabia. Between 1877 and 1880 she traveled throughout the country with her husband Wilfrid Scawen Blunt, a retired diplomat, and became close to the Bedouins who lived there. She wrote two books about her experiences, *Bedouin Tribes of the Euphrates* (1879) and *A Pilgrimage to Nejd, The Cradle of the Arab Race. A Visit to the Court of the Arab Emir, and 'Our Persian Campaign'* (1881).

A more exotic traveler to the region was Lady Hester Stanhope, an Englishwoman who became the first white woman to enter the holy city of Palmyra in western Syria in 1813. She dressed in the clothing of Turkish men and claimed that the indigenous people at Palmyra had proclaimed her "Queen of the Desert." Two other explorers who were the first white women to enter holy places in the Middle East were Englishwomen Rosita Forbes and Sarah Hobson. Forbes went into the sacred city of Kufara in the Sahara Desert in present-day Libya in 1920 and Hobson went into an ancient shrine, Qum, in Iran, in 1970 disguised as a man.

The East

Western women also explored Asia. From 1848 to 1853, Englishwoman Lucy Atkinson traveled through Russia with her husband, an artist who wanted to sketch the region. She lived among the nomadic tribes of Russia's mountain regions for a short time after giving birth to a son and also canoed around the Altin Kool, or Golden Lake, and down the Biya and Katun Rivers. In the 1870s, Elizabeth "Nina" Mazuchelli was the first European woman to explore the eastern Himalayan mountains and was accompanied by her husband and 80 servants. Throughout the 1920s, on four expeditions, Dutch mountaineer Jenny Visser-Hooft and her husband explored, mapped, and researched the flora and fauna of the Karakorum Glaciers in Pakistan and India. Visser-Hooft also climbed other mountains in Central Asia, as well as in the Alps and the Caucasus. Swiss explorer Ella Maillart also traveled extensively through Asia, visiting places in China and Russia that no other Europeans had seen before. During one trip from Peking into India in 1935, she and English journalist Peter Fleming traveled 3,500 miles through a desert and over mountains in less than seven months, largely on foot or horseback.

But perhaps the most famous of the Western women to travel to Asia were nineteenth-century British missionaries attempting to bring Christianity to Tibet, a country then closed to foreigners. British missionary Annie Royle Taylor was the first European woman to visit Tibet, entering the country in 1892 in a failed attempt to reach the holy city of Lhasa. Frenchwoman Alexandra David-Neel was the first European woman to enter Lhasa; she did so in 1924, after a year's journey. Meanwhile, English missionary Mildred Cable and Algerians Evangeline and Francesca French traveled throughout China's Gobi Desert to preach the gospel from 1916 to 1931.

The Poles

Whereas Tibet attracted women explorers with strong religious beliefs, the North and South Poles attracted those who wanted a physical challenge. One Arctic traveler was Canadian Agnes Dean Cameron, who, in 1908, was one of the first women to travel through the region; she wrote a book about her experiences, *The New North: An Account of a Woman's Journey Through Canada to the Arctic* (1909). In 1955, American explorer Louise Arner Boyd became the first woman to fly over the North Pole. During the 1920s and 1930s, she also made several land expeditions to the Arctic, collecting detailed information about regions that had never been studied. However, Boyd was accompanied by many scientists in her explorations. In contrast, American explorer Helen Thayer was the first woman to ski to the magnetic North Pole alone in 1988. She spent 27 days traveling 345 miles across a polar ice cap to reach the site. American explorer Shirley Metz was the first woman to ski to the South Pole—a feat she accomplished in 1988. In 1993, American explorer Ann Bancroft led the first all-woman skiing expedition to the South Pole, during which expedition members participated in scientific tests that examined the effects of a polar environment on the human body and mind. Bancroft has also participated in other polar expeditions.

Other Regions

Other women explorers associated with specific regions of the world include Frances Hornby Barkley, Jeanne Carr, Mena Benson Hubbard, Beatrice Grimshaw, and Lilian Brown. Englishwoman Frances Barkley accompanied her fur-trader husband during her eight-year journeys beginning in 1786 to western Canada and was the first white woman to see some of its regions; Barkley Sound, Frances Island, and Hornby Peak were named for her. American botanist Jeanne Carr explored unseen regions of what is now California's Yosemite National Park during the mid-1800s. Canadian Mena Benson Hubbard was the first white woman to cross Labrador; during her 1905 43-day expedition she canoed along the Nascaupee and George Rivers to map their course and recorded information on flora and fauna via scientific instruments and photographs. Irish journalist Beatrice Grimshaw was the first white woman to travel down the Sepik River in New Guinea, exploring the South Seas between 1906 and 1936. Englishwoman Lilian Brown accompanied anthropologist Frederick Mitchell-Hedges on his expeditions to Central America from 1921 to 1931, studying indigenous tribes and collecting material for the British Museum.

Some women explorers, however, were not associated with one particular region, but instead traveled throughout the world. One of the most famous of these was Englishwoman Isabella Bird Bishop, although she was not technically an explorer because she only visited places that had already been discovered by other Europeans. However, she discovered new aspects of foreign cultures and wrote about them in numerous books, which earned her recognition from geographic societies during the nineteenth century. Another world explorer during this period was Lady Isabel Burton, the wife of British explorer and diplomat Sir Richard Burton. She accompanied her husband on some of his travels to Africa, South America, Syria, India, and Egypt and was particularly drawn to the people of the Middle East.

American Fanny Bullock Workman also traveled the world with her husband. The two took several bicycle tours and were also prominent mountaineers. During several mountaineering expeditions from 1898 to 1906, Fanny Workman became the first woman to explore a number of peaks and glaciers. Her ascent of Mount Koser Gunge in the Karakoram mountain range in 1899 gave her the record for the highest climb by a woman at that time, as did her ascent of Nun Kun in 1906. She and her husband also provided the first maps to some of the peaks they reached. In 1912 they embarked on a major expedition to the Srachen Glacier that was sponsored by the Royal Geographical Society.

Geographic Societies

These societies were important in the history of exploration, and one such society was founded by women for women. In 1925, American explorer Harriet Chalmers Adams, who is primarily known for her travels through Spain and Portugal, helped found the Society of Women Geographers and served as the organization's first president until 1933. The society encourages all forms of geographical exploration and research, interpreting "geographer" to include such allied disciplines as anthropology, archaeology, biology, ecology, geology, and oceanography. Its members have included hunter American Delia Denning Akeley, round-the-world traveler American Blair Niles, American mountaineer Annie Smith Peck, American anthropologist Margaret Mead, and British archaeologist Mary Leakey.

One of the most recent women explorers supported by the National Geographic Society is American undersea explorer Sylvia Earle, who is preparing to launch a five-year series of expeditions designed to explore the deepest parts of 12 U.S. National Marine Sanctuaries. As leader of the National Geographic Society's Sustainable Seas Expeditions (SSE), she will pilot a one-person submarine into each sanctuary and broadcast information about her discoveries on the

Internet. *See also* Adams, Harriet Chalmers; Africa; Akeley, Delia Denning; anthropologists and archaeologists; Asia; Atkinson, Lucy; Baker, Florence; Barkley, Frances; Bell, Gertrude; Bishop, Isabella Lucy Bird; Blunt, Anne; Boyd, Louise Arner; Burton, Isabel; Cable, Mildred; Cameron, Agnes Dean; David-Neel, Alexandra; Earle, Sylvia; French, Evangeline and Francesca; Grimshaw, (Ethel) Beatrice; Hubbard, Mena Benson; Johnson, Osa; Kingsley, Mary; Leakey, Mary; Maillart, Ella; Mazuchelli, Elizabeth Sarah; Mead, Margaret; missionaries; Niles, Blair; organizations and associations; Peck, Annie Smith; Poles, North and South; Sheldon, May French; Stanhope, Hester; Taylor, Annie Royle; Thayer, Helen; Tinne, "Alexine" (Alexandrine); Visser-Hooft, Jenny; Workman, Fanny Bullock.

Further Reading: Adams, W. H. Davenport. *Celebrated Women Travellers of the Nineteenth Century*. London: W. S. Sonnenschein, 1883; Birkett, Dea. *Spinsters Abroad: Victorian Lady Explorers*. Oxford: Basil Blackwell, 1989; Foster, Shirley. *Across New Worlds: Nineteenth-Century Women Explorers and Their Writings*. New York: Harvester Wheatsheaf, 1990; McCarry, Charles, ed. *From the Field: The Best of National Geographic*. Washington, DC: National Geographic Society, 1997; Middleton, Dorothy. *Victorian Lady Travellers*. London: Routledge & Kegan Paul, 1965; Miller, Luree. *On Top of the World: Five Women Explorers in Tibet*. London: Paddington Press, 1976; Tinling, Marion. *Women Into the Unknown: A Sourcebook on Women Travelers*. Westport, CT: Greenwood Press, 1989.

F

Family Travel. *See* CHILDREN, TRAVELING WITH; SPOUSES.

Farnham, Eliza W. (1815–1864)

Born in 1815, American author and social reformer Eliza W. Farnham traveled from New York to Illinois at the age of 20, accompanied by her brother Henry. Her book *Life in Prairie Land*, which was published in 1846, describes her journey and subsequent experiences in what was then an uncivilized land where her sister Mary had settled five years earlier.

Farnham's book was popular among city dwellers to the east, to whom a primitive life was unfamiliar. However, in a preface the author warns her readers:

> It must not be forgotten . . . that a large class of minds have no adaptation to the conditions of life in the West. This is more especially true of my own sex. Very many ladies are so unfortunate as to have had their minds thoroughly distorted from all true and natural modes of action by an artificial and pernicious course of education, or the influence of a false social position. They cannot endure the sudden and complete transition which is forced upon them by emigration to the West. Hence a class may always be found who dislike the country; who see and feel only its disadvantages; who endure the self-denial it imposes without enjoying any of the freedom it confers; who suffer the loss of artificial luxuries, but never appreciate what is offered in exchange for them. Persons so constituted ought never to entertain for a moment the project of emigration. They destroy their own happiness, and materially diminish that of others. Their discontent and pining are tolerated with much impatience, because those who do not sympathize with them, see so much to enjoy and so little to endure, that their griefs command little or no respect. (pp. iv-v)

However, in describing life on the prairie, Farnham does make her situation sound superior to those of city dwellers. For example, she quotes Mary as saying:

> Social and physical freedom exist here . . . in their most enlarged forms. In less favored portions of the earth, man is more or less enslaved. Want, custom, artificial desires, or some of the thousand phantoms which tread upon the heels of human enjoyment, restrain his freedom. They limit his action, give complexion to his feeling, oppress his thought, cut off his communion with the primal sources of truth. His necessities have each an individual voice, and call loudly for effort. He may not rest till this is made. Here, it is to a great extent otherwise. Our genial climate and exhaustless soil afford an abundant and ready return for his labors. He is soon released from want, and his faculties, rebounding from their depressed condition, go leisurely forth in quest of happiness. There is just enough of ease in his outward circumstances to excite, instead of enervating the tone of his

energies; and with enlarged capacities for enjoyment, he finds himself surrounded by the most propitious array of facts and objects for promoting it. Nature in her loveliest and benignest aspect is spread before him. She invites him to her acquaintance; and while he courts it, the jarring selfishness in which his life has been spent softens into greater harmony with the good, the true, and the beautiful in creation. He becomes a better, wiser, happier man. . . . [He has found] freedom from want, purchased with a moderate use of his physical powers; freedom from social trammels; freedom from the struggles of an emulation founded in vanity, or other vitiated desires; from the myriad forms of ruinous and slavish excess which feeling takes in more populous regions. . . . (pp. 89–90)

Farnham extols the virtues of nature throughout her work, often to excess. At one point, after a long passage on nature's beauty, she echoes Mary's comments by saying: "My worship [of nature] is kindled. . .into far more intense life than by the displays of human power. Living much with nature, makes me wiser, better, purer, and therefore happier!" (p. 209) Consequently her work is not only an example of travel literature but also of environmental literature, which promotes a closer relationship between humans and the environment.

In her later years, Farnham wrote novels and essays. Married to a travel writer, she had the opportunity to travel throughout the American West. She also worked as a prison matron and became involved in the plight of lower-class women in California Gold Rush towns. Farnham died in 1864. *See also* ECOTOURISM; TRAVEL WRITERS.

Further Reading: Farnham, Eliza W. *Life in Prairie Land*. New York: Harper & Brothers, 1846.

Fearn, Anne Walter (1865–1939)

Born in 1865, Anne Walter Fearn is representative of American missionaries who brought medical knowledge to other countries. Fearn grew up in Mississippi as the daughter of a wealthy attorney and attended medical school in Philadelphia, earning her medical degree in 1893. She then left for China as part of a missionary effort run by the Methodist Episcopal Church, South. In 1896, after marrying a fellow missionary, Fearn began her medical practice in China. She lived most of her life in the city of Shanghai, where she established the Fearn Sanitorium. However, she also traveled to the United States to call attention to the problems of China. In addition, she wrote an autobiography about her experiences in China, *My Days of Strength* (1939). Fearn continued to live in Shanghai after her husband's death in 1926. She died in 1939. *See also* MISSIONARIES.

Further Reading: Mueller, John Theodore. *Great Missionaries to China*. Freeport, NY: Books for Libraries Press, 1972.

Ferber, Edna (1885–1968)

Born in 1885, American novelist Edna Ferber helped popularize the idea of women as business travelers through a series of stories about a traveling saleswoman, Emma McChesney, who crossed the country peddling underskirts. Ferber wrote 30 of these stories,

Edna Ferber. *Library of Congress.*

which were published in national magazines in the early twentieth century, before abandoning the series. She then went on to write plays and novels, including *Showboat* (1926), *Cimarron* (1929), *Giant* (1952), and *Ice Palace* (1958). She also wrote two autobiographies, *A Peculiar Treasure* (1939) and *A Kind of Magic* (1963). She died in 1968.

Further Reading: Gilbert, Julie. *Ferber: A Biography of Edna Ferber and Her Circle*. New York: Applause Theatre Books, 1998; Shaughnessy, Mary Rose. *Women and Success in American Society in the Works of Edna Ferber*. New York: Gordon Press, 1977.

Field, Kate (1838–1896)

Born in 1838, American journalist Kate Field provided one of the few firsthand accounts of the political situation in Spain during the Carlist Wars (1873–1876). Her work was published in 1875 as *Ten Days in Spain*. Field was also one of the first American women to travel alone through Spain and to openly establish her own newspaper, *Kate Field's Washington* (D.C.). She began this weekly paper in 1890, by which time she was already an experienced journalist. Her father had been a St. Louis, Missouri, newspaper publisher; after his death in 1856 and a few years as the ward of her uncle, Field decided to support herself by writing articles and acting in plays. Some of these articles appeared in *Atlantic Monthly*. She also wrote a book ridiculing American tourists abroad, *Hap-Hazard* (1873). Field traveled extensively throughout the world, both for pleasure and as a lecturer but did not view herself as being an ordinary tourist. She died in 1896.

Further Reading: Field, Kate. *Kate Field: Selected Letters, 1838–1896*. Edited and with an introduction by Carolyn J. Moss. Carbondale: Southern Illinois University Press, 1996; Whiting, Lilian. *Kate Field, a Record*. Boston: Little, Brown, 1899.

Finch, Linda (1951–)

Born in 1951, American aviator Linda Finch made a flight around the globe in 1997, duplicating and completing the voyage begun by Amelia Earhart in 1937. In order to make her re-creation more exact, Finch flew in the same type of plane as Earhart, a 1935 Lockheed Electra 10E that Finch herself painstakingly restored with the help of her sponsor, aircraft company Pratt and Whitney. Finch's plane, however, had some additional features unknown in Earhart's day. It included a satellite terminal so Finch could send and receive e-mails, faxes, digital images, and weather information while in flight, as well as Global Positioning System navigation equipment so that she would not get lost at sea as Earhart apparently had. Finch's plan was also equipped to carry more fuel than Earhart's: 1,800 gallons as opposed to 800. In all, Finch took 10 weeks to go across the Atlantic and Pacific Oceans, traveling in 8- to 12-hour stretches.

By the time she embarked on her journey, Finch was already an experienced aviator, having logged more than 8,000 hours in the air as a pilot. She received her license to fly at age 29 after becoming interested in vintage aircraft. Today she is a well-known restorer of such craft and has flown them in numerous air shows over the past 10 years. *See also* AIR TRAVEL; CIRCUMNAVIGATORS AND ROUND-THE-WORLD TRAVELERS; EARHART, AMELIA.

Further Reading: Allstar Network. Aeronautics Learning Laboratory for Science, Technology, and Research. "Linda Finch." http://www.allstar.fiu.edu/aero/finch.html (July 1999); Lefevre, Greg. "Pilot Attempts Earhart Flight 60 Years Later." CNN News. http://www.cnnsf.com/newsvault/output/earhart.html (May 2000).

Forbes, Rosita (1893–1967)

Born in 1893 in England, Rosita Forbes was the first white woman to enter the sacred city of Kufara in the Sahara (in present-day Libya), a feat she accomplished in 1920. On this and many other expeditions, Forbes recorded her observations not only in writing but also with a camera. She produced numerous photographs and books from her adventures, including *The Secret of the Sahara: Kufara* (1921), *From Red Sea to Blue Nile: Abyssinian Adventures* (1925), *India of the Princes* (1939), and *A Unicorn in the Bahamas* (1939). She also wrote books about the famous political leaders she met and interviewed on her travels through the Middle East, Asia, and South America, including *These Are Real People* (1937) and *These Men I Knew* (1940). Forbes soon became famous herself and was active in London's Royal Geographical Society. In 1944 she published an autobiographical work entitled *Gypsy in the Sun,* which recounts her travels of 1920 to 1934. She died in 1967. *See also* AFRICA; ASIA.

Further Reading: Forbes, Rosita Torr. *Gypsy in the Sun*. New York: E. P. Dutton, 1944.

Forced Relocation. *See* MIGRATION.

Fossey, Dian (1932–1985)

Born in 1932, Dian Fossey is representative of American women who have traveled for the sake of science. To study rare mountain gorillas, she established the Karisoke Research Center in Rwanda, Africa; her book *Gorillas in the Mist,* published in 1983, describes her experiences there.

Fossey received a degree in occupational therapy in 1954 and began working with disabled children in the United States. However, in 1963, after reading about the gorilla research of American zoologist George Schaller, she decided to travel to Africa where she met British anthropologist Louis Leakey. In 1966 Leakey hired Fossey to undertake her own study of the gorillas. She set up her Karisoke research site the following year.

Awarded a Ph.D. in zoology from Cambridge University in 1974, Fossey provided a great deal of new information about gorilla behavior. However, some of the gorillas in Fossey's group were killed by poachers during her 22-year study. In 1985, Fossey herself was murdered, also probably by poachers. Today an environmental group called the Dian Fossey Gorilla Fund continues to work on gorilla conservation issues. *See also* ADVENTURE TRAVEL; ECO-TOURISM; SCIENTISTS.

Further Reading: Fossey, Dian. *Gorillas in the Mist.* Boston: Houghton Mifflin, 1983; Kevles, Bettyann. *Watching the Wild Apes: The Primate*

Dian Fossey in 1985. © *Yann Arthus-Bertrand/CORBIS.*

Studies of Goodall, Fossey, and Galdikas. New York: Dutton, 1976; Mowat, Farley. *Woman in the Mists: The Story of Dian Fossey and the Mountain Gorillas of Africa.* New York: Warner Books, 1987.

Fountaine, Margaret (1862–1940)

Born in 1862, English entomologist Margaret Elizabeth Fountaine was a butterfly collector who traveled throughout the world in search of specimens. Beginning in 1890, she visited every continent and most major countries, including Syria, Turkey, Algeria, Spain, Africa, the Caribbean, India, New Zealand, Fiji, South America, the United States, and the Far East. She collected over 20,000 butterflies and kept detailed diaries of her adventures, which were published posthumously as *Love Among the Butterflies: The Travels and Adventures of a Victorian Lady* (1982). Her writings describe her romantic escapades with many men, as well as her relationship with the natural world. In describing one of her expeditions, she writes:

> The way was long, the broad, dry riverbed with its burning hot sands often made me foot-sore, and the ascent up the rocky side of the mountain was steep and arduous. . . . And then . . . such a world of flowers and butterflies into which we presently descended on the other side! Tall orange marigolds grew in rank profusion beneath the slender shades of the umbrella pines, while the hot winds would murmur through their branches. . . . And by and by we would descend by another way to which we had come up, hot and thirsty with our day's chase, and longing to reach the spot where we would stop and drink from a mountain spring. (Morris, p. 116)

Fountaine also wrote scholarly articles about butterflies, and in her last years she drew up a will that ordered her diaries remain sealed until 1978. She chose this date because it was one 100 years from the time of her first diary entry in 1878. Fountaine died in 1940. *See also* SCIENTISTS.

Further Reading: Fountaine, Margaret. *Love Among the Butterflies: The Travels and Adventures of a Victorian Lady.* New York: Penguin Books, 1982; Morris, Mary. *Maiden Voyages: Writings of Women Travelers.* New York: Vintage, 1993.

Franklin, Lady Jane (1792–1875)

Born in 1792, Englishwoman Lady Jane Franklin was one of the most notable women travelers of the Arctic. She was an avid traveler from childhood, and after marrying Arctic explorer John Franklin in 1828, she became even more of an adventurer. She helped him plan his expeditions, joined him on them occasionally, and traveled alone when he was away, going to such places as North America, Europe, and the West Indies.

When her husband's ship disappeared in the Arctic in the late 1840s, Franklin arranged a series of rescue missions, but although they lasted from 1850 to 1857 he was never found. However, in the course of these missions, explorers learned many new things about the Arctic. As a result, Franklin was honored by the Royal Geographical Society in 1860. During all of her travels, both before and after her husband's death, Franklin kept diaries of her experiences; excerpts from them were later published in *The Life, Diaries and Correspondence of Jane, Lady Franklin, 1792–1875* (1923). *See also* POLES, NORTH AND SOUTH.

Further Reading: Robinson, Jane. *Wayward Women: A Guide to Women Travellers*. Oxford: Oxford University Press, 1990.

French, Evangeline (1869-1960) and Francesca (1871–1960)

Born in 1869 and 1871, respectively, Algerian missionaries Evangeline and Francesca French are representative of Victorian women who traveled because of their religious beliefs. They worked for the China Inland Mission, spreading Christianity through China, and in 1900 they were almost killed during a massacre of foreign missionaries in that country. In 1902 they joined fellow missionary Mildred Cable in her efforts to establish a Christian school, but eventually the three women decided to go into the Gobi Desert to preach. They traveled throughout the region for over 15 years, spreading Christianity and learning about Buddhism in return. Periodically they would return to civilization for supplies, then set out again.

Their freedom to travel ended, however, in 1931 when a political-religious conflict led to military rule. The new government restricted the women to the city of Tunhwang, but after several months they managed to escape across the desert. They left China shortly thereafter and did not return until 1935, when they could only travel in the company of an official Chinese guide. In their later years the women made two more trips: in 1946 they traveled to Africa, Australia, New Zealand, and Europe, and in 1947 they visited South America. Both Evangeline and Francesca died in 1960. *See also* ASIA; MISSIONARIES.

Further Reading: Tinling, Marion. *Women into the Unknown: A Sourcebook on Women Travelers.* Westport, CT: Greenwood Press, 1989.

From the Field

Published in 1997, *From the Field: The Best of National Geographic* is the only book-length collection of travel articles from *National Geographic* magazine, which since its first issue in 1888, has sponsored both female and male explorers and travelers throughout the world. The women authors in the volume include American travel writer and photographer Eliza Ruhamah Scidmore, American poet Anne Morrow Lindbergh, American aviator Amelia Earhardt, American anthropologists Jane Goodall and Dian Fossey, and travelers Donna K. Grosvenor, Valerie Taylor, and Patricia Jones-Jackson. Scidmore writes about seeing the results of a tsunami in Japan in 1896 and about the visiting the Ganges River and climbing a rock in Sigiri, Sri Lanka, in 1907; Lindbergh contributed an article in 1934 about flying with her husband, famed aviator Charles Lindbergh, around the northern Atlantic Ocean. Earhart discussed how she would handle an emergency while flying over the Pacific Ocean. Goodall contributed articles about her experiences with African chimpanzees. Grosvenor, wrote about a visit to Bali with her husband Gilbert in 1969. Fossey contributed an article about Rwandan mountain gorillas in 1970. Taylor wrote about diving with sharks. Jones-Jackson died in an automobile accident in 1986 while on assignment for *National Geographic* in the Sea Island of Georgia and the Carolinas, where she had gone to study linguistics.

In an introduction to the work, editor Charles McCarry reports that all of the articles in the anthology were actually written by the authors—a practice that he says was uncommon. He explains:

> Until 1990, most adventure stories and many natural-history articles were ghost-written. That is why there are so few of them in this anthology: it is reserved for the real thing. One real writer, William Graves, who retired in 1994 as the seventh editor of the *Geographic*, is the actual author of most first-person tales about expeditions and adventures published in the last quarter of a century. (p. 7)

Alhough women are not represented in this volume in as great a number as the men, their unique perspective on world exploration comes through in their articles. Moreover, in his introduction McCarry

Anne Morrow Lindbergh in 1929. © *Bettmann/CORBIS.*

names Eliza Ruhamah Scidmore as the best author the *Geographic* ever had. *See also* EARHART, AMELIA; FOSSEY, DIAN; GOODALL, JANE; SCIDMORE, ELIZA RUHAMAH.

Further Reading: McCarry, Charles, ed. *From the Field: The Best of National Geographic*. Washington, DC: National Geographic Society, 1997.

Fry, Elizabeth (1780–1845)

Born in 1780, Englishwoman Elizabeth Fry is representative of women who traveled for social reform. A Quaker minister in London, she began to speak out against inhumane prison conditions in 1813. Within a short time she had launched a campaign to improve the lot of women convicts, both in England and in a British penal colony in Australia. She particularly fought to have women convicts housed separately from men to reduce incidents of rape.

In 1817 Fry founded the Prisoners Aid Society and the following year she testified before a Royal Commission on prison conditions. During the 1820s she toured British prisons and established more reform groups; in 1838 King Louis Phillipe of France invited her to tour French prisons. Her visit and subsequent evaluation of French prison conditions led to reforms. She traveled to various European countries and met with other heads of state, encouraging reform efforts throughout the continent. She was also involved in projects to improve women's educational facilities and housing for the poor. Fry also made improvements to nursing schools. She died in 1845. *See also* AUSTRALIA.

Further Reading: Graham, Maureen. *Women of Power and Presence: The Spiritual Formation of Four Quaker Women Ministers*. Wallingford, PA: Pendle Hill Publications, 1990; Rose, June. *Elizabeth Fry*. New York: St. Martin's Press, 1981.

Fuller, Margaret (1810–1850)

Born in 1810, feminist Margaret Fuller was one of few nineteenth-century American women to write about their experiences in continental Europe. As Leo Hamalian points out in his book *Ladies on the Loose: Women Travellers of the 18th and 19th Centuries*: "A host of European women, especially from England, travelled in America and wrote about their travels during the eighteenth and nineteenth centuries. By contrast, few American women who went abroad during the same period apparently thought it was worth recording the impressions of the Old World." (p. 27)

Prior to embarking on her trip to Europe, Fuller was the editor of *The Dial* magazine and wrote critical articles for the *New York Tribune*. In 1846 she and two wealthy friends went to Italy, where she fell in love with a marquis, and remained in the country until 1850. On her way home, her ship sank and everyone on board was drowned. Her *Memoirs,* in two volumes, were published posthumously in 1852. Her other works include *Summer on the Lakes* (1844) and *Women in the Nineteenth Century* (1846).

Further Reading: Hamalian, Leo, ed. *Ladies on the Loose: Women Travellers of the 18th and 19th Centuries*. New York: Dodd, Mead, 1981.

G

Galdikas, Birute (1946–)

Born in Germany in 1946, anthropologist Birute Galdikas is representative of women who travel back and forth between two locations for most of their lives. Galdikas was raised in Canada but traveled to the rain forests of Borneo in the late 1960s to study orangutans and earn her Ph.D. from the University of California at Los Angeles (UCLA). Galdikas's mentor in establishing her career was anthropologist Louis Leakey, who also supported the work of Jane Goodall and Dian Fossey. She has spent over 20 years dividing her time between Canada and the jungles of Borneo, and she is now the world's foremost expert on orangutans, directing the Orangutan Education and Care Center in Borneo. In particular, Galdikas discovered that orangutans are extremely solitary animals unable to work in strong social groups. She argues that this lack of cooperation and teamwork has hindered orangutans' evolution, suggesting that humans have evolved and thrived specifically because of thier family groups. *See also* FOSSEY, DIAN; GOODALL, JANE.

Further Reading: Kevles, Bettyann. *Watching the Wild Apes: The Primate Studies of Goodall, Fossey, and Galdikas*. New York: Dutton, 1976.

Garrod, Dorothy Annie Elizabeth (1892–1969)

Born in 1892, English archaeologist Dorothy Annie Elizabeth Garrod was the first woman to study prehistoric people from the Paleolithic era, or Old Stone Age. She began this research in Gibraltar in 1925, but her excavations of Old Stone Age sites also took her to such places as France, Palestine, Lebanon, Kurdistan, and Bulgaria.

Garrod made many important discoveries during her career. For example, while leading a joint U.S.-British excavation at Mount Carmel in Palestine, she found the first evidence of Old Stone Age and Middle Stone Age culture in the region. She also established that a primitive form of early man and an advanced one existed concurrently. In addition, Garrod was among the first to use aerial photography to document archaeological expeditions. As a result of her work, she received many awards and was the first woman to hold a professorship at Cambridge University in England, where she was a professor of archaeology from 1939 to 1952. Garrod died in 1969. *See also* ANTHROPOLOGISTS AND ARCHAEOLOGISTS.

Further Reading: Garrod, D. A. E. "The Middle Palaeolithic of the Near East and the Problem of Mount Carmel Man." *Journal of the Royal Anthropological Institute*, 92, part 2 (1962).

Gaunt, Mary (1861–1942)

Born in 1861, Australian author Mary Eliza Bakewell Gaunt traveled to Europe, India, Africa, China, Siberia, and the West Indies and wrote several books about her experiences specifically to gather material for her adventure novels. Therefore, according to scholar Jane Robinson, "Mary Gaunt's journeys were all business trips." (p. 181) These novels, which were published during the early 1900s, introduced her readers to new and exotic lands. But Gaunt also wrote travel books based on her experiences which encouraged other adventurous women to follow in her footsteps. Her first such book, *Alone in West Africa* (1912), describes two trips to Africa in 1908 and 1910.

Gaunt was a student at Melbourne University when she decided to become a writer. Her first works were newspaper articles, followed by a romance novel entitled *Dave's Sweetheart* (1894). After the death of her husband in 1900, Gaunt moved to London and began to travel in earnest. In her later years she lived in Italy, but in World War II was forced to relocate to France. She died there in 1942. *See also* AFRICA.

Further Reading: Gaunt, Mary. *Alone in West Africa*. London: W. T. Laurie, 1912; Robinson, Jane. *Wayward Women: A Guide to Women Travellers*. Oxford: Oxford University Press, 1990.

Gear. *See* CHALLENGES FOR THE MODERN TRAVELLER; CLOTHING.

Gellhorn, Martha (1908–1998)

Born in 1908 in Saint Louis, Missouri, Martha Gellhorn was one of the first female war correspondents, traveling throughout the world to report news stories. She got her first assignment in 1937, when she covered the Spanish Civil War for *Collier's Weekly*. While in Spain she met and married author Ernest Hemingway, whom she left after five years.

Gellhorn soon became *Collier's* most widely read author. She also wrote for *The Atlantic Monthly* for over three decades, covering many important stories firsthand. She was in Czechoslovakia in 1939 to report on the Nazi occupation there; she rode with British pilots on bombing raids over Germany during World War II; and she was the only woman to report on the Russian bombing of Helsinki, Finland. One of her last important stories was the 1989 American invasion of Panama. Gellhorn also interviewed many world leaders during her career, and, in addition to her articles, she wrote five novels, 14 novellas, and numerous short stories. She also published a popular book of travel stories, *Travels with Myself and Another* (1979).

Travels with Myself and Another is unusual in that it describes the author's worst travel experiences. In an introduction to the work, Gellhorn says:

> We can't all be Marco Polo or Freya Stark but millions of us are travellers nevertheless. The great travellers, living and dead, are in a class by themselves, unequalled professionals. We are amateurs and though we too have our moments of glory we also tire, our spirits sag, we have our moments of rancor. . . .The fact is, we cherish our disasters and here we are one up on the great travellers who have every impressive qualification for the job but lack jokes. . . . All amateur travellers have experienced horror journeys, long or short, sooner or later, one way or another. As a student of disaster, I note that we react alike to our tribulations: frayed and bitter at the time, proud afterwards. Nothing is better for self-esteem than survival. (pp. 11–12)

Gellhorn had no trouble coming up with her disaster stories. At the time her book was written, she had visited 53 countries and every state in the United States except Alaska, as well as numerous islands throughout the world. She had also visited several countries more than once. After describing some of her worst travel experiences, she notes that such experiences are more common now among the population at large because more people can afford to travel. She says:

> Amateur travel always used to be a pastime for the privileged; now it is a pastime for everyone. Perhaps the greatest social change since the Second World War is the way citizens of the free nations travel as never before in history. We have become a vast floating population and an industry; we are essential to many national economies not that we are therefore treated with loving gratitude, more as if we were gold-bearing locusts. . . . I have seen many people who looked as if they were on their own kind of horror journey. Men with life-

> less eyes carrying parcels for voracious wives; how cheap these leather wallets are in Florence, this pottery in Oaxaca, these cuckoo clocks in Berne. Groups, in museums and palaces, cowed by guides, their shoulders drooping, their feet swollen. Friends and lovers in shrieking quarrels on that dreamed-of visit to a romantic city . . . Weary queues in railway stations, pushing their luggage ahead inch by inch. Couples grey and silent with melancholia in any foreign hotel dining room. Young parents, laden with small children. . . . They were all pleasure-bent but seek and ye shall find does not necessarily apply to travel. Once safe again at home they could forget how awful some, much or most of it had been, bring out their souvenirs, their photos, their edited memories, and plan another holiday. (pp. 282–283)

Gellhorn's bleak picture of tourism is not unknown in travel literature of the period. However, there are few books that focus entirely on bad travel experiences.

Further Reading: Gellhorn, Martha. *Travels with Myself and Another*. New York: Dodd, Mead, 1979.

Gifts of the Wild

Published in 1998, *Gifts of the Wild: A Woman's Book of Adventure* is a popular collection of contemporary women's travel essays compiled by the editors of Adventura Books: Faith Conlon, Ingrid Emerick, and Jennie Goode. The essays are organized into 12 sections according to the type of travel experience, or "gift"; the gifts are ones of serenity, joy, courage, wild creatures, independence, friendship, companionship, mothering, body, solace, wild places, and age. Most of these essays focus on the authors' emotions; for example, "Gift of Age" has essays by women in their late seventies describing their own adventure travel experiences. A few essays, however, provide information about travelers other than the author. For example, in "Gift of the Body" Linda Lewis's essay "Etes-Vous Prêts?" describes the career of Ernestine Bayer, who helped established women's competitive rowing as a sport in America. *Gifts of the Wild* also offers biographical information about the author of each essay.

Further Reading: Conlon, Faith, Ingrid Emerick, and Jennie Goode, eds. *Gifts of the Wild: A Woman's Book of Adventure.* Seattle, WA: Seal Press, 1998.

Gill, Isobel (1845–?)

Lady Isobel Gill is representative of nineteenth-century women who accompanied their husbands on scientific missions that involved research but not geographical exploration. The wife of Scottish astronomer David Gill, she traveled with him on a 1877 voyage to Ascension Island, in the South Atlantic Ocean approximately midway between Africa and South America, to find a better view of the stars and planets. The Gills spent six months on the volcanic island, and Isobel Gill subsequently wrote a detailed account of their experiences, *Six Months in Ascension: An Unscientific Account of a Scientific Expedition* (1878). She was the first woman to write about the island. *See also* SPOUSES.

Further Reading: Gill, Isobel. *Six Months in Ascension: An Unscientific Account of a Scientific Expedition.* London: J. Murray, 1878.

Girl Scouts of the U.S.A.

Created in 1912 in Savannah, Georgia, by outdoorswoman Juliette Low, the Girl Scouts of America is an organization dedicated to encouraging girls to participate in outdoor activities. As such it offers girls unique opportunities to travel and camp. The organization is affiliated with the Girl Guides, which was founded in England in 1910. Beginning in 1963, members were organized into troops according to girls' ages.

The first Girl Scout troop had only 18 members who primarily spent their time hiking and camping. In 1913 the organization provided them with the first official Girl Scouts handbook entitled *How Girls Can Help Their Country*, which featured instructions on knot-tying, first aid, and outdoor cooking. In World War I, Girl Scouts became active in volunteer work, helping at local hospitals and selling war bonds. By 1926, the scouts had 137,000 members. The Girl

Scouts had its national headquarters in Washington, D.C., and a national training center for Girl Scout leaders, Camp Edith Macy, in upstate New York. The group was also extremely diversified for its time, having troops for disabled girls, Native American girls, and African American girls. In 1934, it began financing activities by selling cookies, a practice the organization continues today.

The current membership of the Girl Scouts is approximately 3.5 million, with more than 2,670,000 girls in five program levels: Senior Girl Scouts, ages 14–17 or grades 9–12; Cadette Girl Scouts, ages 11–14 or grades 6–9; Junior Girl Scouts, ages 8–11 or grades 3–6; Brownie Girl Scouts, ages 6–8 or grades 1–3; and Daisy Girl Scouts, ages 5–6 or grades kindergarten and 1. The remaining members are adults, both women and men, who act as volunteer leaders, board members, and specialists in various aspects of child development, adult education, administration, and wilderness experiences. Girls Scouts make more than 223,000 trips both in the United States and abroad; members of overseas troops are girls from American military families and their civilian support personnel. The national headquarters of the Girl Scouts is now located at 420 Fifth Avenue, New York, NY 10018-2798, and the organization maintains an Internet Web site at http://www.gsusa.org. *See also* LOW, JULIETTE.

Further Reading: Girl Scouts of the USA. *75 Years of Girl Scouting*. New York: Girl Scouts of the USA, 1986.

Gold Rush

The term "Gold Rush," when referring to events in the United States, is generally used to refer to an influx of people into California a year after the discovery of gold there in 1848. This event is highly significant in terms of women's travel, because it caused a wave of migration that brought many women from the eastern United States to the West. The first wave of California immigrants was composed mostly of men who hoped to strike it rich, either by mining or by selling goods and services to miners, but it wasn't long before women followed. According to Elizabeth Margo in *Women at the Gold Rush,*

> Within a few months [of the gold strike] the women came streaming in. The great majority of these first women were adventuresses, female Argonauts [gold miners], having many of the characteristics of their male counterparts: seeking excitement, craving easy money, wanting to cut themselves a big slice of the golden pie. Some of them left by request of their governments—France and Australia, particularly. Some of them indentured their services as a means of earning passage money and woke up to find that service was to be given in the dives and fandango cellars [i.e., bordellos] of San Francisco. By 1852, women were no longer notably scarce except in the higher mountain camps. But respectable women, though they were arriving in increasing numbers, were still a scarcity item. (pp. 8-9). One of the few responsible women was Louise Clappe. A collection of Clappe's letters, published in 1933 as *The Shirley Letters*, provides insights into how such women dealt with their travel experiences.

By 1854 enough wives had migrated to the gold fields to have an impact on San Francisco schools; that year there were over 2,500 children enrolled—or 70 percent of all children in the city over the age of four. But an even greater impact was made by prostitutes. In 1850, approximately 2,000 unmarried women entered California, and most of them made their living through prostitution. In the two years following the discovery of gold, the majority of prostitutes migrating to California came from Mexico, France, and Peru. The Mexicans were typically white slaves sold into prostitution by the ship captains who brought them north. Theoretically they could earn their freedom by working to pay off their cost to bordello owners, but few did. Many French prostitutes, however, had freedom from the outset. Rarely were they slaves; some worked in partnership with French gamblers who protected them from abuse by clients, while others worked "freelance." The French government provided these women—and their protectors, if they had them—with free ship's passage to California in order to lessen the number of prostitutes in France. South American prostitutes had to arrange their own passage; like the French they were often free agents and in addition to prostitution they performed as dancers and entertainers. The most beautiful of these women usually chose to become the mistresses of powerful men. Some Frenchwomen did this as well. One example was Fanny Bendixen. She arrived in the mining region in 1850, and when she discovered that finding gold was difficult she settled in San Francisco, where she became the mistress of a wealthy criminal. When she left him to marry another man, he tried to kill her fiancé. Nonetheless, the wedding proceeded, and the couple moved to Victoria, British Columbia, to establish a hotel. In 1866 they divorced and Bendixen went to Barkerville, British Columbia, which was having its own gold rush. There she ran a saloon for a time before moving from town to town to establish saloons, boarding houses, and brothels in order to capitalize on the Canadian gold rush profits. When the rush ended, she settled in Barkerville and ran a saloon until her death in 1899.

As the Gold Rush wore on, Chinese prostitutes also began to arrive in California. Like Mexican prostitutes, these women were slaves who had to earn their freedom. One of the lucky ones was Polly Bemis. Sent to a California mining town as a young girl to work in a brothel, she was either given her freedom or earned it by making and selling jewelry. In either case, once she was free, Bemis ran a boarding house, married, and in 1894 settled in Idaho.

In the hierarchy of prostitution, the Chinese were at the bottom; above them were the American and English prostitutes, then the South American and Mexican prostitutes, and at the top the very expensive French prostitutes. Many of the latter lived quite extravagantly. In fact, miners sometimes referred to French prostitutes as California's aristocracy. But even many lesser prostitutes were treated well. The reason for this, according to Elizabeth Margo, had to do with the ratio of women to men during the Gold Rush. She reports:

> At least one women in five, in 1850, must have been viable merchandise [i.e., desirable as a prostitute, mistress, or wife]. But while the overall proportion of women and children to men is usually stated at one to seven, the proportion of salable females to wifeless forty-niners could not have been much more than one to 25. This situation still made of women a premium item: to be granted extreme chivalry if she were another Argonaut's wife, and to be battled over if she were frankly on the market. (p. 79)

Women who were neither wives nor prostitutes nor mistresses—and who had no interest in mining gold—were in the minority. Of these women, some had run away from bad marriages. Others were widowed and came to California for a fresh start. An independent woman in California typically earned her living as a laundress, cook, or housekeeper if she were not inclined towards prostitution. As more families became established in the region, there was also work for teachers and governesses. As Margo reports, the encroachment of civilization "increased the influence of Victorian womanhood" during the 1850s, bringing with it a decline in prostitution and female entrepreneurship and an increase in wives who adhered to a rigid social and moral code. (p. 88). *See also* CLAPPE LOUISE

Further Reading: Harding, Les. *The Journeys of Remarkable Women: Their Travels on the Canadian Frontier.* Waterloo, ON: Escart Press, 1994; Levy, Jo Ann. *They Saw the Elephant: Women in the California Gold Rush.* Norman: University of Oklahoma Press, 1992; Margo, Elizabeth. *Women of the Gold Rush.* 1955. Reprint. New York: Indian Head Books, 1995. Mayer, Melanie J. *Klondike Women: True Tales of the 1897–98 Gold Rush.* Athens: Swallow Press/Ohio University Press, 1989; Stefoff, Rebecca. *Women Pioneers.* New York: Facts On File, 1995.

Goldman, Hetty (1881–1972)

Born in 1881, American archaeologist Hetty Goldman traveled more extensively than many in her field. She was also the first woman to head archaeological excavations sanctioned by government officials. Her first excavation was in Halae, Greece, in 1911, while she was studying at the American School of Classical Studies in Athens, Greece. This was followed by excavations in Ionia, Asia Minor, in 1921, in Yugoslavia in 1932, and in Cilicia Tarsus, Turkey, in 1934 and 1939. (The interruption was caused, in part, by the instability of the region in the years preceding World War II.) Goldman received her Ph.D. in archaeology from Radcliffe College (based on her excavation work in Halae) in 1916, after receiving her master's degree there in 1911. During World War I she worked as a Red Cross volunteer in New York, and during World War II she was involved in efforts to help German Jews enter the United States to escape the Nazis. In the 1950s and 1960s, she wrote extensively about her archaeological expeditions in books and articles. Goldman also received several prestigious awards for her contributions to her field. She died in 1972. *See also* ANTHROPOLOGISTS AND ARCHAEOLOGISTS.

Further Reading: Goldman, Hetty. *Excavations at Eutresis in Boeotia*. Cambridge, MA: Harvard University Press, 1931.

Goodall, Jane (1934–)

Born in 1934, English anthropologist Jane Goodall is one of the best-known women to travel for the sake of science. As a young woman she went to Kenya in Africa to work as a secretary; in 1957 she became assistant secretary to anthropologist Louis B. Leakey, the curator of the National Museum in the city of Nairobi. Goodall accompanied Leakey and his wife Mary on their archaeological expeditions, and eventually he asked her to undertake a study of chimpanzees in the wild.

Goodall began her research in 1960, establishing the Gombe Stream Chimpanzee Reserve beside Lake Tanganyika in Tanzania with the help of funding from the Wilkie Foundation and the National Geographic Society. At the start, she was accompanied in Tanzania by her mother, Vanne Goodall, because the African authorities would not allow a young unmarried woman to live without a female companion; however, five months later they permitted Vanne to return to England. In 1961 Goodall also went to London, where Leakey had arranged for her to attend Cambridge University to take up ethology, the study of animal behavior. She returned to Gombe each summer.

Goodall received her Ph.D. in 1965. That same year she married wildlife photographer Hugo van Lawick. Goodall had met him four years earlier, when he was filming her chimpanzees for a National Geographic documentary. Goodall's book *In the Shadow of Man* (1971), which documents the first 10 years of her research at Gombe, includes some of van Lawick's photographs.

After van Lawick and Goodall had a child, they restricted their time at Gombe because they considered the wilderness too dangerous for their infant. Now, however, Goodall is again active in chimpanzee research and conservation, and her study program at Gombe continues to attract students from all over the world. In 1977 she founded the Jane Goodall Institute, which is dedicated to the preservation of wild chimpanzees. She also continues to lecture and write about her experiences in Africa. Her works include *The Chimpanzees of Gombe: Patterns of Behavior* (1986) and *Through a Window: My Thirty Years with the Chimpanzees of Gombe* (1990). *See also* AFRICA.

Further Reading: Coerr, Eleanor. *Jane Goodall*. New York: Putnam, 1976; Goodall, Jane. *Africa in My Blood: An Autobiography in Letters: The Early Years, 1934–1966*. Edited by Dale Peterson. Boston: Houghton Mifflin, 2000; ———. *The Chimpanzees of Gombe: Patterns of Behavior.* Boston: Houghton Mifflin, 1986. ———. *In the Shadow of Man*. Boston: Houghton Mifflin, 1988; ———. *Through a Window: My Thirty Years With The Chimpanzees Of Gombe.* Boston: Houghton Mifflin, 1990. Kevles, Bettyann. *Watching the Wild Apes: The Primate Studies of Goodall, Fossey, and Galdikas*. New York: Dutton, 1976.

Governesses

A governess is a woman employed to teach a family's children in their home; her lessons generally address behavior as well as academics. Families who hire governesses are usually wealthy and many of them travel abroad, taking their governess along to watch the children. Some governesses, however, have sought more than a vacation experience as part of their job. Interested in learning about foreign cultures in depth, they have traveled abroad to teach children living in other countries.

Of these women, the most famous is Anna Leonowens, an Englishwoman who became the governess for the royal family in Siam (present-day Thailand) in 1862. She is the subject of a 1945 book, *Anna and the King of Siam* by Margaret Landon, on which a musical and several movies were based. Leonowens taught over 60 royal children and their mothers, educating them in Western knowledge, manners, and customs, and worked as the king's secretary, perhaps influencing some of his decisions. She remained in her position until 1867, when she retired to the United States because of poor health. Three years later, she published a book about her experiences, *The English Governess at the Siamese Court: Recollections of Six Years in the Royal Palace at Bangkok* (1870).

The Siamese king was not the only foreign leader to want an English governess for his children. During the Victorian era many monarchs and noblemen believed that it would be beneficial for their offspring to be educated in Western ways. However, not every governess had as positive an experience as Leonowens. For example, when British governess Emmeline Lott traveled to Egypt to teach the five-year-old heir to the throne in 1863, she found her charge difficult and King Ismael Pacha lewd. She remained in her position for two years but complained about it in such books as *The English Governess in Egypt: Harem Life in Egypt and Constantinople* (1866; 2 volumes) and *Nights in the Harem: Or, the Mohaddetyn in the Palace of Ghezire* (1867; 2 volumes).

Some Victorian women had an even more difficult time in the role of governess. Emigration societies in Great Britain convinced many educated women to travel abroad in order to work as governesses—not for royalty but for colonists—in such places as Australia and Canada. These women often found themselves cooking and cleaning as well as teaching and were extremely unhappy with their lot. Moreover, because these emigration societies were successful promoters, in some places there were more governesses than there were positions for them. Yet without jobs, these women could not afford to pay for passage back to England. *See also* CHILDREN, TRAVELING WITH; EMIGRATION SOCIETIES (BRITISH); LEONOWENS, ANNA HARRIETTE; LOTT, EMMELINE.

Further Reading: Broughton, Trev, and Ruth Symes. *The Governess: An Anthology*. New York: St. Martin's Press, 1997; Leonowens, Anna Harriette. *The English Governess at the Siamese Court: Recollections of Six Years in the Royal Palace at Bangkok*. 1870. Reprint, London: Folio Society, 1980.

Graham, Maria (1785–1842)

Born in 1785, Maria Graham was a popular English author who wrote several travel books, including *Journal of a Residence in India* (1812) and *Journal of a Voyage to Brazil and Residence There During Part of the Years 1821, 1822, 1823* (1824). She was the wife of a sea captain and accompanied him on his voyages until 1822, when he died en route to Chile. Graham remained in South America—first in Chile, then in Brazil, where she worked for a few years as governess to the daughter of Brazil's emperor—until 1827, whereupon she went to London. There she married an artist who traveled with her through Europe, sketching illustrations for some of her books. Graham died in 1842. *See also* SEA TRAVEL.

Further Reading: Graham, Maria. *Journal of a Voyage to Brazil*. 1824. Reprint. London: Longman, Hurst, 1984.

Graham Bower, Ursula (1914–1988)

Born in 1914, Ursula Graham Bower traveled to India in the 1930s and 1940s, part of the time as a photographer for the Royal Geographical Society. She became close to the Indian people, particularly in the Naga hills in the Assam region, and by World War II was considered to be one of the foremost experts on the Naga culture. During the war she ran a refugee camp in the area. From

1948 to 1954 she lived in Kenya, and at the end of her life she lived on the Isle of Mull in Scotland. She also wrote two books about her travel experiences, *Naga Path* (1950) and *The Hidden Land* (1953). Graham Bower died in 1988.

Further Reading: Graham Bower, Ursula. *The Hidden Path*. London: John Murray, 1953; ——. *Naga Path*. London: John Murray, 1950.

Grand Tours

During the eighteenth and nineteenth centuries, the Grand Tour was a popular travel experience for wealthy British young men and, eventually, young women. The latter were typically accompanied by their mothers or older brothers. Travelers went to Europe ostensibly for educational purposes, visiting museums, historical sights, and similar attractions in order to become more "cultured." However, they were also taking advantage of one of the few ways that upper-class society allowed young single men and women to have adventures.

Women making Grand Tours typically went to the same places in Europe; in the sixteenth and seventeenth centuries the most popular destination was France, whereas in the eighteenth century the traditional Grand Tour comprised visits to key cities in France, Italy, Germany, Switzerland, and the Netherlands. As the eighteenth century progressed, Italy became the most popular destination, and travelers began spending a disproportionate amount of their time there.

With the addition of Italy into the typical itinerary, the purpose of the Grand Tour began to shift from self-improvement to enjoyment, particularly for men. Historian Lynne Withey reports:

> For Western Europeans, Italy was the source of all that was important in their culture, both ancient and modern: home of the Romans, whose language formed the core of upper-class British education and whose government and art remained models for emulation, and of the great Renaissance artists, considered by the British to be the finest exemplars of modern aesthetic taste. . . . [But] Italy was gaining popularity for less exalted reasons as well. The Italians' reputation as a gregarious and uninhibited people, the presumed easy availability of women, even the brilliant colors of the landscape and the warm climate exercised a powerful attraction for northern Europeans. Certainly, the pleasures of the senses were always part of the grand tourists' motives—at least if the cautionary literature warning of the temptations facing young travelers is any indication—but the traditional purpose of the grand tour was educational, focused on visiting historical and cultural sites and observing what were loosely known as the manners and customs of foreign nations. Toward the end of the eighteenth century, however, the sensual pleasures of the Mediterranean climate and culture were more readily acknowledged as one of the main reasons to visit Italy. (pp. 7–8)

The most popular cities to visit in Italy were Rome, Florence, Naples, and Venice. According to eighteenth-century British traveler Lady Mary Wortley Montague, Rome alone required a full month to properly visit all of its historical sites, and other writers noted that spending six weeks there was not uncommon for people on a Grand Tour. Other countries held varying degrees of appeal. Many travelers spent little time in Germany, which lacked the type of great art and monuments that people on a Grand Tour typically sought. Switzerland was also relatively unpopular, except for the city of Geneva—because it was famous as the birthplace of religious leader John Calvin—and certain alpine slopes that attracted those inclined to mountaineering. In France, the main attraction was Paris. As with Rome, many tourists spent several weeks there, not only sightseeing but also shopping. Lynn Withey explains:

> Once suitably lodged [in Paris], the British visitor's first task was acquiring a Parisian wardrobe. Despite deep ambivalence about French culture, British visitors conceded (implicitly, at least) Paris's place as the fashion capital of Europe, and, not wanting to appear conspicuous as foreigners, they took pains to dress in the French style. Preliminaries accomplished, visitors

> typically spent their days visiting churches and other public buildings, royal palaces, and homes of noblemen, paying particular attention to their art collections. In the eighteenth century, works of art were nearly all in private homes (apart from the paintings and sculpture that were an integral part of churches and other public buildings), although they were generally accessible to people with the appropriate credentials who requested permission to see them. (p. 16)

In Paris as in most other places, tourists stayed in small hotels or as guests of people who shared their social standing. In the eighteenth and early nineteenth centuries, Grand Tours were expensive and therefore commonly embarked upon only by the wealthiest in society. This changed in the mid-nineteenth century, when English cabinetmaker Thomas Cook set out to make tourism affordable to the masses. He first got the idea of helping travelers when he chartered a train to take members of his temperance group to a regional meeting. Shortly thereafter, Cook started escorting people on low-cost group tours in Great Britain, and in 1855 he offered his first budget tour to France. Later he added Switzerland and then Italy to his itineraries. Many of Cook's tourists were single women who felt pressured by society not to travel alone. These women were interested in the freedom that travel provided and therefore were just as pleased to be enjoying the scenery as they were to be studying great art.

But while middle-class women were enjoying the benefits of travel, the upper classes were dismayed to find hoards of working-class tourists invading their favorite European destinations. They consequently sought travel experiences far from the crowds, in exotic places like Egypt and the Middle East, but within a short time Cook began offering tours to those regions too. By 1875 he had added Germany, Austria, Belgium, Holland, the Middle East, Egypt, and Scandinavia to his offerings; Spain, India, the United States, Australia, and New Zealand soon followed. Even worse as far as upper-class Britains were concerned, beginning in 1867 he aggressively promoted his tours to Americans, and other tour companies soon sprung up to do the same. For the next few years, approximately 40,000 U.S. sightseers visited Europe each year, further crowding tourist attractions.

As wealthy travelers tried to find new ways to set themselves apart from the lower classes, it became fashionable to embark on world tours, visiting not just one or two countries but a whole host of them. However, in 1872 Cook launched a world tour. Personally conducted by Cook himself, it lasted for 222 days and included stops in several major U.S. and Japanese cities as well as visits to India and Egypt. Upper classes would now have to distinguish themselves not by the countries they visited but by the expense of their accommodations.

Meanwhile, middle-class women travelers were accused of using Grand Tours solely as a way to improve their social standing and perhaps attract a wealthy suitor. In fact, this concept was so prevalent that it became common in early nineteenth-century fiction. Several novels of the period feature middle-class women who spend their Grand Tours attempting to acquire upper-class husbands, either for themselves or for their daughters.

For many women, the Grand Tour of Europe became the Great Equalizer. These women treated their travels like a social event, rarely venturing off the well-trod path. Those who went to the Middle East were somewhat more adventurous, but even then the majority stayed in large cities like Cairo, and when they went afield were accompanied by guides, servants, and plenty of supplies; their food was shipped from England. According to Withey, Cook's tours to the Holy Land was "luxury camping," whereby travelers had as many comforts as possible. (p. 258)

So comfortable were Cook's tours, in fact, that by the end of the nineteenth century his Middle East and Egypt excursions were attracting the upper classes as well, who suffered the presence of the middle classes in order to benefit from Cook's good care. In addition, areas that experienced an influx of tourists because of Cook's efforts became sites for luxury hotels and resorts. This was true not only in the Middle East and Egypt but also in other parts of the world. For example, nineteenth-century Ceylon had several luxury hotels built specifically to attract European tourists, whose presence strengthened the economy.

In the United States, however, the emphasis was on budget rather than luxury accommodations. When the American Grand Tour became popular in the 1870s, many new hotels were built for travelers on a budget. There were also inexpensive tourist "packages" that combined accommodations with sightseeing opportunities. For example, a seven-day tour of Yellowstone, which included a stay in a tent camp, cost just $35.

Today the concept of the Grand Tour has entirely given way to that of the packaged tour. Thanks to the efforts of Thomas Cook, modern travel agents offer people a variety of guided tour opportunities throughout the world. *See also* CLASS DISTINCTIONS.

Further Reading: Anderson, Patrick. *Over the Alps: Reflections on Travel and Travel Writing with Special Reference to the Grand Tours of Boswell, Beckford and Byron*. London: Hart-Davis, 1969; Withey, Lynne. *Grand Tours and Cook's Tours: A History of Leisure Travel, 1750–1915*. New York: W. Morrow, 1997.

Great Britain

Great Britain has produced a large number of women travelers and explorers, some of them the most notable in the world. In part this is because since the time of Queen Elizabeth I (1558–1603), its government has long encouraged exploration and travel.

The largest category of women traveler to come out of England is the colonist. Englishwomen began settling in North America in the early 1600s in what is now Massachusetts and Virginia, and many other women subsequently went to Canada, Australia, New Zealand, South Africa, and India to start new lives. All of these countries were at one time colonies of Great Britain, and several private organizations encouraged educated women to settle in one of them.

Emigration Societies

One of the first such organizations was the Female Middle Class Emigration Society, established in London in 1862. The organization provided interest-free loans to women—particularly governesses—who could not afford their passage to British colonies, with a particular emphasis on Australia. In 1908, these loans were taken over by the British Women's Emigration Association, which was established in 1901 to promote emigration among educated Englishwomen.

It offered emigrants protection and connections as they traveled and initial lodging with good families once they arrived at their destinations. This group included Scottish and Irish women as well as Englishwomen in its efforts, and it primarily focused on South African emigration. In 1902 part of the organization split off to become the South African Colonization society; once this society was created, the British Women's Emigration Association shifted its emphasis to encouraging emigration to Canada, New Zealand, and Australia. Meanwhile another emigration society, the Colonial Intelligence League, also promoted travel to Canada. In 1919 the League merged with the British Women's Emigration Association and the South African Colonial Society; the newly united trio was named the Society for the Overseas Settlement of British Women. This group continued to promote women's emigration to British colonies until the 1960s.

Settlers

In addition to single women helped by emigration societies, many married women left England with their husbands to settle new lands. These include pioneers like Rebecca Burlend, who sailed to America in 1831 with her husband and their five youngest children. The Burlends made their way to Illinois and had lived there for 15 years when Rebecca Burlend decided to visit her eldest son, Edward, back in England. During this visit she told him about her experiences as a pioneer, and he used her recollections to write *A True Picture of Emigration* (1848), which inspired other English families to emigrate to America.

Another notable English colonist who wrote about her experiences was Lady Mary Anne Barker. She traveled to New Zealand in 1865 to establish a sheep farm with her husband, remaining there until her husband received a government post on a nearby island in 1868. Like Burlend, Barker's books about her experiences, which include *Station Life in New Zealand* (1870) and *A Christmas Cake in Four Quarters* (1872), increased interest in emigration to New Zealand among her peers in England.

One colonist who did more than write about her life was Deborah Moody, who traveled from England to America in 1639 to settle in the Massachusetts Bay Colony. Unhappy with its restrictions regarding religion, Moody, an Anabaptist, established her own colony, Gravesend, where she could practice her faith. It was the only permanent settlement of its time created by a woman. Similarly, Anna Leonowens influenced political policies after she traveled to Siam (present-day Thailand) in 1862 to become governess to its royal family. She taught more than 60 royal children and their mothers, educating them in Western knowledge, manners, and customs, and worked as the king's secretary until 1867.

Some colonists, however, went on their voyages unwillingly. During the 1700s England began shipping convicts to Australia, and the conditions there were poor. One woman convict who gained notoriety was Mary Bryant, who escaped from a penal colony at Botany Bay in 1792. Sentenced to prison for stealing a woman's coat, she was among the first group of prisoners ever sent to Botany Bay, and some historians suspect she was raped en route. After reaching the colony she married a fellow convict, William Bryant, and in March 1791 she and her husband escaped with several others in an open rowboat. The group headed north along the coastlines of Australia and New Guinea for 69 days, ultimately reaching the island of Timor in the East Indies. There they were soon arrested and shipped back to England, where in 1792 Bryant was again imprisoned. However, the public was so taken with the story of her travels that they called for her release, and she was freed in 1793.

Missionaries and Pilgrims

Another type of woman traveler to come out of England was the missionary. Like emigration societies, British missionary societies such as the China Inland Mission arranged for single women to travel abroad, but their charges were given the task of spreading Christianity in non-Christian regions of Asia and Africa. Of these missionaries, the most notable are Mildred Cable and Anne Royle Taylor. Cable traveled throughout the Gobi Desert in China for 15 years during the early 1900s, along with Algerian missionaries Evangeline and Francesca French. In the 1940s the three women made several more trips, traveling to Africa, Australia, New Zealand, Europe, and South America. British missionary Annie Royle Taylor was the first European woman to visit Tibet. She first went to China in 1884 and set out for Tibet's holy city of Lhasa in 1892, traveling more than 1,000 miles into the country's interior dressed as a Tibetan nun. Three days before she was to have reached Lhasa, she was betrayed to authorities by a disgruntled servant, arrested, and told to leave the country. For approximately 20 years she remained near the Tibetan border doing missionary work.

Religion also motivated some of the earliest Englishwomen to travel long distances: those making pilgrimages to holy sites in the Middle East. One such pilgrim was Margery Kempe, who in 1413 traveled from England to the Middle East to visit places mentioned in the New Testament of the Bible. She was gone about a year, doing most of her traveling on foot or by donkey. This tradition of religious pilgrimages continued into the modern era, although to a lesser extent, and in the 1930s, Lady Evelyn Cobbold became the first Englishwoman to make a religious pilgrimage based on the Islamic faith rather than Christianity, traveling to the holy city of Mecca.

On the High Seas

Other long-distance British travelers were the wives or daughters of sea captains. Some of them, such as Maria Graham and Alice Rowe Snow, wrote books based on their travel adventures. A more colorful category of sea traveler was the female pirate, including Lady Killigrew, Maria Lindsey Cobham, and Mary Read. Lady Killigrew sailed along the coast of England with her husband, Sir John, during the sixteenth century; in 1582, she led her own attack on a ship in port, killing its crew and stealing its goods. During the eighteenth century, an era sometimes called the Golden Age of Piracy, Maria Lindsey Cobham and her husband terrorized ships on the Atlantic Ocean while Mary Read plundered vessels in the Caribbean. All of these women were eventually sentenced to death for their crimes, although some were spared execution because of their gender.

Explorers and Scientists

Britain also produced several women explorers and scientists who traveled the world in search of new discoveries. One such woman was Mary Kingsley, who explored West Africa in 1892 and again between 1894 and 1895, and wrote a very popular book about her experiences, *Travels in West Africa* (1897). During her second expedition, she studied indigenous cultures, collected specimens of plants and fish (particularly freshwater ones) for the Royal Botanical Gardens, and made a pioneering trek through the Gabon Mountains and ascended Mount Cameroon. Those who encountered her during her travels usually assumed she was a missionary.

Other notable British explorers were Elizabeth Mazuchelli, Lady Hester Stanhope, Ella Constance Sykes, Lady Isabel Burton, and Charlotte Cameron. Mazuchelli was the first European woman to explore the eastern Himalaya Mountains; in 1871 she and her husband spent two months traveling through the mountain range with approximately 80 servants. Lady Hester Stanhope traveled through the Middle East in the 1810s, exploring the desert in western Syria, while Ella Constance Sykes became the first Englishwoman to visit certain parts of Persia, Russia, and China—usually accompanied by her brother Percy, a British diplomat—during the late nineteenth and early twentieth centuries. Lady Isabel Burton accompanied her husband, British explorer and diplomat Richard Burton, on some of his travels to Africa, South America, Syria, India, and Egypt between the years 1869 and 1890. Explorer Charlotte Cameron went on expeditions throughout the world between 1910 and 1925, visiting such places as Africa, Polynesia, Borneo, Asia, India, South America, the Yukon, and Alaska.

Scientists

Women scientists who traveled far from England include archaeologists Gertrude Bell, Margaret Murray, Gertrude Caton-Thompson, Dorothy Garrod, and Mary Leakey, and anthropologists Lilian Brown and Jane Goodall. Bell traveled throughout the world during the late nineteenth and early twentieth centuries, exploring various cities and archaeological sites in the Middle East; in 1914 she went deeper into the Arabian Desert than any previous female British explorer had ever traveled. Murray excavated in Egypt in the late 1890s; her work on Egyptian archaeology, *The Splendor That Was Egypt* (1931), is still considered a classic in the field. Caton-Thompson also worked on archaeological sites in Egypt, between 1921 and 1928, and in subsequent years in Africa and southern Arabia. Garrod excavated prehistoric villages in Gibraltar, France, Palestine, Lebanon, Kurdistan, and Bulgaria in the 1920s. Brown accompanied anthropologist Frederick Mitchell-Hedges on his expeditions to Central America from 1921 to 1931, making several trips to study indigenous tribes and collect material for the British Museum. Leakey excavated prehistoric sites in Africa from the 1930s through the 1970s. Goodall went to Kenya, Africa, in the 1960s to study chimpanzees in the wild; her research there continues today.

A more active scientific traveler was entomologist Margaret Fountaine, a butterfly collector who went throughout the world in search of specimens. Beginning in 1890, she visited most major countries, including Syria, Turkey, Algeria, Spain, Africa, the Caribbean, India, New Zealand, Fiji, South America, the United States, and the Far East. Along the way she collected over 20,000 butterflies. Lucy Cheesman was another insect collector who traveled extensively to find specimens. Between 1923 and 1955 she collected thousands of insects in the South Pacific, traveling to islands in the New Hebrides, New Guinea, and New Caledonia.

Other women from Britain who traveled for scientific reasons include Lady Isobel Gill, filmmaker Caroline Hamilton, Marika Hanbury-Tenison, and Amelia Murray. Gill accompanied her husband on a 1877 voyage to Ascension Island to find a better view of the stars and planets, while Hamilton led the first all-woman expedition to the geographical North Pole in 1997 as part of a scientific study. Journalist Marika Hanbury-Tenison went to South America, Indonesia, and Malaysia during the 1970s to study tribal cultures; Amelia Murray was an amateur botanist who toured North America and Cuba between 1854 and 1855.

Traveling Writers

Many of these women wrote books about their experiences, but they were not known as travel writers. British women in that category include Louisa Costello, Isabella Bird Bishop, Constance Cumming, Amelia Edwards, Mariana Stark, Frances Trollope, and Ethel Tweedie. Costello is considered by many historians to be the first woman to make a living writing travel books; her works include *Bearn and the Pyrenees: A Legendary Tour to the Country of Henri Quatre* (1844) and *A Tour to and from Venice, by the Vaudois and the Tyrol* (1846). However, Isabella Bird Bishop is more famous; she wrote a total of nine books, including *A Lady's Life in the Rocky Mountains* (1879) and *Unbeaten Tracks in Japan* (1880), and traveled not only to North America and Japan but to the South Pacific, Hawaii, Asia, and the Middle East.

Stark wrote one of the most important travel guidebooks of the nineteenth century, *Travels on the Continent Written for the Use and Particular Information of Travellers* (1820), which provided detailed information for tourists heading for the European continent—later editions included information on Russia and Scandinavia as well. Cumming made her first overseas journey in 1868, touring India with her sister and brother-in-law, and subsequently went to such exotic places as Fiji, Ceylon, Tonga, Samoa, Tahiti, Japan, and China, as well as the coasts of California and the Pacific Northwest. Her books include *Wanderings in China* (1881) and *At Home in Fiji* (1886). Amelia Edwards's books include *Untrodden Peaks and Unfrequented Valleys: A Midsummer Ramble in the Dolomites* (1873), an account of her journey through the Dolomite Mountains in northern Italy in 1872, and *A Thousand Miles Up the Nile* (1877), which concerns a 1873 trip up Egypt's Nile River and provided the fullest account of Egyptian ruins up until that time.

Trollope and Tweedie each wrote a series of travel books. Trollope's include the controversial *Domestic Manners of the Americans* (1832), which was based on her experiences visiting the United States from 1827 to 1830, while Tweedie—one of the most prolific female travel writers of the late nineteenth and early twentieth centuries—wrote *Through Finland in Carts* (1897), *Mexico As I Saw It* (1901), *America As I Saw It* (1913), *An Adventurous Journey (Russia-Siberia-China)* (1926), and *My Legacy Cruise (The Peak Year of My Life)* (1936).

Two other notable British travel writers were Lady Dorothy Mills and Christina Dodwell. Mills was the first white woman to visit Timbuktu, in present-day Mali, (1923) and parts of West Africa and South America (1926 and 1930, respectively); her books include *The Road to Timbuktu* (1924), *Through Liberia* (1926), and *The Country of the Orinoco* (1931). Dodwell's books include *Travels with Fortune: An African Adventure*

(1979), which concerns three years she spent crossing Africa, *In Papau New Guinea* (1983), *Traveller in China* (1985), *Traveller on Horseback: In Eastern Turkey and Iran* (1987), *Travels with Pegasus: A Microlight Journey Across West Africa* (1990), *Beyond Siberia* (1993), and *Madagascar Travels* (1995). Dodwell also wrote a guidebook for fellow travelers entitled *An Explorer's Handbook: An Unconventional Guide for Travelers to Remote Regions* (1986).

Politics and Culture

Some British women chose to write about political situations in the places they visited. For example, Mary Wollstonecraft was an English feminist who visited France in 1792 to study women's rights there. She became famous for a feminist book published the same year, *A Vindication of the Rights of Women*. Nineteenth-century social reformer Frances Wright traveled from England to America in 1818 to study democratic government, touring the United States and Canada for two years before returning to London. Mary Edith Durham visited the Balkans during the early 1900s and became involved in political causes there. Elspeth Josceline Huxley traveled to Africa in the 1940s to write about politics, as did Florence Dixie and Margery Freda Perham in the 1880s and 1920s, respectively.

Medical and Other Quests

Nurse Kate Marsden embarked on her travels to discover a cure for leprosy. Having heard that Siberian shamans had discovered an herb that would end the disease, in 1890 she petitioned the Queen of England and the Empress of Russia for permission to travel through Siberia, which was at the time closed to foreigners. They granted her request, and she went to Russia by way of the Middle East, where she studied some advanced cases of leprosy. She then traveled into Siberia, but after three months she fell ill and had to return home empty-handed.

In contrast, Lucy Atkinson, Frances Ann Hopkins, and Marianne North traveled because of art. Lucy Atkinson also traveled in the company of her husband, an architect who wanted to make detailed sketches of the Siberian landscape, villages, and people. The two spent from 1848 to 1853 in Russia. Artist Frances Hopkins went to Canada in 1838 after her husband went to work for the Hudson Bay Company there, and she went with the company's fur traders on at least two voyages, one in 1869 and the other in 1870, to sketch and paint scenes from the region. She continued to travel and paint after her husband's death in 1893. Marianne North searched throughout the world for subjects for her oil paintings, which featured plants from around the world. Beginning in 1870 she traveled through Europe, the United States, Jamaica, Haiti, South America, Japan, Singapore, the Far East, several Pacific and Atlantic islands, Australia, and South Africa. Her last journey was to Chile in 1884.

Mrs. F. D. Bridges and her husband were sightseers, traveling the world between 1877 to 1880, as were Gwendolen Dorrien Smith and Lady C. C. Vyvyan, who took a trip across Canada in 1926. Lady Mary Wortley Montague traveled from England to Turkey in 1716 to be with her husband, who had been named ambassador to Turkey, while Sophia Poole lived with her two small children in Egypt from 1842 to 1849 to experience Egyptian culture. Other sightseers include Emmeline Stuart Wortley and her daughter Victoria, who went to North and South America, Spain, and Egypt as well as Africa during the 1840s and 1850s; "Vita" Sackville-West, who went through the Middle East and Russia in 1925; and, filmmaker Sarah Hobson, who explored the Middle East dressed as a man in 1970.

Adventure Seekers

British women have also traveled for adventure. For example, mountaineer Anna Pigeon traveled throughout the world looking for new climbing challenges. She was among Great Britain's most accomplished mountaineers, ascending over 63 major peaks in the late nineteenth and early twentieth centuries. British mountaineers Nea Morin and Dorothy Pilley also participated in several important climbing expeditions throughout the world during that time. Aviator Amy Johnson flew from England to Australia in 1930, becoming the first woman pilot to make this journey solo, and Beryl Markham made a solo flight across the Atlantic Ocean in 1936. Great Britain continues to produce women who travel throughout the world in search of adventure or to further science. *See also* AFRICA; ANTHROPOLOGISTS, AND ARCHEOLOGISTS; ATKINSON, LUCY; AUSTRALIA AND NEW ZEALAND; BARKER, MARY ANNE; BELL, GERTRUDE; BISHOP, ISABELLA BIRD; BRYANT, MARY; BURLEND, REBECCA LUCY; BURTON, ISABEL; CABLE, MILDRED; CAMERON, CHARLOTTE; CATON-THOMPSON, GERTRUDE; CHEESMEN, LUCY EVELYN; COBBOLD, EVELYN; COBHAM, MARIA LINDSEY; COSTELLO, LOUISA STUART; CUMMING, CONSTANCE; DIXIE, FLORENCE; DODWELL, CHRISTINA; DURHAM, MARY EDITH; EDWARDS, AMELIA; EMIGRATION SOCIETIES (BRITISH); EXPLORERS; FOUNTAINE, MARGARET; FRENCH, EVANGELINE AND FRANCESCA; GARROD, DOROTHY ANNIE ELIZABETH; GILL, ISOBEL; GOODALL, JANE; GRAHAM, MARIA; HAMILTON, CAROLINE; HANBURY-TENISON, MARIKA; HOBSON, SARAH; HOPKINS, FRANCES ANN; HUXLEY, ELSPETH JOSCELINE; JOHNSON, AMY; KEMPE, MARGERY; KINGSLEY, MARY; LATIN AMERICA; LEAKEY, MARY; LEONOWENS, ANNA HARRIETTE; MARKHAM, BERYL; MARSDEN, KATE; MAZUCHELLI, ELIZABETH SARAH; MILLS, DOROTHY; MISSIONARIES; MONTAGUE, MARY WORTLEY; MOODY, DEBORAH; MORIN, NEA; MURRAY, AMELIA; MURRAY, MARGARET ALICE; NORTH AMERICA; NORTH, MARIANNE; PERHAM, MARGERY FREDA; PIGEON, ANNA; PILGRIMS; PILLEY, DOROTHY; PIRATES; POLES, NORTH AND SOUTH; POOLE, SOPHIA; QUEENS; SCIENTISTS; SEA TRAVEL; SMITH, GWENDOLEN DORRIEN; STANHOPE, HESTER; STARK, MARIANA; SYKES, ELLA CONSTANCE; TAYLOR, ANNE ROYLE; TRAVEL WRITERS; TROLLOPE, FRANCES; TWEEDIE, ETHEL BRILLIANA; VYVYAN, C. C.; WOLLSTONECRAFT, MARY; WORTLEY, EMMELINE STUART AND VICTORIA; WRIGHT, FRANCES.

Further reading: Baker, Daniel. *Explorers and Discoverers of the World.* New York: Gale, 1993; Geniesse, Jane Fletcher. *The Passionate Nomad: The Life of Freya Stark.* New York: Random House, 1999; Lovell, Mary S. *A Rage to Live: A Biography of Richard and Isabel Burton.* New York: W.W. Norton & Co., 1998; Wallach, Janet. *Desert Queen: The Extraordinary Life of Gertrude Bell, Adventurer, Advisor to Kings, Ally of Lawrence of Arabia.* New York: Anchor Books, 1999.

Griffin, Maud (1880–1971)

Born in 1880, American Maud Griffin was the only licensed female ship pilot of the 1920s. She operated a tugboat on the waterways of Houston, Texas, at first with her husband George, but alone after his death in 1924. Eventually she became known as "Tugboat Annie." She retired in 1932 and died in 1971.

Further Reading: Raine, Norman Reilly. *Tugboat Annie: Great Stories from the Saturday Evening Post.* Indianapolis, IN: Curtis Publishing, 1977.

Grimshaw, (Ethel) Beatrice (?–1953)

Irish journalist and fiction writer (Ethel) Beatrice Grimshaw was the first white woman to travel down the Sepik River in New Guinea. Between 1906 and 1936 she explored the South Seas in order to gather material for romances and adventure stories and wrote about her experiences in such books as *In the Strange South Seas* (1907) and *Isles of Adventure* (1930). She was also an accomplished bicyclist, hard-hat ocean diver, and diamond prospector. She spent the last 14 years of her life in Australia, where she died in 1953.

Further Reading: Robinson, Jane. *Wayward Women: A Guide to Women Travellers.* Oxford: Oxford University Press, 1990.

Group Tours

The majority of women who travel as tourists rather than for business purposes prefer not to travel alone, and those who cannot find their own travel companions often sign up for group tours.

In fact, the first organization dedicated to promoting group tours was created with women in mind. When Thomas Cook, the first travel agent and tour guide, started his tourism business in the 1850s, he did so in part because it was considered improper and unsafe for women to travel alone. Even in groups, women were expected to be accompanied by a male, and Cook decided to fill that role. He gathered together large groups of single and married women and acted as their chaperone, escorting them to various sites first in Europe and then elsewhere throughout the world.

Today women continue to seek out group excursions, not because of societal pressures but because group travel offers pleasant companionship and is usually less expensive than traveling alone. A particularly popular choice among women is a tour that involves travel by cruise ship. Such vessels carry thousands of passengers each year on trips lasting from two days to several months.

Both land and sea tour packages are usually purchased through travel agents, 80 percent of which in the United States are women. Approximately 80 percent of travel agency managers are also women. According to a 1998 survey by *Travel Weekly* magazine, the average salary of a travel agent is $25,614 a year. However, travel agents and managers also receive discounts on travel-related services and therefore have greater opportunity to travel. They are also part of a lucrative business; tourism generates over $100 billion in revenues in the United States alone, with group tours and cruises the most profitable aspect of the business for travel agents. *See also* GRAND TOURS.

Further Reading: Maxtone-Graham, John. *Crossing & Cruising: From the Golden Era of Ocean Liners to the Luxury of Cruise Ships of Today.* New York: Macmillan Library Reference, 1979; ———. *Liners to the Sun.* Dobbs Ferry, NY: Sheridan House, 1985; Withey, Lynne. *Grand Tours and Cook's Tours: A History of Leisure Travel, 1750–1915.* New York: W. Morrow, 1997; The Women's Travel Club. http://www.womenstravelclub.com (May 2000).

Guidebooks, Travel

Travel guidebooks help travelers find places to visit, stay, eat, and shop that provide, in the opinion of their authors, the optimal travel experience. These books also offer information regarding customs and general travel tips. Consequently old guidebooks offer insights into the past, as well as into the mindsets of the people who wrote them and read them.

The first person who could truly be called a professional guidebook author, Mariana Stark, was concerned in her early works with describing scenery. In the early 1800s, most of her readers had never been away from home and had no access to photographs that could show them what places such as Italy looked like. As time passed, there were more pictures of exotic lands and more women willing to travel to see them; therefore Stark's works began to include fewer descriptive passages and more tips on how to have an enjoyable travel experience. In an 1820 guidebook, *Travels on the Continent: Written for the Use and Particular Information of Travellers,* Stark also reassured her readers that traveling through continental Europe was safe despite political unrest. In addition, this work provided detailed information on art galleries and artwork, because many tourists of the period were interested in "bettering" themselves through exposure to great art and architecture. Stark rated paintings according to their level of importance and interest.

Many modern travelers are more interested in leisure travel; therefore, modern guidebooks typically tell travelers how to find the best hotels and excellent food. Early guidebooks did this too, but not nearly to the same degree. Inn and hotel information in eighteenth- and nineteenth-century guidebooks, if any was provided, was generally brief, and the food of a region was more often discussed in general terms rather than with references to specific restaurants as in modern guidebooks. Similarly, while early guidebooks concentrated on scenic attractions and sometimes provided maps on how to reach them, modern guidebooks often focus on experiences and activities, telling tourists what places provide the most enjoyable pastimes.

Both early and modern guidebooks come in two varieties: those that concentrate on a particular country or type of travel experience, and those that provide more general travel information. Mariana Stark, for example, wrote several editions of a popular guidebook that discussed Italy alone. In modern times, many guidebooks, such as Frommer's and Fodor's, are organized in separate volumes according to country.

But whereas early guidebooks were always either for men or for both genders, in modern times there have been many guides written specifically with women in mind. For example, Thalia Zepatos's books, *Adventures in Good Company: The Complete Guide to Woman's Tours and Outdoor Trips* (1994) and *A Journey of One's Own: Uncommon Advice for the Independent Woman Traveler* (1996), are intended for women seeking general information about travel. The former gives detailed information on more than 75 tour companies that offer trips specifically designed for women, as well as advice on how to select a tour. The latter volume combines tips on how to handle travel problems such as sexual harassment with stories told by women travelers.

This blend of tips and stories is very popular in modern travel literature written for women. Other examples in the genre include *More Women Travel: Adventures and Advice for More than 60 Countries* (1995) edited by Natania Jansz and Miranda Davies, and *Gutsy Women: Travel Tips and Wisdom for the Road* (1996) edited by Marybeth Bond, just one of a series of books published by Travelers' Tales. Other titles in the series are collections of travel essays organized around a particular country or city, or around a central theme like food or women's issues. The latter include *A Woman's World: True Stories of Life on the Road* (1995) edited by Marybeth Bond, *Women in the Wild: True Stories of Adventure and Connection* (1998) edited by Lucy McCauley, and *A Mother's World: Journeys of the Heart* (1998) edited by Marybeth Bond and Pamela Michael.

Another book that combines travel essays with travel tips is *Go Girl!: The Black Woman's Guide to Travel and Adventure* (1997) edited by Elaine Lee. It includes 52 travel essays written by black women for black women. This book is therefore an example of a work targeted for a very specific group of women. Similarly, *Damron Women's Traveller 1998* (1997) edited by Gina M. Gatta offers information specifically for lesbian travelers, including facts about lesbian tours, festivals, conferences, campsites, city clubs, and women-run businesses. The *Damron Road Atlas* has maps pinpointing gay and lesbian accommodations, bars, and bookstores in dozens of major cities and resort destinations throughout the world and *Damron Accommodations* provides reviews of gay- and lesbian-friendly hotels, inns, bed-and-breakfasts, and other accommodations around the world. Other guidebooks targeted for specific groups include *The Active Woman Vacation Guide* (1997) by Evelyn Kaye, which is designed for women seeking adventure travel experiences; *Traveling Solo: Advice and Ideas for More than 250 Vacations* (1999)by Eleanor Berman, which is for women who want to travel without a partner; *A Foxy Old Woman's Guide to Traveling Alone: Around Town and Around the World* (1995) by Jay Ben-Lesser, which offers advice to the older traveler. Other guidebooks abound, as do collections of travel essays that offer stories without including travel tips. *See also* TRAVEL WRITERS.

Further Reading: Ben-Lesser, Jay. *A Foxy Old Woman's Guide to Traveling Alone: Around Town and Around the World*. Freedom, CA: Crossing Press, 1995; Berman, Eleanor. *Traveling Solo: Advice and Ideas for More than 250 Vacations*. Old Saybrook, CT: Globe Pequot Press, 1999; Bond, Marybeth, ed. *Gutsy Women: Travel Tips and Wisdom for the Road*. San Francisco: Travelers' Tales, 1996; ———. *A Woman's World: True Stories of Life on the Road*. San Francisco: Travelers' Tales, 1995; Bond, Marybeth, and Pamela Michael, eds. *A Mother's World: Journeys of the Heart*. San Francisco: Travelers' Tales, 1998; Gatta, Gina M., ed. *Damron Women's Traveller 1998*. San Francisco: Damron, 1997; Jansz, Natania, and Miranda Davies, eds. *More Women Travel: Adventures and Advice for More than 60 Countries*. London: Rough Guides, 1995; Kaye, Evelyn. *The Active Woman Vacation Guide*. Boulder, CO: Blue Panda Publications, 1997; Lee, Elaine, ed. *Go Girl!: The Black Woman's Guide to Travel and Adventure*. Portland, OR: Eighth Mountain Press, 1997; McCauley, Lucy, ed. *Women in the Wild: True Stories of Adventure and Connection*. San Francisco: Travelers' Tales, 1998; Zepatos, Thalia. *Adventures in Good Com-*

pany: The Complete Guide to Woman's Tours and Outdoor Trips. Portland, OR: Eighth Mountain Press, 1994; ———. *A Journey of One's Own: Uncommon Advice for the Independent Woman Traveler*. Portland, OR: Eighth Mountain Press, 1996.

Gun, Isabel

In 1806 Isabel Gun sailed to Canada from the Orkney Islands off Scotland disguised as a man, "John Fubbister," having accepted a contract to work for the Hudson's Bay Company, a mercantile corporation established by the British in Canada in 1670. The Hudson's Bay Company engaged in fur trading with Native Americans and encouraged Canadian exploration and settlement. Under their employ, Gun participated in fur trapping expeditions until 1807, when she revealed her gender by giving birth to a boy. Since she had a work contract, her employers retained her services as a washerwoman, but as soon as the contract expired they sent her back to Scotland. Her fate is uncertain, as is the identity of her baby's father, although many suspect it was a company employee who had accompanied her from the Orkney Islands.

Further Reading: Bryce, George. *The Remarkable History of the Hudson's Bay Company, including that of the French Traders of North-western Canada and of the North-west, XY, and Astor Fur Companies*. New York: B. Franklin, 1968 (reprint of 1900 edition); O'Meara, Walter. *Daughters of the Country: The Women of the Fur Traders and Mountain Men*. New York: Harcourt, Brace & World, 1968; Tharp, Louise Hall. *Company of Adventurers; The Story of the Hudson's Bay Company*. Boston: Little, Brown, 1946.

H

Hadley, Leila (1925–)

Born in 1925 to wealthy New York socialites who traveled extensively, Leila Hadley is one of the foremost female travel writers in America. She has visited many exotic places, including remote regions of India, Tibet, Africa, the Far and Middle East, Central America, the Caribbean, and Southeast Asia. Among her most widely read works are a series of guidebooks on traveling with children, but she is known for her travel memoirs as well. These include *Give Me the World* (1958), which describes a sailing trip she took with her husband and four-year-old son from Singapore to Naples, and *A Journey with Elsa Cloud*, Hadley's best selling book to date.

Published in 1997, *A Journey with Elsa Cloud* describes one of Hadley's trips to India in the 1970s, which not only involved sightseeing but also an attempt to reconnect with her estranged daughter, Veronica. Her daughter had been in Dharamsala, India, for two years, studying Buddhism as one of the followers of Tibet's exiled Dalai Lama. As Hadley writes about her daughter, India, and Buddhism, she also brings up memories of her own childhood relationship with her mother. The book therefore combines a travelogue with a personal story of connection and dawning self-awareness—an external journey with an internal one—as do many other works of women's travel literature.

For much of the first half of the book, Hadley and her daughter connect via simple pursuits: shopping and eating. Hadley offers detailed descriptions of bazaars and the bargaining process. Shortly after arriving in India she says:

> The prospect of an alien, hot country where I can bargain for hours for the fine work of artisans and craftsmen always sings a siren song for me, and nowhere more impassioned than in India. . . . I want to see belled toe-rings and wide ankle bracelets weighed in a pan scale for their value by their total weight. I want to watch merchants fling lengths of billowing golden silk cloth around me for the pleasure of my selection. I want to look for treasures and curiosities and objects of everyday life whose purpose I will not know until told. I want to steep myself like a teabag in fantasy and reality, a brew of romance and the genial trickeries of bargaining. . . . Oriental bargaining . . . is a way of life and a ubiquitous custom, so shopping and bargaining become a rite of passage, an initiation into the local society and aesthetic idiom. Instead of feeling self-indulgent and extravagant, I feel virtuous; it's like doing research, an exercise in selective acquisition, a way to become instantly connected and accepted. (pp. 83–84)

Hadley also speaks of bargaining for food items and describes some of her meals, saying: "When one travels, one eats, thinks of food, needs food, wants food, looks for food in a way one never seems to do at home." (p. 247) Midway through the book, she turns her attention to the people she meets and to the sites she visits, particularly those devoted to Buddhism. She begins to understand her daughter's beliefs, and, at the end of the book, she meets with the Dalai Lama himself. Afterward she says: "I feel free and light, disencumbered, as if the restrictive wrappings of worry, anxiety, depression, tiresome bindings of twentieth-century acedia, have been stripped away to reveal another self—not an entirely new self, but one that is better, more comfortable." (p. 600)

In addition to sharing her personal travel experiences in books like *A Journey With Elsa Cloud*, Hadley has made significant contributions to the field of women's exploration. She co-founded Wings Trust, which offers support to women explorers and archives their literature. She also belongs to other organizations devoted to exploration, including the Society of Women Geographers and the Explorers Club. In addition, in 1992 she became a board member of and contributing editor to *Tricycle: The Buddhist Review*. She currently lives in New York. *See also* Asia.

Further Reading: Hadley, Leila. *A Journey with Elsa Cloud*. Paperback edition. New York: Penguin Books, 1998.

Hahn, Emily (1905–)

Born in 1905, American author Emily Hahn traveled to the Belgian Congo in Africa in 1930 as a Red Cross worker, offering aid to Pygmy tribes. She also worked as a courier in Santa Fe, New Mexico, and as a mining engineer in St. Louis, Missouri, having been one of the first women in the United States to graduate with a degree in mining. In 1935 she moved to China, where she lived until 1944. She wrote several books based on her experiences there, including *China to Me* (1944) and *Times and Places* (1970). In the latter, which was a compilation of articles that originally appeared in the *New Yorker* magazine, she writes about her addiction to opium, explaining that when she first arrived in China she had no knowledge of the drug. She writes:

> As a newcomer, I couldn't have known that a lot of the drug was being used here, there, and everywhere in town. I had no way of recognizing the smell, though it pervaded the poorer district. I assumed that the odor, something like burning caramel or those herbal cigarettes . . . was just part of the mysterious effluvia produced in Chinese cookhouses. Walking happily through the side streets and alleys, pausing here and there to let a rickshaw or a cart trundle by, I would sniff and move on, unaware that someone close at hand was indulging in what the books called that vile, accursed drug. (p. 270)

Eventually Hahn shed her addiction to the drug and went on to write many more books, including novels and children's books. She currently lives in New York City.

Further Reading: Hahn, Emily. *China to Me: A Partial Autobiography*, 1944. Reprint. New York: Da Capo Press, 1975; ———. *Times and Places*. New York: Cromwell, 1970; Morris, Mary. *Maiden Voyages: Writings of Women Travelers*. New York: Vintage, 1993.

Hall, Mary (1857–1912)

Born in 1857 in England, Mary Hall was the first woman to journey longitudinally through Africa. She began her adventure in South Africa and went north to Cairo, Egypt, by railway, steamer, rickshaw, foot travel, and being carried by porters. Although Hall traveled with servants and guides, her peers considered her "unaccompanied" during her seven-month trip because she lacked a European companion. She wrote about her experiences in a book, *A Woman's Trek from the Cape to Cairo* (1907). Hall's trip through Africa began in 1905, after she had spent a year sightseeing in South Africa. But she became a traveler years earlier, as a young woman who had been advised that travel would improve her poor health. It did, and she continued to travel throughout her life, visiting places all over the world. Her second book, *A Woman in the Antipodes and in the Far East* (1914), describes a trip to New Zealand, Australia, the South Pacific, China, and Siberia. Hall died in 1912. *See also* Africa.

Further Reading: Hall, Mary. *A Woman in the Antipodes and in the Far East*. London: Methuen, 1914; ———. *A Woman's Trek from the Cape to Cairo*. London: Methuen, 1907.

Hamilton, Caroline (1964–)

Born in 1964, British filmmaker Caroline Hamilton led the first all-woman expedition to the geographical North Pole in 1997. There were five relay teams in the expedition, each with 20 women, and each team walked and/or skied to a predesignated point where an airplane picked them up and left the next team. Each team traveled for approximately two weeks, although two guides—Matty McNair from the United States and Denise Martin of Canada—went the entire 625 miles from Ward Hunt Island in northern Canada to the North Pole.

The idea for the expedition came to Hamilton in 1995, when she met Pen Hadow, a polar explorer and head of the Polar Travel Company. The two discussed possible travel experiences in the Arctic, and Hadow suggested the relay. Hamilton then advertised internationally for other participants, and more than 200 women applied. Team members were selected via an endurance test that involved such exercises as hiking 40 miles with a heavy pack. They then went through nine months of training, which included learning survival and navigational skills, and spent the 10 days prior to the expedition in a remote northern Canadian village practicing with sleds and skis. Hamilton was part of the final team that reached the North Pole, after a 13-hour day of traveling across ice. Once there she planted the British flag and sang that country's national anthem. As a result of this success, Hamilton and five other women from her North Pole expedition are currently planning an all-woman South Pole expedition, which will be run in a similar fashion. *See also* POLES, NORTH AND SOUTH.

Further Reading: BBC Online. "Polar Women Begin Marathon." http://news6.thdo.bbc.co.uk/hi/english/uk/newsid_500000/500719.stm, (June 2000).; McVitie's Penguin Polar Expedition. http://www.stuff.co.uk/media/polar-relay/expedition.html (May 2000).

Hanbury-Tenison, Marika (1938–1982)

Born in 1938, English journalist Marika Hanbury-Tenison went to South America, Indonesia, and Malaysia during the 1970s to study tribal cultures. She was the wife of noted Cornish explorer Robin Hanbury-Tenison, and her first significant trip was to accompany him to the Brazilian jungle in 1971. At that time, indigenous tribes were being forced from their lands by developers, and many of them had congregated in encampments established by the British government or private charitable groups. The Hanbury-Tenisons visited these camps to speak with local people and study their living conditions. After returning to Great Britain in 1972, Marika Hanbury-Tenison wrote about their findings in *For Better, For Worse: To the Brazilian Jungles and Back Again* (1972), published in some countries as *Tagging Along* (1972). Meanwhile, her husband published a more scholarly work on the same subject, *A Question of Survival for the Indians of Brazil* (1973).

In 1973, the Hanbury-Tenisons again visited tribes, this time in Indonesia. As with Brazil, this was a region where indigenous people were being pushed from their lands, and, in many cases, they were reacting with violence. Nonetheless, the Hanbury-Tenisons managed to visit about a dozen tribes without being attacked, although Marika did become ill during her travels. She wrote about her Indonesian adventures in *A Slice of Spice* (1974), a popular work in contrast to her husband's more scholarly *A Pattern of Peoples: A Journey among the Tribes of Outer Indonesian Islands* (1975). Marika Hanbury-Tenison made her last research trip with her husband in 1979, when the couple visited Malaysia as part of a Royal Geographical Society scientific expedition. Shortly thereafter she was diagnosed with cancer. She died in 1982. *See also* SPOUSES.

Further Reading: Hanbury-Tenison, Marika. *For Better, For Worse: To the Brazilian Jungles And Back Again*. London, Hutchinson, 1972; ———. *A Slice of Spice: Travels to the Indonesian Islands*. London: Hutchinson, 1974.

Harassment. *See* CHALLENGES FOR THE MODERN WOMAN TRAVELER.

Harrison, Barbara Grizzuti (1934–)

Born in 1934, American author Barbara Grizzuti Harrison has written numerous articles, novels, and nonfiction books, but is perhaps best known for a work of modern travel literature entitled *Italian Days*. Published in 1989, the book is typical of its genre in that it is not just an exploration of place but of self.

Italian Days describes Harrison's journey to Italy in 1985 to learn about her heritage as the child of Italian-American parents. At the beginning of her work, she spends much of her time describing various sights and their historical significance, talking about her tourist activities in such cities as Milan, Venice, Florence, and Rome and comparing Italian and American cultures. But as her book progresses, her journey becomes more personal as she describes meetings with various family relatives and tells stories of her childhood.

Throughout *Italian Days,* Harrison also talks about the emotional side of travel. For example, in describing her arrival in Italy she says:

> One feels a quickening of the pulse when one crosses a border. It is strange: What difference is there (for example) between the landscape of Canada and that of the northern United States; or, in North Africa, what difference exists between the stretch of land on the coastal road of Tunisia and that of Libya? None, one would say; and yet one knows—I have always known—when one crosses these or any other borders. Are the trees different? the sky? It is mysterious; but one senses subliminally on these long and lonely stretches of road that one is in another country, and it is only after the subliminal awareness reaches one's consciousness that one looks for tangible proofs (in Canada, a maple leaf on a mailbox; in Tunisia, a road sign in French). This phenomenon stems from a change in character so subtle it is indescribable. (p. 7)

Harrison is excited by travel, but she believes that its purpose involves more than just pleasure. She says: "What's the point of traveling if one isn't changed—I don't say uplifted—by the experience?" (p. 13) This philosophy is tied with her need to learn about her Italian heritage rather than just visit Italian historical sites and museums.

In addition to her trip to Italy, Harrison has traveled elsewhere throughout the world, and she has lived in Libya, India, and Guatemala. She now lives in New York City, where she was raised. *See also* TRAVEL WRITERS.

Further Reading: Harrison, Barbara Grizzuti. *Italian Days*. New York: Weidenfeld and Nicholson, 1989.

Harrison, Marguerite Baker (1879–1967)

Born in 1879, American author Marguerite Baker Harrison ostensibly was a journalist who traveled to report stories and make documentaries. However, these activities eventually became a cover for her work as a spy, a job she accepted after incurring debts during the illness and death of her husband.

By that time she had already traveled a great deal; when she was a child her family often went abroad and she learned to speak both French and German. After her husband died of cancer in 1915, she took a job with the *Baltimore Sun*, first as a society editor and then as columnist. In 1918 she went to Germany under the pretense of reporting for the *Sun* on the events of World War I. In reality, she transmitted information from the United States government to operatives abroad.

After this assignment ended, Harrison was able to pay her debts, but she soon requested another spy mission. She was sent to Russia, again appearing to be a reporter, but after a month the Russian government found out her true purpose and imprisoned her. She was freed only after she agreed to work as a double agent, spying for Russia as she met with Americans living in Russia while pretending to continue her work for the United States. But Harrison's loyalty remained with the United States, and the information she provided was false or worthless. When the Russian government learned this, Harrison was again imprisoned. This time she was freed because the United States government agreed to exchange food shipments to Russia, which was experiencing a famine, for the release of all American prisoners in Russian prisons.

When she returned to the United States, Harrison wrote a book about her experiences, *Marooned in Moscow: The Story of an American Woman Imprisoned in Russia* (1921). She also gave lectures about Russia. However, she longed to travel abroad once more, and in 1922 she went to Japan, Korea, and Manchuria to write a series of magazine articles about Asia. She then decided to revisit Russia, where she was again arrested as a spy. After several weeks she was released but warned never to return.

Back in the United States, Harrison wrote a book comparing and contrasting Russian and Japanese politics, *Red Bear or Yellow Dragon* (1924). But, as before, she soon became restless and accepted a job as co-director of a film, *Grass* (1925), documenting the travels of a Persian tribe. In 1925 she helped found the Society of Women Geographers, and she continued to travel throughout her later years, not only in Europe and Asia but in other parts of the world as well. She also continued to write; her works include a book on Asian politics, *Asia Reborn*, first published in 1928 and revised in 1938. Harrison died in 1967.

Further Reading: Harrison, Marguerite. *There's Always Tomorrow: The Story of a Checkered Life*. New York: Farrar & Rinehart, 1935.

Hasell, Eva (1886–1974)

Born in 1886, Frances Hatton Eva Hassell is representative of British missionaries of the early twentieth century. She preached the gospel throughout Canada during the 1920s, organizing motoring caravans that brought Christian literature to remote areas, and wrote three books about her experiences: *Across the Prairie in a Motor Caravan: A 3,000 Miles Tour by Two Englishwomen on Behalf of a Religious Education* (1922); *Through Western Canada in a Caravan* (1925); and *Canyons, Cans and Caravans* (1930). As the years passed, she added travel anecdotes to her sermons, telling her audiences about her adventures. One popular tale was about Hasell's encounter with a grizzly bear who stole her breakfast before moving on. After entertaining the crowd, Hasell would ask for donations, which she later used to maintain her caravan and establish permanent church schools. She retired after World War II and died in 1974. *See also* MISSIONARIES.

Further Reading: Hassell, Eva. *Across the Prairie in a Motor Caravan*. London: Society for Promoting Christian Knowledge, 1922.

Hathorn, Susan (born ca. 1830)

Susan Hathorn is representative of nineteenth-century American women who married sea captains and joined their voyages. Hathorn's journal of her adventures, which occurred during her first year of marriage in 1855, are particularly detailed. They describe Hathorn's travels from the Eastern seaboard of the United States to Cuba, the Caribbean, London, and Savannah, Georgia, and her decision to settle in Maine to await the birth of her child. Hathorn's writings and her life are the subject of *A Bride's Passage: Susan Hathorn's Year Under Sail* (1997) by Catherine Petroski. *See also* SEA TRAVEL; SPOUSES.

Further Reading: Petroski, Catherine. *A Bride's Passage: Susan Hathorn's Year Under Sail*. Boston: Northeastern University Press, 1997.

Hawes, Harriet Boyd (1871–1945)

Born in 1871, American archaeologist Harriet Boyd Hawes was the first person to discover a settlement of the Minoans, a people of ancient Greece. She traveled to the Greek island of Crete in 1901 while on leave from her job teaching Greek and Greek archaeology at Smith College in Northhampton, Massachusetts. Shortly after her arrival she found a Minoan town dating from the Early Bronze Age, and, as the leader of its excavation, she became the first woman to head a major archaeological project in the field. This had long been her goal. While at Smith, she had been denied inclusion in the fieldwork activities of the American School of Classical Studies in Athens, Greece, because of her gender. After her discovery, however, she was considered one of the foremost experts in archaeology. In 1902 she became the first woman to address the Archaeological Institute of America; she lectured throughout the country before returning to Greece to discover and excavate more settlements. In 1901 Smith College awarded her master's degree in archaeology.

Hawes's travels to Greece were particularly significant because that country was in turmoil, first because of the Greco-Turkish War and then because of World War I. During the former conflict, Hawes served as a volunteer nurse in Thessaly, Greece, and during World War I she carried supplies from the United States to Corfu, Greece. She also engaged in fundraising activities. After World War I she taught at Wellesley College in Wellesley, Massachusetts. Hawes retired in 1936 and died in 1945. *See also* ANTHROPOLOGISTS AND ARCHAEOLOGISTS.

Further Reading: Allesbrook, Mary. *Born to Rebel: The Life of Harriet Boyd Hawes*. Bloomington, IN: Oxbow Books, 1992.

Helena, Saint (ca. 250–ca. 330 C.E.)

Born in approximately 250 C.E., Saint Helena is one of the best known women travelers of ancient times. She made a pilgrimage from Rome to Jerusalem in approximately 324 C.E., where she found pieces of wood that she believed had come from Christ's cross. On the site of her discovery she founded the Church of the Holy Sepulchre. She then went to Bethlehem and established the Church of the Nativity on the spot where she believed Christ had been born. As the mother of Emperor Constantine the Great (ca. 274–337 C.E.), Helena had a powerful influence on the people of the Roman Empire, and many made their own pilgrimages to the churches Helena had established. Because of her contributions to Christian spirituality, she was declared a saint immediately after her death in approximately 330 C.E. *See also* PILGRIMS.

Further Reading: Baker, G. P. *Constantine the Great and the Christian Revolution*. New York: Barnes & Noble, 1967.

Hobson, Sarah (1947–)

Born in 1947, British filmmaker Sarah Hobson explored the Middle East dressed as a man, reporting on her experiences in a book that was titled *Through Persia in Disguise* in London and *Masquerade: An Adventure in Iran* in the United States. Published in 1973 and 1979, respectively, it describes Hobson's travels to parts of Iran that had never before been seen by Western women; her disguise gave her the freedom to visit many places normally forbidden to those of her gender. But Hobson explains that her original reason for adopting her male disguise was not curiosity about these forbidden places but concerns for her safety. She says:

> From what I had heard, it seemed that as an unveiled Christian girl in a Muslim country, I would experience pestering and unpleasantness. And the danger would probably be greater if I were to visit remote areas in Iran, and holy places where unveiled women had rarely been seen. I felt I could cause trouble, to myself and to others. [Therefore] I decided to dress as a boy, for though such disguise might not always convince, I felt it would give some protection. (p. 2)

Hobson's disguise was so convincing that she fooled people who gave her a ride as she hitchhiked through Spain, Italy, and Turkey to Istanbul, where she got a bus for Tehran, Iran. There she toured the city, studying art and architecture, before going to the holy city of Qum, 90 miles to the south. In Qum she visited a shrine—becoming the first white woman to enter it—and learned about the spiritual beliefs of the Iranian people. Throughout her visit she was afraid that her deception would be discovered; she believed that if she was revealed as a woman, she would be attacked by an angry mob for going places where foreign women were not allowed. Eventually she became so anxious about being unmasked that she left Qum for other, less spiritually significant, parts of Iran. However, before leaving the country she returned to Qum to see friends she had made there, and they finally saw through her disguise. Nonetheless, they did not turn her in to authorities.

In writing about her experiences in the 1979 version of her book, Hobson says that she learned as much from her masquerade as she did from her sightseeing. She reports:

> The experience changed me in many ways: it made me understand more about myself, particularly as a woman; it made me reconsider my values, and my behavior within relationships; it set me on a course of . . . research relating to other cultures, and on a quest for the meaning of human development. . . . (p. i)

In an epilogue she adds that she went back to Iran in 1978 during a time of political unrest. On this occasion, she wore women's clothing, including a veil, and had a very different travel experience. She says:

> I found the veil an enormous benefit, though as a garment it seemed inefficient for it required the mouth or a hand to prevent it slipping away from the head. It encompassed my whole body, and I found it gave me my own territory which strangers could not intrude on. It also prevented propositioning, because the way I learnt to wear it, plus the colour and weight of the veil, signalled me as a modest woman uninterested in flirtation. . . . And the veil, in a funny way, made me feel more secure in a climate that was volatile, not just because of the dominant male culture but because of political instability, social uncertainties, and geographical rawness—affecting women and men alike. (pp. 178–179)

Hobson suggests that the veil, though seemingly restrictive, in many ways gave her more freedom, because as a woman she was allowed to converse with other women at will, and learned about aspects of Iranian life that she had neglected during her previous visit.

Hobson also traveled to India and Africa without being in disguise. In the early 1980s she produced a children's television series on life in other countries. *See also* AFRICA; ASIA; DISGUISES.

Further Reading: Hobson, Sarah. *Masquerade: An Adventure in Iran*. Chicago: Academy Chicago, 1979.

Homesteading

Homesteading was the claiming and settling of federal lands in the United States under the Homestead Act of 1862, a practice that encouraged thousands of American settlers

Family with covered wagon in 1886. © *CORBIS*

to migrate west in the nineteenth century. The act was passed as a result of settlers' complaints that it was unfair for them to have to pay for virgin land, which only became valuable after they developed it. The Homestead Act awarded 160-acre plots to individuals who paid a small registration fee, built a homestead, and both lived on and cultivated the land for five years. Women were allowed to homestead their own claims as well as men, and many benefited from this policy. One such woman was Elinore Pruitt Stewart, who wrote about her homesteading experiences in *Letters of a Woman Homesteader*. Published in 1914, this book is a collection of letters that Stewart wrote from 1909 to 1913 to describe life on her homestead in Burnt Fork, Wyoming. Stewart initially went to Wyoming to work as a housekeeper for a rancher there, but after marrying her boss she established her own homesteading claim on property adjoining his. To satisfy the residency requirement of the Homestead Act, the couple built a house straddling their two pieces of lands.

In her letters, Stewart promoted homesteading it as a way for women to better themselves financially. She says:

> To me, homesteading is the solution of all poverty's problems . . . any woman who can stand her own company, can see the beauty of the sunset, loves growing things, and is willing to put in as much time at careful labor as she does over the washtub, will certainly succeed; will have independence, plenty to eat all the time, and a home of her own in the end. . . . I am only thinking of the troops of tired, worried women, sometimes even cold and hungry, scared to death of losing their places to work, who could have plenty to eat, who could have good fires by gathering the wood, and comfortable homes of their own, if they but had the courage and determination to get them. (pp. 214–217)

In all, 250 million acres were homesteaded by the 1950s, but loopholes in the law allowed homesteading not only by individuals but also by representatives of larger groups such as railroad companies and land speculators. This meant that not all of the 160-acre plots homesteaded were held by individuals. *See also* STEWART, ELINORE PRUITT.

Further Reading: Jones-Eddy, Julie. *Homesteading Women: An Oral History of Colorado, 1890–1950*. New York: Twayne Publishers, 1992; Stewart, Elinore Pruitt. *Letters of a Woman Homesteader*. 1914. Reprint. Boston: Houghton Mifflin, 1982.

Hopkins, Frances Ann (1818–1918)

Born in 1818, English artist Frances Ann Hopkins traveled extensively in Canada, drawing the attention of other women travelers to the region by virtue of her artwork. She first went to Canada with her husband and stepchildren in 1838 after he went to work for the Hudson Bay Company in Quebec. She then accompanied the company's fur traders on at least two voyages, one in 1869 and the other in 1870, and sketched and painted scenes from the region. In 1869 her painting *Canoes in a Fog, Lake Superior*, was exhibited at the Royal Academy in London, and several more exhibitions of her work followed, both in England and in France. In all, 85 of her works were shown, many depicting canoe travel related to the fur trade. Hopkins returned to England with her husband shortly before his death in 1893, then resumed her traveling. She also continued to paint. She died in 1918.

Further Reading: Clark, Janet. *Frances Anne Hopkins, 1838–1919: Canadian Scenery*. Thunder Bay, Ontario: Thunder Bay Art Gallery, 1990.

Hotels. *See* ACCOMMODATIONS; YOUNG WOMEN'S CHRISTIAN ASSOCIATION (YWCA).

Hubbard, Mena Benson (1870–1956)

Born in 1870, Mena Benson Hubbard was the first white woman to cross Labrador. She was also the first person to publish a detailed description of the region, writing *A Woman's Way through Unknown Labrador: An Account of the Exploration of the Nascaupee and George Rivers* (1908). During her 1905 expedition, Hubbard canoed along the

Nascaupee and George Rivers to map their course and record information on flora and fauna via scientific instruments and photographs.

Accompanying Hubbard on her 43-day trek was explorer George Elson. Elson had been on a failed expedition in 1903 with Hubbard's husband, who died during the journey, and Hubbard undertook her own crossing to honor her husband's memory. She also wanted to steal fame from explorer Dillon Wallace, who was planning his own Labrador expedition. Like Elson, Wallace had been on the failed 1903 expedition, but afterward he dismissed the contributions of Hubbard's husband as insignificant.

Hubbard was honored in her native Canada for her success but never undertook another such journey. She died in 1956.

Further Reading: Hubbard, Mena Benson. *A Woman's Way through Unknown Labrador: An Account of the Exploration of the Nascaupee and George Rivers*. New York: McClure, 1908.

Hurston, Zora Neale (1891–1960)

Born in 1891 in the United States, African American anthropologist and novelist Zora Neale Hurston is representative of women who traveled for research purposes. During the early 1930s she went throughout the American South and the West Indies in search of folk literature, producing some of the most important collections of African-American folklore. They include *Mules and Men* (1935), which was based on her field research in the South, and *Tell My Horse* (1938), which was based on her visits to Haiti and Jamaica. Hurston also wrote several novels, including *Their Eyes Were Watching God* (1937) and *Seraph on the Suwanne* (1948), as well as an autobiography, *Dust Tracks on a Road* (1942). Hurston died in 1960.

Further Reading: Hemenway, Robert E. *Zora Neale Hurston: A Literary Biography*. With a foreword by Alice Walker. Urbana: University of Illinois Press, 1977; Howard, Lillie P. *Zora Neale Hurston*. Boston: Twayne Publishers, 1980.

Zora Neal Hurston in the 1950s. © *CORBIS.*

Hutchings, Florence (1865–1880) and Gertrude (1868–?)

Born in 1865 and 1868, respectively, Florence and Gertrude Hutchings were the first white women born in what is now Yosemite National Park in California; their father opened the first hotel there in 1864. They spent their childhoods exploring the region and guiding visitors to various scenic spots, and they were the first women to climb Half Dome Mountain. In 1880, Florence was killed in a mountain-climbing accident; Gertrude moved to Vermont but in her later years returned to Yosemite to camp.

Further Reading: Kaufman, Polly Welts. *National Parks and the Woman's Voice*. Albuquerque: University of New Mexico Press, 1996.

Hutchison, Isobel Wylie (1889–1982)

Born in 1889, Scots botanist Isobel Wylie Hutchison traveled to Greenland, Alaska, Canada, and the Aleutian Islands to collect specimens of Arctic plants for the British Museum. During her research expeditions she sometimes lived with indigenous people and sometimes camped alone under brutal conditions. For example, during one trip to Herschel Island near the Arctic Circle, she had to travel across the ice by dogsled and endure sub-freezing temperatures while camping. In Greenland she hunted her own food while living in an Eskimo village. Hutchison wrote many articles and books about her experiences, including *On Greenland's Closed Shore* (1930) and *North to the Rime-Ringed Sun: Being the Record of an Alaska-Canadian Journey Made in 1933–34* (1934). She died in 1982.

Further Reading: Robinson, Jane. *Wayward Women: A Guide to Women Travellers*. Oxford: Oxford University Press, 1990.

Huxley, Elspeth Josceline (1907–)

Born in 1907, English author Elspeth Josceline Huxley wrote extensively about East Africa, traveling throughout the region to do research for her books and articles. She wrote about a variety of subjects related to Africa, including politics and tribal cultures. However, her best known work is an autobiographical novel, *The Flame Trees of Thika: Memories of an African Childhood* (1959), which describes her childhood in Thika, Kenya. Huxley's parents had left England after buying a 500-acre African plantation, and Huxley lived with them there from 1913 to 1925, except for a brief period in England during World War I.

As an adult, Huxley lived in London, where she met and married a sales and marketing expert. The representative of a tea company, he traveled throughout Europe and the British colonies, and Huxley accompanied him. She also frequently traveled to Africa to visit her parents, and eventually she decided to write about her childhood memories and current observations of Africa. Her works include *Settlers of Kenya* (1948), *Four Guineas: A Journey through West Africa* (1954), *The Challenge of Africa* (1971), and *Out in the Midday Sun: My Kenya* (1985). In addition, a collection of her correspondence with fellow African traveler Margery Perham, *Race and Politics in Kenya: A Correspondence between Elspeth Huxley and Margery Perham*, was published in 1944. *See also* AFRICA; PERHAM, MARGERY FREDA.

Further Reading: Huxley, Elspeth Joscelin Grant. *Out in the Midday Sun: My Kenya*. U.S. edition. New York: Viking, 1987; ———. *Race and Politics in Kenya: A Correspondence Between Elspeth Huxley and Margery Perham*. 1944. Reprint, with an introduction by Lord Lugard. Westport, CT: Greenwood Press, 1975; ———. *Settlers of Kenya*. 1948. Reprint. Westport, CT: Greenwood Press, 1975; ———. *White Man's Country: Lord Delamere and the Making of Kenya*. New York: Praeger, 1968.

I

Iditarod Trail Dog Sled Race

The Iditarod Trail Dog Sled Race is a challenging dogsled race across frozen Alaskan terrain amid the threat of thin ice, 100 mph winds, avalanches, wild animals, and other dangers. It originated in 1967 to commemorate the travels of dog sled mushers who rushed to deliver serum from Nenana to Nome—a distance of 674 miles—during a diphtheria epidemic in 1927; the serum had been sent by train from Anchorage but became stranded in Nenana after a snowstorm blocked the tracks. The 1967 race covered only 56 miles; since 1973, however, it has covered 1,160 miles from Anchorage to Nome along an old dogsled mail route. At one point the trail splits; in odd-numbered years, competitors take the southern route and in even-numbered years the northern route. In its entirety, the course is known as

Mushers Dee Dee Jonrowe and Susan Butcher take a break from the Iditarod race at the Mountain checkpoint. Alaska, USA. 1963. © *Paul A. Souders/CORBIS.*

the Iditarod Trail and crosses two mountain ranges, the Alaska and the Kuskokwim, as well as frozen waterways.

Men and women compete against each other in the race under equal circumstances. The first women to enter the Iditarod were Mary Shields and Lolly Medley in 1974. At that event, Shields became the first woman to finish the race, placing twenty-third; Medley arrived at the finish line 26 minutes later. The first woman to win the Iditarod was Libby Riddles in 1985. Another woman, Susan Butcher, won the race in 1986 and again in 1987, 1988, and 1990. Another notable female Iditarod competitor is DeeDee Jonrowe, who placed second in the race in 1993. She is the only woman to have competed in both the Iditarod and another major dogsledding race, the Alpirod, for three years straight, in 1992, 1993, and 1994. Jonrowe competed in the Iditarod for the first time in 1980 and has co-authored a book about the competition, *Iditarod Dreams: A Year in the Life of Alaskan Sled Dog Racer DeeDee Jonrowe* (1995). Libby Riddles has also written about the Iditarod in *Race Across Alaska: First Woman to Win the Iditarod Tells Her Story* (1988). *See also* Butcher, Susan; Jonrowe, DeeDee; Riddles, Libby; Shields, Mary.

Further Reading: Dolan, Ellen M. *Susan Butcher and the Iditarod Trail.* New York: Walker, 1993; Freedman, Lew. *Iditarod Classics: Tales of the Trail from the Men and Women Who Race Across Alaska.* Fairbanks, AK: Epicenter Press, 1992; Jones, Tim. *The Last Great Race: The Iditarod.* Harrisburg, PA: Stackpole Books, 1988; Jonrowe, DeeDee, with Lewis Freedman. *Iditarod Dreams: A Year in the Life of Alaskan Sled Dog Racer DeeDee Jonrowe.* Fairbanks AK: Epicenter Press, 1995; Nielsen, Nicki J. *The Iditarod: Women on the Trail.* Anchorage, AK: Wolfdog Publications, 1986; Riddles, Libby, and Tim Jones. *Race Across Alaska: First Woman to Win the Iditarod Tells her Story.* Harrisburg, PA: Stackpole Books, 1988.

Industrial Revolution

The Industrial Revolution of the nineteenth century encouraged millions of people, many of them young women, to leave their rural homes and travel to cities in search of employment. During this period new machines were invented that made jobs easier to perform, and men—who typically demanded higher wages than women—were not the only ones who could operate them. Therefore as more and more new factories were built to accommodate these new machines, factory owners turned to women and children for their workforce. This was particularly true in the textile industry.

The majority of women who traveled to industrial regions to work were young. For example, in Stockport, England, in 1841, the average age of women weavers was 20; in 1861, it was 24. In Lowell, Massachusetts, in the 1830s and '40s, 80 percent of the female workers were under 30; many were as young as 15. Since many of these women were away from home for the first time, various societies sprung up to offer them support. Their efforts included providing safe places for these women to live and social events at which they could meet reputable young men while chaperoned. But in traveling to their new lives, many of these women were without chaperones or support; they trusted on social conventions to keep their virtue safe. The same was true for women workers in the pre-industrial period. At this time many young women from England traveled across the Atlantic Ocean to America, where they became servants or plantation or factory workers. By 1870, 50 percent were servants. This changed, however, with the advent of the twentieth century, when women increasingly became office workers instead of servants and factory workers; by 1920, roughly 40 percent of American women who worked were teachers, shop clerks, or clerical workers. These women also often traveled far from home, leaving rural communities for cities, and experienced the same problems associated with being separated from their families. However, relocation for work purposes offered them many opportunities—including independence—that they would not have had at home.

Although the Industrial Revolution gave women an opportunity to travel far from home, once they arrived in factory towns they were often exploited. Factory workers labored long hours under poor conditions, and were usually housed together in dormitories, separated by gender, that were unsanitary and/or overly crowded. By the twentieth century, labor laws and unions had improved working conditions. However, these laws did nothing to address the social problems brought about by the mass separation of women from their families.

Further Reading: Henderson, W. O. *The Industrial Revolution in Europe, 1815–1914*. Chicago: Quadrangle Books, 1961.

J

Jackson, Monica

In 1955, mountaineer Monica Jackson set off with two other women climbers, Elizabeth Stark and Evelyn Camrass, on the First Women's Himalayan Expedition. All of the women were members of the Ladies Scottish Climbing Club, and they intended to find a new path through a series of glaciers and to validate the accuracy of a map of the region. In addition to achieving their goals within only two months, the women were the first Europeans to climb Mount Gyalgen, which they named for their Nepalese Sherpa guide. Jackson later wrote about their experiences in *Tents in the Clouds: The First Women's Himalayan Expedition* (1956). In subsequent years she continued to climb, ascending several peaks in Turkey and surrounding regions in 1965. *See also* MOUNTAINEERING.

Further Reading: Jackson, Monica, with Elizabeth Stark. *Tents in the Clouds: The First Women's Himalayan Expedition*, London: Collins, 1956.

James, Naomi (1949–)

Born in 1949, Naomi James of Ireland was the first woman to circumnavigate the globe alone via Cape Horn. She embarked on her adventure in a 53-foot yacht, traveling 252 days from September 1977 to June 1978. Her only companion was a kitten that was washed overboard during her journey.

James wrote two books based on her experiences, *Woman Alone* (1978) and *Alone Around the World: At One With the Sea* (1979). She also wrote another autobiographical work, *At Sea On Land* (1981), and a book about historic sailing expeditions entitled *Courage at Sea: Tales of Heroic Voyages* (1987). *Courage at Sea* includes an

Naomi James checks the rigging on the yacht *Koiter Lady* before setting off in the Transatlantic Race from Plymouth in 1980. © *Hulton-Deutsch Collection/ CORBIS.*

account of James's victory, with her husband, in a 1982 trimaran race around Great Britain.

James also won the Ladies' Division of the 1980 Observer Transatlantic Race; in fact, she was the only woman to finish the race. James decided to stop competing in such events shortly after her husband was killed in a sailing accident in 1983. *See also* SEA TRAVEL.

Further Reading: James, Naomi. *Alone Around the World*. New York: Coward, McCann & Geoghegan, 1979; ———. *Courage at Sea: Tales of Heroic Voyages*. U.S. edition. Topsfield, MA: Salem House Publishers, 1988; ———. *Woman Alone*. London: Daily Express, 1978.

Jameson, Anna Brownell (1794–1860)

Born in 1794, Irish author Anna Brownell Jameson was the first European woman to shoot the rapids at Sault St. Marie in Canada. She traveled throughout eastern Canada as well as in the United States. Her first trip was to Europe as a governess, during which she wrote a book, *Diary of an Ennuyeé* (1834).

While in England she also met and married Robert J. Jameson, who was named attorney general of Upper Canada in 1833. She joined him there in 1836, primarily because it was important for his political career to have a wife by his side; after her arrival he was appointed vice chancellor. Anna Jameson then traveled alone to various Canadian towns, as well as to Mackinac Island, Manitoulin Island, and Lake Huron. She left Canada in 1837, and she subsequently wrote more books about her travel experiences, including *Sketches in Canada, and Rambles Among the Red Men* (1852). In her later years she became an art historian in Europe and wrote books about historical subjects. Jameson died in 1860. *See also* NORTH AMERICA.

Further Reading: Jameson, Anna. *Sketches in Canada, and Rambles Among the Red Men*. London: Longman, Brown, Green, and Longmans, 1852; ———.*Visits and Sketches at Home and Abroad. With Tales and Miscellanies Now First Collected, and a New Edition of the "Diary of an ennuyée."* New York: Harper & Brothers, 1834.

Johnson, Amy (1903–1941)

Born in 1903 in Hull, England, Amy Johnson was the first woman to fly solo from England to Australia, a journey of approximately 11,000 miles in a single-engine Gypsy Moth airplane. Her journey took place in 1930, two years after Bert Hinkler became the first solo aviator to fly the same route. (Hinkler died during a second such flight in 1933.) Johnson hoped to beat Hinkler's record of 16 days, but her own flight took 19 days, including both planned stops and forced landings.

Johnson did set records in several other flights, including a 1931 trip from England to Japan, a 1932 trip from England to Capetown, South Africa, and a 1936 trip from England to Capetown; the two South African flights were made solo. Johnson and her husband set a record together when they flew nonstop from England to India in 1934 as part of an England-to-Australia air race. Her marriage to the fellow pilot was short-lived because of the pressures of Johnson's fame. She was celebrated by the press throughout her career and was the subject of several books and articles. She also wrote her own book based on her experiences, *Sky Roads of the World* (1939).

During World War II, Johnson volunteered to serve in the Air Transport Auxiliary Force, delivering planes from the factories where they were built to various Royal Air Force bases. During one such flight in 1941 Johnson became disoriented in bad weather, ran out of fuel, and crashed into the River Thames in London. A ship immediately came to her rescue, but she was swept under its bow and drowned. *See also* AIR TRAVEL.

Further Reading: Johnson, Amy. *Sky Roads of the World*. London: W. & R. Chambers, 1939.

Johnson, Osa (1894–1953)

Born in 1894, American explorer and photographer Osa Johnson was the first woman to take a movie camera to Africa, traveling throughout the continent as well as in the South Pacific with her husband, fellow photographer Martin Johnson. When they mar-

Osa Johnson during safari in Africa, probably 1926. *Library of Congress.*

ried, he had just returned from accompanying author Jack London to the South Seas; shortly thereafter he wrote *Through the South Seas with Jack London* (1913). Once the book was published, the couple embarked on a speaking tour through the United States, during which Martin told stories and showed photographers while Osa provided piano accompaniment.

Four years later the two went on their first overseas trip together. They visited several South Pacific islands specifically looking for cannibals, and Johnson was nearly taken captive by one tribe on the island of Malekula; the chief had just grabbed her and was about to carry her off when a British war ship appeared on the horizon, frightening him into letting her go. Johnson and her husband also filmed the indigenous people they encountered. When they returned to the United States, they made a movie from their footage, *Among the Cannibal Isles of the South Seas* (1917). Over the next few years, they continued to travel to various islands to film indigenous people, producing more movies and a book, *Cannibal-Land* (1922).

During this time they met African explorer Carl Akeley (husband of explorer Delia Denning Akeley), who encouraged them to photograph wild animals in Africa. Acting on his advice, they sought funding from a photographic company, Kodak, and the American Museum of Natural History in New York City, and in 1923 they embarked on a major expedition to Africa. There they found a heavily used watering hole and built a house nearby, where they lived for four years photographing big game. They then visited several other sites in Africa, creating movies from their experiences.

The Johnsons returned to the United States in 1932 but left for Africa again almost immediately, this time to fly from one remote region to another. In 1937 their plane crashed; Martin was killed while Osa survived and later remarried. In 1940 she published a book about her life with Martin, *I Married Adventure*, which was her most popular work. Two more autobiographical works followed: *Four Years in Paradise* (1941) and *Bride in the Solomons* (1944). Johnson died in 1953. *See also* AFRICA; PHOTOGRAPHERS AND ARTISTS.

Further Reading: Johnson, Osa. *Bride in the Solomons*. Boston: Houghton Mifflin, 1944; ———. *Four Years in Paradise*. Philadelphia: J. B. Lippincott, 1941; ———. *I Married Adventure: The Lives and Adventures of Martin and Osa Johnson*. Philadelphia: J. B. Lippincott, 1940.

Jonrowe, DeeDee (1953–)

DeeDee Jonrowe is the only woman to have competed in two major dogsledding races, the Iditarod and the Alpirod, for three years straight, in 1992, 1993, and 1994. She began racing in 1978, driving a friend's dogs in the Anchorage Fur Rendezvous Woman's World Championship, but did not perform very well. In 1977, after marrying a man with an interest in dogsledding, she began buying and training her own dogs, and by 1979 she had a team of 25. She competed in the Iditarod for the first time in 1980, and in 1993 she finished second in the race. She has co-authored a book about the competition, *Iditarod Dreams: A Year in the Life of Alaskan Sled Dog Racer DeeDee Jonrowe* (1995). She currently lives in Willow, Alaska, with her husband. *See also* IDITAROD TRAIL DOG SLED RACE.

Further Reading: Dolan, Ellen M. *Susan Butcher and the Iditarod Trail*. New York: Walker, 1993; Jonrowe, DeeDee, with Lewis Freedman. *Iditarod Dreams: A Year in the Life of Alaskan Sled Dog Racer DeeDee Jonrowe*. Fairbanks, AK: Epicenter Press, 1995; Nielsen, Nicki J. *The Iditarod: Women on the Trail.* Anchorage, AK: Wolfdog Publications, 1986; Jones, Tim. *The Last Great Race: The Iditarod.* Harrisburg, PA: Stackpole Books, 1988.

A Journey of One's Own

Published in 1996, *A Journey of One's Own: Uncommon Advice for the Independent Woman Traveler* by Thalia Zepatos is one of the foremost guidebooks related to solo women's travel, although it also includes information on traveling with a companion. It alternates essays by woman travelers with chapters on subjects such as planning an itinerary and dealing with health and safety issues while abroad. It also includes a list of resources for travelers, including travel books and magazines, Internet sites, and travel organizations. *See also* GUIDEBOOKS, TRAVEL.

Further Reading: Zepatos, Thalia. *A Journey of One's Own: Uncommon Advice for the Independent Woman Traveler*. Portland, OR: Eighth Mountain Press, 1996.

K

Keen, Dora

In 1912, American mountaineer Dora Keen led the first expedition up Alaska's Mount Blackburn. Although this expedition failed, a 1913 ascent was successful, making Keen the first woman to reach the mountain's peak. The 1913 expedition, which included seven male climbers, was also the first to climb a significant distance at night. The group experienced serious avalanches, but no one was killed. *See also* MOUNTAINEERING.

Further Reading: Birkett, Bill, and Bill Peascod. *Women Climbing: 200 Years of Achievement*. Bronson, MO: Mountaineer Books, 1990.

Kemble, Fanny (1809–1893)

Born in 1809, Fanny Kemble was an English actress who traveled internationally to perform during the 1830s. After touring the east coast of America in 1832, she wrote a book about her travels entitled *Journal* (1835). By the time it was published, she had abandoned acting to marry a plantation owner from Georgia, an experience she documented in *Journal of a Residence on a Georgian Plantation in 1838–1839* (1863). When she divorced him six years later, she took an extended trip to Europe, writing about it in *A Year of Consolation* (1847). In her later years she continued to travel and to write autobiographical material. Kemble died in 1893.

Further Reading: Armstrong, Margaret. *Fanny Kemble, A Passionate Victorian*. New York: Macmillan, 1938; Rushmore, Robert. *Fanny Kemble*. New York: Crowell-Collier Press, 1970.

Kempe, Margery (ca. 1373–ca. 1440)

Born in approximately 1373, Margery Kempe is one of the best-known religious pilgrims of the Middle Ages. In 1413, she left on a religious pilgrimage from England to the Middle East, where she visited many sites mentioned in the New Testament of the Bible. She then traveled to Rome, Italy, and Damascus, Syria, before sailing home. She was gone about a year and did most of her traveling on foot or by donkey.

Kempe made several subsequent religious pilgrimages, although primarily to local shrines, and visited Germany and the Baltic as well. However, her pilgrimage to the Holy Land has received the most attention because someone wrote down her stories about this trip. The manuscript survives as *The Book of Margery Kempe*, first published in 1936. Historians believe that Kempe died in approximately 1440. *See also* PILGRIMS.

Further Reading: Atkinson, Clarissa W. *Mystic and Pilgrim: The Book and the World of Margery Kempe*. Ithaca, NY: Cornell University Press, 1983; Collis, Louise. *Memoirs of a Medieval woman: The Life and Times of Margery Kempe*. New York: Crowell, 1964; Kempe, Margery. *The Book of Margery Kempe*. 1936. Reprint, translated by B. A. Windeatt. New York: Viking Penguin, 1985; Thornton, Martin. *Margery Kempe:*

An Example in the English Pastoral Tradition. London: S.P.C.K., 1960.

Kingsley, Mary (1862–1900)

Born in 1862, English explorer Mary Kingsley traveled throughout West Africa and wrote a very popular book about her experiences, *Travels in West Africa* (1897), which inspired other travelers to turn their attention to the African continent. The book actually describes two trips that Kingsley made to the region. She embarked on her first trip to Africa, which lasted nine months, after the death of her parents in 1892. Kingsley had been the caretaker for her physically and mentally ill mother, and once free of her responsibility she immediately left for the Canary Islands, where she met some people bound for West Africa and decided to go there too. In writing of this decision in an introduction to *Travels in West Africa*, scholar Elizabeth Claridge says:

> Few women of any period have stepped so abruptly from conventional circumstances into the unknown. Mary Kingsley is often compared to Marianne North, Constance Gorden Cumming, and Isabella Bird, those other redoubtable travellers who left the confines of Victorian England to explore remote and hazardous parts of the world, but they progressed to their more difficult journeys from preliminary excursions abroad. When Mary Kingsley boarded *S.S. Lagos*, the only woman apart from a stewardess . . . , her experience of foreign travel amounted to a week in Paris with a woman friend in 1888, and a trip to the Canaries by Castle liner to restore her spirits after her parents' deaths. (p. ix)

Kingsley's second African trip lasted for 11 months in 1894–1895, and it provides the bulk of material in *Travels in West Africa*. However, the book opens with the author's brief explanation of why she decided to go to Africa for the first time, despite the objections of friends who thought such a journey was too dangerous for a woman. Kingsley left Liverpool, England, for Sierra Leone, a West African shipping port controlled by the British. From there she traveled down the African coast to Angola and then into the continent's interior.

In describing Kingsley's second expedition, *Travels in West Africa* includes excerpts from Kingsley's diaries, even though the author says that "publishing a diary is a form of literary crime." (p. 100) However, she adds that "no one expects literature in a book of travel," continuing:

> [T]here are things to be said in favour of the diary form, particularly when it is kept in a little known and wild region, for the reader gets therein notice of things that, although unimportant in themselves, yet go to make up the conditions of life under which men and things exist. The worst of it is these things are not often presented in their due and proper proportion in diaries. Many pages in my journals that I will spare you display this crime to perfection. For example: "Awful turn up with crocodile about ten—paraffin good for over-oiled boots—Evil spirits crawl on ground, hence high lintel—Odeaka cheese is made thus:—" Then comes half a yard on Odeaka cheese making. (p. 100)

Travels in West Africa also includes detailed descriptions, observations, and stories of adventure related to Kingsley's travels, as well as commentaries about African politics and policies. As an example of the latter, Kingsley says:

> All the training the boys [in mission schools] get is religious and scholastic. . . . It is strange that no technical instruction is given by any government out here. All of the governments support mission schools by grants: but the natives turned out by the schools are at the best only fit for clerks, and the rest of the world seems to have got a glut of clerks already, and Africa does not want clerks yet, it wants planters—I do not say only plantation hands, for I am sure from what I have seen in Cameroons of the self-taught native planters there, that intelligent Africans could do an immense amount to develop the resources of the country. (p. 206–207)

Kingsley also goes into depth about African culture. She studied several tribes of the interior and made detailed notes regarding their customs. As an example, she says:

> These customs [regarding the birth of twins in one region] . . . amounted to the mother of twins being kept in her hut for a year after the birth. Then there was a great dance and certain ceremonies, during which the lady and the doctor, not the husband, had their legs painted white. When the ceremonials were over the woman returned to her ordinary avocations.
>
> There is always a sense of there being something uncanny regarding twins in West Africa, and in those tribes where they are not killed they are regarded as requiring great care to prevent them from dying on their own account. I remember once among the Tschwi trying to amuse a sickly child with an image which was near it and which I thought was its doll. . . . I found out that the image was not a doll at all but an image of the child's dead twin which was being kept near it as a habitation for the deceased twin's soul, so that it might not have to wander about, and, feeling lonely, call its companion after it. (p. 473)

In addition to such details in her text, Kingsley offers appendixes that report on African trade and labor issues, diseases, reptiles, fish, and plants. During her travels she collected animal and plant specimens for British museums. When she returned to England she also gave public lectures on Africa. In both her lectures and her book, she endeavored to portray Africa and its people as realistically as possible. As she explains to British readers in her preface:

> "I have never accepted an explanation of a native custom from one person alone, nor have I set down things as being prevalent customs from having seen a single instance. I have endeavored to give you as honest an account of the general state and manner of life . . . and some description of the various types of country. . . . Your superior culture-instincts may militate against your enjoying West Africa, but if you go there you will find things as I have said." (pp. xx–xxi)

Interestingly, those who encountered Kingsley during her travels usually assumed she was a missionary, because most British women who went to Africa at that time were there to spread Christianity. However, Kingsley was an agnostic and had no interest in converting Africans to Western religions. She was, however, interested in African politics. Back in England she lectured about this subject as well as about her adventures, and her book *West African Studies* (1899) was largely political. She disagreed with the prevailing viewpoint that Africans were incapable of managing their own affairs, and while she never suggested that England should give up its governing of West Africa, she did argue that more respect should be afforded to indigenous cultures.

Kingsley's last work was *The Story of West Africa* (1900). Shortly before it was published, she went to South Africa to work as a nurse and newspaper correspondent during the Boer War, where she fell ill and died. *See also* AFRICA; BISHOP, ISABELLA LUCY BIRD; CUMMING, CONSTANCE; EXPLORERS; MISSIONARIES; NORTH, MARIANNE; VICTORIAN ERA.

Further Reading: Frank, Katherine. *A Voyager Out: The Life of Mary Kingsley*. Boston: Houghton Mifflin, 1986; Hughes, Jean Gordon. *Invincible Miss: The Adventures of Mary Kingsley*. London: Macmillan, 1968; Kingsley, Mary H. *Travels in West Africa*. 1897. Reprint, with introduction by Elizabeth Claridge. London: Virago Press, 1982; Myer, Valerie Grosvenor. *A Victorian Lady in Africa: The Story of Mary Kingsley*. Southampton, England: Ashford Press Pub., 1989.

Knight, Sarah Kemble (1666–1727)

Born in 1666, Sarah Kemble Knight was a Boston schoolteacher who traveled alone on horseback from Boston to New York in 1704, keeping a detailed account en route of places, people, inns, and local customs she encountered. Her diary was published in 1825 as *The Journal of Madame Knight* and again in 1865 with biographical information about Knight. In 1713 Knight moved to Norwalk, Connecticut, where she lived the rest of her life. She died on Christmas Day in 1727. *See also* EQUESTRIENNES.

Further Reading: Knight, Sarah Kemble. *The Journal of Madame Knight*. 1825. Reprint. Chester, CT: Applewood Books, 1992.

Kroeber, Theodora (1897–1979)

Born in 1897, American anthropologist Theodora Kroeber (also known as Kroeber-Quinn) was a member of the Society of Women Geographers who traveled extensively during the 1960s and 1970s to study Native American tribes. She also held a master's degree in clinical psychology and was the mother of well-known author Ursula LeGuin. Kroeber was also a writer, producing several books based on her field research. They include *An American Anthropologist Looks At History* (1963), *Almost Ancestors: The First Californians* (1968), and *Ishi, the Last Yahi: A Documentary History* (1979). She died in 1979. *See also* ANTHROPOLOGISTS AND ARCHAEOLOGISTS.

Further Reading: Kroeber, Theodora, and Robert F. Heizer. edited by F. David Hales. *Almost Ancestors: The First Californians*. San Francisco: The Sierra Club, 1968; Kroeber, Theodora, and Robert F. Heizer, eds. *Ishi, the Last Yahi: A Documentary History*. Berkeley: University of California Press, 1979.

Kuhn, Isobel (1901–1957)

In 1928, Canadian Isobel Kuhn went to China as part of a missionary effort by the China Inland Mission, one of several organizations that sent men and women into Asia to spread Christianity. Shortly after arriving in China she married fellow missionary John Kuhn, and for 20 years the two worked together among the Lisu people of the western province of Yunnan. In 1950 the area became the site of guerilla activity, so they left there for Thailand. Kuhn wrote many articles and books based on her experiences, including *Nests Above the Abyss* (1947), *Green Leaf in Drought-Time: The Story of the Escape of the Last C.I.M. Missionaries from Communist China* (1957), and *Ascent to the Tribes: Pioneering in North Thailand* (1956). She died in 1957. *See also* MISSIONARIES.

Further Reading: Dick, Lois Hoadley. *Isobel Kuhn*. Minneapolis, MN: Bethany House Publishers, 1987.

L

Ladies on the Loose

Published in 1981, *Ladies on the Loose: Women Travellers of the 18th and 19th Centuries* by Leo Hamalian is an anthology of writings by 17 women travelers of the eighteenth and nineteenth centuries. The women represented in the volume are the best known of their type; obscure travelers are not included. But fame was not Hamalian's main criterion for including them. Instead he says the women were selected because their writings are the most readable, adding: "I simply chose passages that struck me as the most interesting and assumed that they would seem the same to my readers." [p. xii] This would suggest that the most famous travelers in history were also the best writers. *See also* Bishop, Isabella Lucy Bird; Blunt, Anne; Kingsley, Mary; Martineau, Harriet; Montague, Mary Wortley; Pfeiffer, Ida Reyer; Piozzi, Hester Lynch; Stanhope, Hester; Tinne, "Alexine" (Alexandrine); Wollstonecraft, Mary; Workman, Fanny Bullock.

Further Reading: Hamalian, Leo. *Ladies on the Loose: Women Travellers of the 18th and 19th Centuries*. New York: Dodd, Mead, 1981.

Lagemodière, Marie-Anne (1780–1875)

Born in 1780, Canadian traveler Marie-Anne Lagemodière is believed to have given birth to the first white child in western Canada. She was a housekeeper in Quebec when she married a fur trader and decided to accompany him on his journeys. Her first trip was an 1806 expedition to an outpost on the Red River in western Canada. There her new husband's Native American mistress tried to poison her. Nonetheless, she remained with him and gave birth to a child in January 1807; in August 1808 she gave birth to a second child.

During these years she lived as a Native American would, hunting beaver and buffalo with her husband, his friends, and their Native American women. However, in 1812 her husband took a job delivering messages back and forth from western to eastern Canada, and Lagemodière settled among a group of pioneers. In 1816, her settlement was destroyed by a hostile Indian tribe, and she was forced to relocate; her husband had some difficulty finding her again. After they reunited, they built a home on a land grant beside the Red River, where Lagemodière remained until her death in 1875. She bore eight children in all. *See also* North America.

Further Reading: Harding, Les. *The Journeys of Remarkable Women: Their Travels on the Canadian Frontier*. Waterloo, Ontario: Escart Press, 1994.

Latin America

Latin America, which comprises both Central and South America, was first explored by nonindigenous people in the late fifteenth century after being claimed by both Spain and Portugal, but, unlike Africa, no famous women travelers are associated with its early exploration. Most women visited the region as immigrants while on voyages elsewhere, particularly during the California Gold Rush that began in 1849. At that time, three routes of travel were available from the eastern United States to California. One was to travel by wagon train across the United States; another was to take a ship to the Isthmus of Panama, then cross over land to take a ship up the west coast of North America; the third was to sail completely around South America via Cape Horn at the southern tip of the continent. Travelers choosing the last two routes might stop for a brief time in Latin American towns along the way, but in their haste to reach California, they typically did not spend a significant time in these places.

However, women travelers not seeking riches and/or a new life in California did sometimes spend time in Latin America. Of these, one of the most notable is Frances Erskine Calderón de la Barca, who traveled in the region from 1839 to 1841 as the wife of a Spanish diplomat and wrote about her experiences. Another notable woman traveler was English author Florence Dixie, who went on a hunting trip to South America with friends in the late 1870s. She too wrote a book, *Across Patagonia* (1880), about her experiences. Several prominent pirates, including Anne Bonney and Mary Read, also visited the region on numerous occasions between 1718 and 1720, and many slaves were transported to the region prior to the twentieth century.

At the beginning of the twentieth century, the focus of women traveling to Latin America turned from adventure and colonization to science. Among the first women scientists to travel in Latin America were American naturalist Ynes Mexia, and English entomologist Margaret Elilzabeth Fontaine. Fountaine went to South America in search of butterfly specimens. Beginning in 1890, she visited every continent and most major countries, including not only South America but also Syria, Turkey, Algeria, Spain, Africa, the Caribbean, India, New Zealand, Fiji, the United States, and the Far East, collecting over 20,000 butterflies during her adventures. Mexia, who made major contributions to the field of botany during the 1920s, began with a first expedition was to Mexico, where from 1925 to 1927 she collected over 1,500 plants, many of them previously unknown. From 1929 to 1932 she traveled in Brazil and Peru. She made another expedition to South America from 1934 to 1936 and another to Mexico from 1936 to 1938.

More recently, sociologists, physicians, and other professionals visited Latin America during the 1960s while volunteering for the Peace Corps, an international aid organization. In the 1970s and 1980s, many women environmentalists traveled to Latin America as well. Today's scientists are as interested in collecting specimens as were Margaret Fountaine and Mexia Ynes in years past; however, modern researchers gather specimens not to classify them but to determine whether they are becoming extinct. Similarly, anthropologists have visited Latin America to study indigenous cultures and preserve them. Several women writers have addressed the problem of threatened cultures in Latin America, including Catherine Caufield and Marika Hanbury-Tenison.

For women who are not involved in scientific or sociological endeavors, tour companies, particularly those offering adventure travel experiences, offer guided tours to Latin America. Many excellent guidebooks offer tips about travel to the region, including Natania Jansz and Miranda Davies's *More Women Travel: Adventures and Advice from More than 60 Countries* (1995). Jansz and Davies also report on cultural aspects of Latin America and discuss women's organizations that are dedicated to feminist activities. *See also* ADVENTURE TRAVEL; CALDERÓN DE LA BARCA, FRANCIS ERSKINE; CAUFIELD, CATHERINE; DIXIE, FLORENCE; FOUNTAINE, MARGARET; HANBURY-TENISON, MARIKA; MEXIA, YNES; PFEIFFER, IDA REYER; PIRATES.

Further Reading: Jansz, Natania, and Miranda Davies, eds. *More Women Travel: Adventures and Advice from More than 60 Countries.* London: Rough Guides, 1995; Tinling, Marion, compiler and ed. *With Women's Eyes: Visitors to the New World, 1775–1918.* Hamden, CT: Archon Books, 1993.

Law, Ruth (1901–1960)

Born in 1901, American aviator Ruth Law set the distance record for nonstop flying in 1916 by traveling 590 miles from Chicago to Hornell, New York, without refueling. She accomplished this six-hour flight, which set the cross-country record for both men and women, by modifying her plane to increase its fuel capacity from 8 to 53 gallons of gas. Her intent was to fly nonstop all the way to New York City, but at Hornell she had to land to refuel. She then took off again, but nightfall forced her to land again, this time in Binghamton, New York. The next morning, she arrived in New York City, which acclaimed her for her exploit.

Ruth Law in 1916. *Library of Congress.*

The public was already familiar with Law by this time because of a publicity stunt she performed the year before. Flying over a baseball stadium in New York, she dropped a grapefruit for Brooklyn Dodger Casey Stengel, a well-known ballplayer, to catch in his mitt. Law was supposed to drop a baseball, but forgot to take it along on her flight and substituted her snack instead.

In 1924, Law became the first woman to earn an international hydroplane license. She was also one of the first members of the Ninety-Nines, an international organization of women pilots. She continued to enjoy flying for most of her life and died in 1960. *See also* AIR TRAVEL.

Further Reading: Brown, Don. *Ruth Law Thrills a Nation.* New York: Houghton Mifflin, 1997. The Ninety-Nines: International Organization of Women Pilots, www.ninety-nines.org (May 2000).

Lawrence, Mary

New Englander Mary Lawrence left behind detailed diaries and letters regarding her life as a whaler's wife during the 1850s. Her husband captained a ship called the *Addison*, and the couple and their young daughter spent months at a time at sea. They also visited exotic places throughout the world, including Hawaii and the South Pacific. *See also* SEA TRAVEL.

Further Reading: De Pauw, Linda Grant. *Seafaring Women.* Boston: Houghton Mifflin, 1982.

Le Blond, Elizabeth (1861–1934)

Born in 1861, Elizabeth Le Blond was a mountaineer who specialized in climbing peaks during winter. She was an expert in snow photography and wrote an important book on the subject in 1895. She also wrote books related to mountaineering; her works include *The High Alps in Winter: Or, Mountaineering in Search of Health* (1883), *My Home in the Alps* (1892), *Adventures on the Roof of the World* (1904), and *Mountaineering in the Land of the Midnight Sun* (1908).

Her autobiography, *Day In, Day Out* (1928), was very popular, as were her guided tours to Spain and Italy.

Le Blond traveled extensively throughout Europe, not only climbing mountains in Norway, Spain, Italy, and Switzerland but also competing in the sport of driving cars up mountainsides. She was also an accomplished skater, tobogganer, and alpine bicyclist, and she served as president of the Ladies' Alpine Club of London. During World War I she served as a nurse and gave lectures on mountain climbing to British troops. *See also* MOUNTAINEERING.

Further Reading: Le Blond, Elizabeth. *Day In, Day Out*. London: John Lane, 1928.

Leakey, Mary (1913–1996)

Born in 1913, English archaeologist Mary Leakey traveled to Africa in the 1930s and worked on several major archaeological excavations there, making several significant fossil discoveries from the 1940s to the 1970s. Her first excavations were in England, but after marrying anthropologist Louis Leakey in 1933, the two concentrated on African sites. Together they reported on new species of early hominids and encouraged fieldwork by other researchers. Mary Leakey wrote several books about their work, including *Olduvai Gorge: My Search for Early Man* (1979). *See also* AFRICA; ANTHROPOLOGISTS AND ARCHAEOLOGISTS.

Further Reading: Leakey, Mary D. *Disclosing the Past*. Boston, MA: G. K. Hall, 1985; ———. *Olduvai Gorge: My Search for Early Man*. London: Collins, 1979.

Leonowens, Anna Harriette (1834–1914)

Born in 1834, Anna Harriette Leonowens is one of the most famous Englishwomen ever to live abroad. A governess for the royal family in Siam (present-day Thailand), she is the subject of a 1945 book, *Anna and the King of Siam* by Margaret Landon. Recently widowed, Leonowens went to Siam to work for its king in 1862. She was already well traveled by then; she and her husband, a military officer, had lived in India, Australia, and Singapore, where Leonowens ran a school. Once in the Siamese court, she taught over 60 royal children and their mothers, educating them in Western knowledge, manners, and customs. She also worked as the king's secretary. She remained in this position until 1867, when she retired to the United States because of poor health. Three years later, she published a book about her experiences: *The English Governess at the Siamese Court: Recollections of Six Years in the Royal Palace at Bangkok* (1870). This work offered Europeans the first thorough description of Siamese life; at the time Leonowens arrived, the country had been open to foreigners for only 12 years, after a closure of over a century. Leonowens encouraged more European involvement in Siam. Her book not only spoke of the country's attempts to modernize its ways, but also suggested that England's guidance was needed, particularly regarding the treatment of slaves. She concludes:

> The capacity of the Siamese race for improvement in any direction has been sufficiently demonstrated, and the government has made fair progress in political and moral reforms; but the conditions of the slaves is such as to excite astonishment and horror. What may be the ultimate fate of Siam under this accursed system, whether she will ever emancipate herself while the world lasts, there is no guessing. The happy examples free intercourse affords, the influence of European ideas, and the compulsion of public opinion, may yet work wonders. (p. 138)

The English Governess at the Siamese Court increased interest in Siam among Europeans, and continued to do so after it was made into a movie, "Anna and the King of Siam" (1946) and a Broadway musical, "The King and I" (1951). These were based not only on Leonowens's book but also on Landon's; the latter suggests that Leonowens and the King were romantically involved, an idea that historians dispute. The 1946 movie was remade in 1999, and the musical version was made as a movie in 1956. Leonowens wrote two other books in addition to *The English Governess*, one on Siamese life and another on her travels through India. Leonowens died in 1914. See also ASIA; GOVERNESSES.

Further Reading: Landon Margaret. *Anna and the King of Siam.* NY: HarperCollins, 2000; Leonowens, Anna Harriette. *The English Governess at the Siamese Court: Recollections of Six Years in the Royal Palace at Bangkok.* 1870. Reprint. London: Folio Society, 1980; ———. *The Romance of the Harem.* Reprint, edited and with an introduction by Susan Morgan. Charlottesville, NC: University Press of Virginia, 1991; ——— . *Siamese Harem Life.* Reprint ed. With an introduction by Freya Stark. New York: Dutton, 1953.

Lewis, Agnes Smith (1843–1926)

Born in 1843, Scots scholar Agnes Smith Lewis and her twin sister Margaret Dunlop Gibson (1843–1920) traveled widely to study ancient cultures, language, and literature. In the 1880s they went to Greece and Egypt, and in 1892 they went to Mount Sinai in the Middle East. At Sinai they visited a monastery housing ancient works and discovered many important manuscripts, including an early version of the Bible's New Testament. Lewis wrote many scholarly articles and books based on her travel experiences, including *In the Shadow of Sinai: A Story of Travel and Research from 1895 to 1897* (1898). She died in 1926.

Further Reading: Lewis, Agnes Smith, and Margaret Dunlop Gibson. *In the Shadow of Sinai.* 1898. Reprint. Portland, OR: Alpha Press, 1999.

Linnea, Ann (1949–)

Born in 1949, American Ann Linnea was the first woman to circumnavigate Lake Superior in a sea kayak, a feat she accomplished during nine weeks in 1992. She began her journey just one day after her forty-third birthday. Embarking in a 17-foot kayak from Duluth, Minnesota, she soon encountered heavy rainstorms, turbulent water, brutally cold temperatures, and fog that was sometimes so thick she could barely see to paddle. Such conditions had often led to shipwrecks on the lake, and it was this challenge that drew Linnea to the adventure. As the wife of a local college teacher and the mother of two teenagers, Linnea used her trip as a time not only to test her physical and mental strength but also to reassess her life. When she returned from her more than 1,200-mile journey, she divorced her husband and changed careers. She also wrote a book about her experiences, *Deep Water Passage: A Spiritual Journey at Midlife* (1995). She currently lives in the Pacific Northwest, where she gives seminars related to human spirituality. *See also* CIRCUMNAVIGATORS AND ROUND-THE-WORLD TRAVELERS.

Further Reading: Linnea, Ann. *Deep Water Passage: A Spiritual Journey at Midlife.* New York: Little, Brown, 1995.

Lott, Emmeline

In 1863, British governess Emmeline Lott traveled to Egypt to teach the five-year-old heir to the throne. Although she found her charge difficult and His Highness Ismael Pacha lewd, she remained in her position for two years. Like another governess abroad, Anna Leonowens, Lott wrote about her experiences living in a foreign culture, and her works were popular in England. Her books include *The English Governess in Egypt: Harem Life in Egypt and Constantinople* (1866; 2 volumes), *Nights in the Harem: Or, the Mohaddetyn in the Palace of Ghezire* (1867; 2 volumes), and *The Grand Pacha's Cruise on the Nile in the Viceroy of Egypt's Yacht* (1869, 2 volumes). *See also* GOVERNESSES.

Further Reading: Lott, Emmeline. *The Grand Pacha's Cruise on the Nile in the Viceroy of Egypt's Yacht.* London: T. C. Newby, 1869.

Low, Juliette Gordon (1860–1927)

Born in 1860, American outdoorswoman Juliette Gordon Low founded the Girl Scouts of America, a group devoted to providing organized outdoor activities for girls. As its first president, Low traveled throughout the country to promote the group. Her idea to start the Girl Scouts came during a trip abroad, when she met the man who had founded the Girl Guides in England. In 1912 Low decided to create an American affiliate of the

Juliette Gordon Low, the founder of Girl Scouts of the United States of America, probably 1900. ©*Hulton-Deutsch Collection/CORBIS.*

Girl Guides in her hometown of Savannah, Georgia. In 1915 she created a national Girl Guides, which she renamed the Girl Scouts of America. The first joint meeting of the Girl Scouts and Britain's Girl Guides was held in 1919. Low died in 1927. *See also* GIRL SCOUTS OF THE U.S.A..

Further Reading: Shultz, Gladys Denny, and Daisy Gordon Lawrence. *Lady from Savannah: The Life of Juliette Low*. New York: Girl Scouts of the U.S.A., 1988.

Ludington, Sybil (1761–1839)

Born in 1761, Sybil Ludington performed a deed similar to that of Paul Revere. On April 26, 1777, she rode to various New York and Connecticut towns warning people that the British had attacked Danbury, Connecticut. She also convinced people to volunteer to fight against the British. Her ride was more than double in distance to Revere's, and she was only 16 at the time. Honored in her own lifetime, she died in 1839. *See also* EQUESTRIENNES; MILIARY SERVICE.

Further Reading: NSDAR (National Society of the Daughters of the American Revolution). "Sybil Ludington—Female Paul Revere." http://www.geocities.com/Heartland/Plains/1789/sybil.html (May 2000).

M

Macaulay, Rose (1881–1958)

Born in 1881, Dame Rose Macaulay was already a noted English novelist when she wrote *The Fabled Shore: From the Pyrenees to Portugal* about her travels in Spain and Portugal in 1948. Published in 1949, it is considered a classic guidebook to the coastal regions of these countries. The author traveled along the coastlines of Catalonia, Valencia, Andalucia, and Algarve from the Pyrenees to Portugal, visiting the cities of Barcelona, Tarragona, Valencia, and Cadiz. She traveled alone by car, and, as scholar Raymond Carr points out in an 1986 introduction to the work, at the time "a woman alone, driving a car, was a phenomenon." (pp. 1–2) Carr adds that "*Fabled Shore* offers something much more valuable than holiday hints. It has a true feel for the Spain that had not been submerged by the tourist avalanche of the 1960s that helped to finance Spain's industrial take-off . . . the Spain of History." (p. 2)

Macaulay includes much historical information in her book, as well as detailed descriptions of the sites she visited. In writing her own travel account she studied those of other writers—"the learned books and the tourist books, the intelligent books and the silly books, the critical books and the gushing books"—and then endeavored to make her own book learned and intelligent. According to Carr, she succeeded. He writes:

> Rose Macaulay is learned. She studied the standard works on Spanish architecture before she set out; she has the geographers and historians of classical antiquity at her finger tips. She remembers that the youths who pester foreign ladies are descendants of Turdentani described by Strabo. It was the sailing book compiled by a Greek mariner in the sixth century B.C., in the version of the late fourth-century Roman poet, Rufus Festus Avienus, that set her out on her travels. It is most refreshing that she expects her readers to understand and enjoy Latin without the benefit of translation.
>
> [But] learning alone does not suffice to create a great travel book. There must be enthusiasm and empathy—the capacity to *feel* the past. This is Rose Macaulay's supreme gift. (p. 3)

As an example of how Macaulay combines travelogue with history, consider her writing on a town along the Andalusian shore:

> In the green sky stars began to candle. I left the warm green Guadalquivir and returned to Sanlucar, where, after some hunting, I got a room for the night. The town was not unattractive; above the lower, newer town where I stayed rose the higher old town, with its square-towered Moorish castle, once the dwelling of Guzman el Bueno, who became Duke of Medina Sidonia, for his war services were rewarded by Alfonso the Wise with the gift of this town just captured from the Moors. The dukes reigned in Sanlucar for several cen-

turies with the firmest tyranny, until it was taken over by the crown in the seventeenth century, having become important as a navigation centre. The magnificent castle is now used as a barracks. English sailors were such frequent visitors to Sanlucar in Henry VIII's reign, and so often far from well, that he founded a hospital for them there in 1517; it looks a fine building, is called the Colegio de San Francisco, and I do not know what it is now used for. The Spanish in some ways more but in others less conservative than ourselves, seldom use buildings for long together for the same purpose, they like to change them about, which all helps to make sightseeing confusing. (pp. 221–222)

Here and elsewhere in the book, Macaulay relates historical information to her sightseeing experiences. After her book was published, the public became captivated by her adventures driving her car through the country, and soon many people were retracing her route. Macaulay died in 1958. *See also* EUROPE, CONTINENTAL; TRANSPORTATION, GROUND; TRAVEL WRITERS.

Further Reading: Bensen, Alice R. *Rose Macaulay*. New York: Twayne Publishers, 1969; Macaulay, Rose. *Fabled Shore: From the Pyrenees to Portugal*. 1949. Reprint. Oxford, NY: Oxford University Press, 1986; ———. *The World My Wilderness*. With a new introduction by Penelope Fitzgerald. London: Virago Press, 1983.

MacEwen, Gwendolyn (1941–1987)

Born in 1941, Canadian poet Gwendolyn MacEwen is representative of women who use their travel experiences to inspire their creative work. Her short story collection *Noman's Land* (1985) is based on a trip she took through the Middle East; the same trip encouraged her to publish *Honey Drum: Seven Tales from Arab Lands* (1984), a collection of Arab folklore. MacEwen's travels through Greece are the subject of *Mermaids and Ikons: A Greek Summer* (1978). She also wrote 10 books of poems as well as poetic dramas for Canadian television, and she received several major awards for her work. MacEwen died in 1987. *See also* TRAVEL WRITERS.

Further Reading: Bartley, Jan. *Invocations, the Poetry and Prose of Gwendolyn MacEwen*. Vancouver: University of British Columbia Press, 1983; MacEwen, Gwendolyn. *Earth-light: Selected Poetry of Gwendolyn MacEwen, 1963–1982*. Toronto: General Publications, 1982; ———. *Mermaids and Ikons: A Greek Summer*. Toronto: Anansi, 1978; ———. *Noman's Land: Stories*. Toronto: Coach House Press, 1985; Sullivan, Rosemary. *Shadow Maker: The Life of Gwendolyn MacEwen*. Toronto: Harper Collins, 1995.

Maillart, Ella (1903–)

Born in 1903, Swiss explorer Ella Maillart traveled extensively through Asia, visiting places in China and Russia that no other European had seen before. During one trip from Peking into India in 1935, she was accompanied by English journalist Peter Fleming; in less than seven months, they traveled 3,500 miles, through a desert and over mountains, largely on foot or horseback. In 1939 Maillart and a female companion went through Afghanistan by car, and later Maillart continued on alone into India and Tibet. She wrote books about all of her travel experiences, including *Turkestan Solo: One Woman's Expedition from the Tien Shan to the Kizul Kum* (1935); *Forbidden Journey* (1937), which concerns her trek with Peter Fleming; *Cruises and Caravans* (1942); *The Cruel Way* (1947), about her Afghanistan experiences, and *The Land of the Sherpas* (1955).

One of Maillart's most popular works, *The Land of the Sherpas* describes her journey through Nepal, an independent kingdom that remained largely isolated from foreigners until 1949, and the Himalayans. The Sherpas are indigenous people of the region who work as guides and porters for mountaineers. Maillart wrote *The Land of the Sherpas* after deciding "to compare the lives of the Sherpas with that of the mountaineers of the Swiss Alps." (p. 15) To this end, she took photographs of each Sherpa she met; there are a total of 78 such illustrations in the book, with one-paragraph descriptions of each individual.

Maillart currently lives in Geneva, Switzerland. *See also* ASIA.

Further Reading: Maillart, Ella. *The Cruel Way*. 1947. Reprint. London: Virago Press, 1986; ———. *The Land of the Sherpas*. London: Hodder and Stoughton, 1955.

Markham, Beryl (1902–1986)

Born in 1902 as Beryl Clutterbuck, British aviator Beryl Markham wrote one of the most popular and well-received travel books of her time, *West With the Night* (1942), about her experiences making a solo flight across the Atlantic Ocean in 1936. Although born in England, Markham grew up in Kenya, Africa, and was one of that continent's first pilots, male or female. She was also the first woman in Africa to receive a license to train

Beryl Markham on the wing of her small plane before beginning her solo flight across the Atlantic Ocean. © *Bettmann/CORBIS.*

racehorses, a certification she earned at age 18. Markham's father was a racehorse breeder and adventurer who taught her to hunt, and as an adult Markham shot big game on African safaris. She was married three times and learned to fly after falling in love with a pilot. In 1931, after only a few months as a pilot, she received a commercial license and began delivering mail, supplies, and passengers to areas in the African bush.

During this time she decided to try for a solo flight record. Her first attempt was a flight from Africa to England, but bad weather and engine trouble thwarted her efforts at several points along the way. Eventually she did reach London, but it took her 23 days. A few years later she attempted to fly solo nonstop from London to New York City; although she succeeded in becoming the first woman aviator to cross the Atlantic she fell short of her mark, crashing into a bog in Nova Scotia.

Markham originally planned to participate in the making of a movie about her Atlantic crossing. When the movie deal fell through, she was asked to write a book about it instead. After *West with the Night*—which tells not only of her famous flight but also her childhood and bush pilot experiences—was published, many people questioned whether Markham wrote the book herself; today it is generally believed that it was the work of Raoul Schumacher, an established author who was Markham's third husband. Nonetheless, it was a bestseller for several years, after which Markham fell into relative obscurity. For a time she lived in California, running an avocado ranch, but in 1952 she went back to Kenya to train racehorses. She died in Kenya in 1986. *See also* AIR TRAVEL.

Further Reading: Gale Group, Women's History. "Beryl Markham." http://www.galegroup.com/freresrc/womenhst/markham.htm (June 2000); Gourley, Catherine. *Beryl Markham: Never Turn Back*. Berkeley, CA: Conari Press, 1997; Lovell, Mary S. *Straight on Till Morning: The Biography of Beryl Markham*. New York: St. Martin's Press, 1987; Markham, Beryl. *West with the Night*. 1942. Reprint. San Francisco: North Point Press, 1983; Trzebinski, Errol. *The Lives of Beryl Markham: Out of Africa's Hidden Free Spirit and Denys Finch Hatton's Last Great Love*. New York: W.W. Norton, 1993.

Marsden, Kate (1859–1931)

Born in 1859, Kate Marsden is representative of women travelers who undertake a quest. In Marsden's case, it was to discover a cure for leprosy. As a British nurse, she had worked with leprosy patients in England, and, in 1890, after hearing that Siberian shamans had discovered an herb that would treat the disease, she petitioned the Queen of England and the Empress of Russia for permission to travel through Siberia, which was at the time closed to foreigners. Once they granted her request, she went to Russia by way of the Middle East, where she studied some advanced cases of leprosy. She then traveled into Siberia, but after three months she fell ill and had to return home empty-handed. Nonetheless, she was made a member of the Royal Geographical Society in honor of her travels. She also wrote a book in which she describes her experience, *On Sledge and Horseback to Outcast Siberian Lepers* (1892). It speaks of long hours on horseback through difficult terrain, as well as nights spent sleeping in unsanitary, insect-plagued camps and villages. She also speaks of the lepers themselves, describing their terrible condition. About one leper settlement she says:

> Twelve men, women, and children, scantily and filthily clothed, were huddled together . . . covered with vermin. The stench was dreadful; one man was dying, two men had lost their toes and half of their feet; they had tied boards from their knees to the ground, so that by this help they could contrive to drag themselves along. One man had no fingers; and the poor stumps, raised to make the sign of the cross, were enough to bring tears to the eyes of the most callous. (Morris, pp. 122–123)

In 1921, she wrote another book about her Siberian experience, *My Mission to Siberia: A Vindication*, to address criticisms about the validity of her trip. Meanwhile, she became ill and was bedridden for the remainder of her life. She died in 1931. *See also* PUBLIC SERVICE.

Further Reading: Marsden, Kate. *On Sledge and Horseback to Outcast Siberian Lepers.* 1892. Reprint. New York: Cassell Publishing, 1982; Morris, Mary. *Maiden Voyages: Writings on Women Travelers.* New York: Vintage, 1993.

Martineau, Harriet (1802–1876)

Born in 1802, English journalist Harriet Martineau traveled internationally to study other cultures and political situations, as well as for pleasure. For example, during a trip to America from 1834–1836, she studied slavery and later wrote about this experience in her 1837 book *Society in America*. The following year she wrote an important guidebook for other Victorian travelers, one of the first books to address the concept—although not by name—of culture shock; entitled *How to Observe Morals and Manners*, it was in three volumes. Her other works were the three-volume *Retrospect of Western Travel* (1838) and *Eastern Life, Present and Past* (1848). Martineau's last major trip was an eight-month journey through Egypt and the Middle East in 1846. She died in 1876. *See also* CULTURE SHOCK.

Further Reading: Martineau, Harriet. *Life in the Wilds. A Tale of the South African Settlement.* New York: J. W. Lovell, 1884; Miller, Florence Fenwick. *Harriet Martineau.* Port Washington, NY: Kennikat Press, 1972; Nevill, John Cranstoun. *Harriet Martineau.* Norwood, PA: Norwood Editions, 1976.

Masterson, Martha Gay (1837–1916)

Born in 1837, Martha Gay Masterson is representative of American pioneers who traveled throughout the country in search of a better life. Masterson grew up in Missouri, but in 1851, when she was 14, her family crossed America to settle in Oregon Territory. Along the way her mother gave birth to her twelfth child, and the wagon train was often attacked by Native Americans.

Once in Oregon, Masterson married a widowed blacksmith with nine children. The couple moved 20 times in 20 years, living in various parts of Oregon, Washington, and Idaho. She had three children of her own but never a permanent home and died in 1916.

Further Reading: Stefoff, Rebecca. *Women Pioneers.* New York: Facts On File, 1995.

Mayes, Frances

American author Frances Mayes is best known for *Under the Tuscan Sun: At Home in Italy*, one of the most popular books in a subgenre of travel writing in which authors are American expatriates living abroad. Published in 1996, *Under the Tuscan Sun* describes Mayes's experiences in Italy, where she settled in 1990 after buying an abandoned house in Tuscany that needed extensive restoration. About her decision to make this purchase and leave America, Mayes writes:

> The house is a metaphor for the self, of course, but it is also totally real. And a *foreign* house exaggerates all the associations houses carry. Because I had ended a long marriage that was not supposed to end and was establishing a new relationship, this house quest felt tied to whatever new identity I would manage to forge. . . . I had the urge to examine my life in another culture and move beyond what I knew. I wanted something of a *physical* dimension that would occupy the mental volume the years of my former life had. . . . The language, history, art, places in Italy are endless—two lifetimes wouldn't be enough. And ah, the foreign self. The new life might shape itself to the contours of the house, which already is at home in the landscape, and to the rhythms around it. (pp. 15–16)

In this way, Mayes's work is typical of modern travel literature, because it involves a personal exploration of the self. Moreover, Mayes reveals that her book was originally a collection of personal writings not intended for publication. She explains:

> [Shortly after moving to Italy] I bought an oversize blank book with Florentine paper covers and blue leather binding. On the first page I wrote ITALY. The book looked as though it should have immortal poetry in it, but I began with lists of wildflowers, lists of projects, new words, sketches of tile in Pompeii. I described rooms, trees, bird calls. I added planting advice. . . . I wrote about the people we met and the food we cooked. . . . Today it

> is stuffed with menus, postcards of paintings, a drawing of a floor plan of an abbey, Italian poems, and diagrams of the garden. Because it is thick, I still have room in it for a few more summers. Now the blue book has become *Under the Tuscan Sun*, a natural outgrowth of my first pleasures here. Restoring, then improving, the house; transforming an overgrown jungle into its proper function as a farm for olives and grapes; exploring the layers and layers of Tuscany and Umbria; cooking in a foreign kitchen and discovering the many links between the food and the culture—these intense joys frame the deeper pleasure of learning to live another kind of life. To bury the grape tendril in such a way that it shoots out new growth I recognize easily as a metaphor for the way life must change from time to time if we are to go forward in our thinking. (p. 2)

Mayes also talks about the significance of travel writing, both for the author and for the reader. In this, too, she reflects a trend in modern travel literature, trying to make the reader feel like the resident of a place rather than a mere observer. Mayes says:

> Writing about this place, our discoveries, wanderings, and daily life, also has been a pleasure. A Chinese poet many centuries ago noticed that to re-create something in words is like being alive twice. . . . My reader, I hope, is like a friend who comes to visit, learned to mound flour on the thick marble counter and work in the egg, a friend who wakes to the four calls of the cuckoo in the linden and walks down the terrace paths singing to the grapes; who picks jars of plums, drives with me to hill towns of round towers and spilling geraniums, who wants to see the olives the first day they are olives. A guest on holiday is intent on pleasure. (p. 3)

Under the Tuscan Sun describes the sights, sounds, and tastes of Mayes's new home, offering recipes along with anecdotes about the Italian countryside. The work has poetic elements, which is understandable given Mayes's background. Her first works were poetry; she is the author of several poetry collections, including *Sunday in Another Country* (1977), *After Such Pleasures* (1979), and *The Arts of Fire* (1982). She has also had many poems published in *Atlantic Monthly, The Iowa Review,* and *New American Writing*. Her autobiographical essays have appeared in *The American Poetry Review, The American Scholar,* and *Ploughshares*. Raised in Georgia, Mayes currently divides her time between homes in Cortona, Italy, and San Francisco. *See also* TRAVEL WRITERS.

Further Reading: Frances Mayes. http://www.websimple.com/utts/aboutauthor.html (May 2000); Mayes, Frances. *After Such Pleasures*. New York: Seven Woods Press, 1979; ———. *The Arts of Fire*. Woodside, CA: Heyeck Press, 1982; ———. *Sunday in Another Country*. Woodside, CA: Heyeck Press, 1977; ———. *Under the Tuscan Sun: At Home in Italy*. San Francisco: Chronicle Books, 1996.

Mazuchelli, Elizabeth Sarah (1832–1914)

Born in 1832, Elizabeth Sarah Mazuchelli, familiarly known as "Nina," was the first European woman to explore the eastern Himalayan mountains. She initially went to the region in 1869, when her husband, a British army chaplain, was stationed in India. In 1871 the couple spent two months traveling with about 80 servants through the mountain range. At first their expedition went well. Then they experienced a string of disasters: the weather turned bad, they got lost, they developed altitude sickness, and their food almost ran out before a delivery of fresh provisions arrived. Nonetheless, they managed to survive their adventures and returned to England the same year. Mazuchelli then wrote a book about her travels, *The Indian Alps and How We Crossed Them: Being a Narrative of Two Years' Residence in the Eastern Himalaya and Two Months' Tour into the Interior* (1876), publishing it under the pseudonym "A Lady Pioneer." She died in 1914.

Further Reading: A Lady Pioneer (Mazuchelli, Elizabeth). *The Indian Alps and How We Crossed Them: Being a Narrative of Two Years' Residence in the Eastern Himalaya and Two Months' Tour into the Interior*. New York: Dodd, Mead, 1876.

McCairen, Patricia (1940–)

Patricia McCairen, an American, was the first woman to travel solo down the Colorado River from northern Arizona to Lake Mead since the Glen Canyon Dam blocked off the route in 1963. It took her 25 days to make the 280-mile journey through the Grand Canyon in her 15-foot raft, which she named the *Sunshine Lady*. She describes her experiences in a book, *Canyon Solitude: A Woman's Solo River Journey Through the Grand Canyon* (1998). McCairen had rafted the river in a group on six previous occasions and has been a rafter for 20 years, once working as a river guide. *See also* ADVENTURE TRAVEL.

For Further Reading: McCairen, Patricia. *Canyon Solitude: A Woman's Solo River Journey through the Grand Canyon*. Seattle, WA: Seal Press, 1998.

McCarthy, Mary (1912–1989)

Born in 1912, American Mary McCarthy is the author of a popular book about her travels in Italy, *Stones of Florence* (1959), which includes historical information along with descriptions of the country and people. She describes the city of Florence at length, saying that most people who have never been there have a false idea of the city. She adds that Florence is inhospitable to tourists, saying:

Mary McCarthy in 1963. © *Dick DeMarsico/Library of Congress.*

> Tourism, in a certain sense, is an accidental by-product of the city—at once profitable and a nuisance, adding to the noise and congestion, raising prices for the population. Florence is a working city, a market centre, a railway junction. . . . The small hotels and restaurants are full of commercial travellers, wine salesmen . . . , textile representatives . . . , dealers in marble. . . . Everyone is on the move, buying, selling, delivering, and tourists get in the way of this diversified commerce. The Florentines, on the whole, would be happy to be rid of them. The shopkeepers . . . , the owners of hotels and restaurants, the thieves . . . might regret their departure, but the tourist is seldom led to suspect this. There is no city in Italy that treats its tourists so summarily, that caters so little to their comfort. (Morris, pp. 264–265)

In addition to books on Italy, McCarthy is the author of several novels and autobiographical works, including *Memories of a Catholic Girlhood* (1957), *How I Grew* (1987), and *A Charmed Life* (1992), which was published posthumously. She died in 1989.

Further Reading: Gelderman, Carol W. *Mary McCarthy: A Life*. New York: St. Martin's Press, 1989; McCarthy, Mary. *A Charmed Life*. San Diego, CA: Harcourt Brace Jovanovich, 1992; ———. *How I Grew*. San Diego, CA: Harcourt Brace Jovanovich, 1987; ———. *Intellectual Memoirs, New York, 1936–1938*. Reprint. San Diego, CA: Harcourt Brace, 1993; ———. *The Stones of Florence*. 1959. Reprint. San Diego, CA: Harcourt Brace Jovanovich, 1987; McKenzie, Barbara. *Mary McCarthy*. New York: Twayne Publishers, 1967; Morris, Mary. *Maiden Voyages: Writings of Women Travelers*. New York: Vintage Press, 1993.

Mead, Margaret (1901–1978)

Born in 1901, American anthropologist Margaret Mead is representative of women who

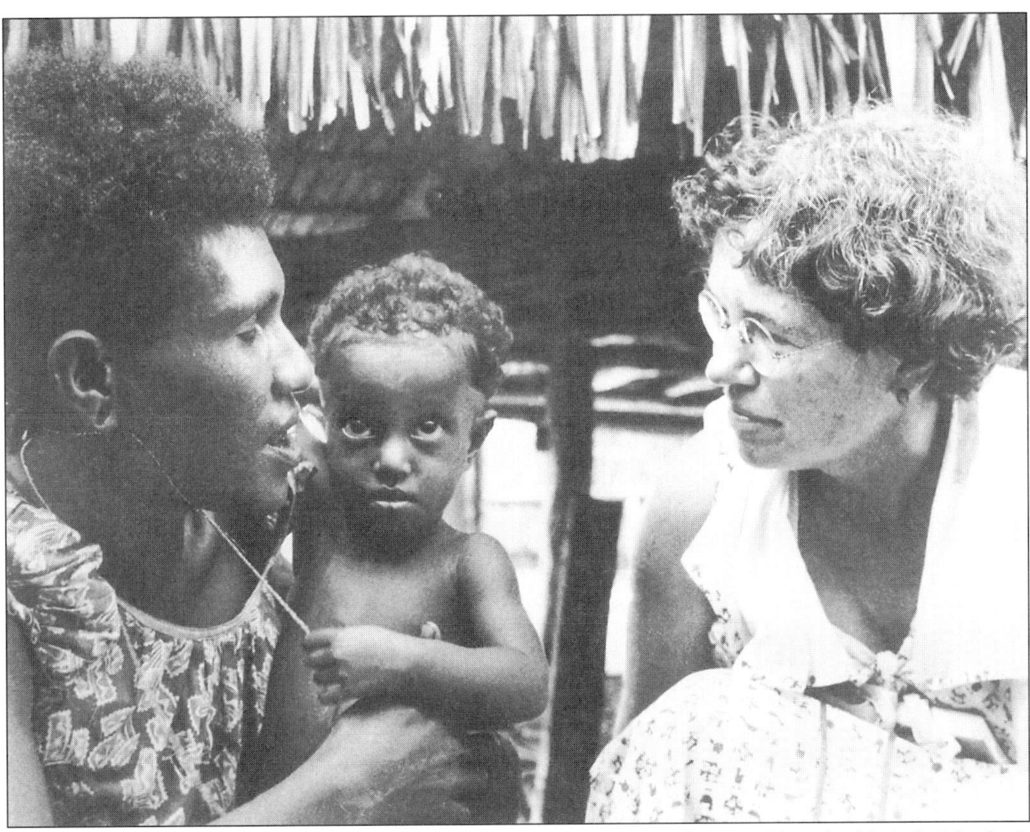

Anthropologist Margaret Mead is shown here with a Manus mother in 1953 during a visit to the Admiralty Islands. © *Bettmann/CORBIS.*

travel to exotic locations to compare other cultures to their own. Her first trip was to the South Pacific in 1925–1926, where she was earning her doctorate in anthropology at Columbia University. Upon her return, she became assistant curator of ethnology at the American Museum of Natural History in New York City, a position she held until 1964, when she became curator. From 1954 until her retirement from the museum in 1969 she also taught anthropology at Columbia University. From 1968 to 1971 she taught at Fordham University as well.

Mead published several books based on her research. Her first, which appeared in 1928, was *Coming of Age in Samoa*. Additional works include *Growing Up in New Guinea* (1930), *A Way of Seeing* (1961), which is a collection of essays related to American culture, and an autobiography, *Blackberry Winter* (1972). She died in 1978.

See also ANTHROPOLOGISTS AND ARCHAEOLOGISTS.

Further Reading: Mead, Margaret. *Letters from the Field, 1925–1975*. New York: Harper & Row, 1977; ———. *World Enough: Rethinking the Future*. Boston: Little, Brown, 1975; Mead, Margaret, and Rhoda Metraux. *A Way of Seeing*. 1961. Reprint. New York: McCall Pub., 1970; Rice, Edward. *Margaret Mead: A Portrait*. New York: Harper & Row, 1979.

Meggers, Betty (1921–)

Born in 1921, American anthropologist Betty Meggers is representative of women who traveled extensively to conduct field research in indigenous cultures. Her focus has been on Latin America, on the regions along the Amazon and the Andes River as well as in Guyana, Ecuador, Venezuela, Peru, and Chile. She has also conducted research in a few other regions outside of Latin America, particularly the Lesser Antilles and Micronesia.

Meggers received her Ph.D. from Columbia University in 1952 and was the executive secretary of the American Anthropological Association from 1959 to 1961. From 1961 to 1963 she was an administrator for a lecture program of the association and the National Science Foundation. Meggers has also worked as a research associate for the Smithsonian Institution in Washington, D.C., and, in 1981, she was named one of the institute's experts. She belongs to, and in some cases serves on the boards of, numerous professional organizations, including the Society for American Archaeology and the Anthropological Association of Washington. She has received many awards for her work and has written many books and articles about Latin American culture and history. *See also* LATIN AMERICA.

Further Reading: Meggers, Betty Jane. *Amazonia: Man and Culture in a Counterfeit Paradise*. Chicago: Aldine, Atherton, 1971; ———. *Prehistoric America: An Ecological Perspective*. New York: Aldine Publishing, 1979.

Mexia, Ynes (1870–1938)

Born in 1870, American naturalist Ynes Mexia made major contributions to the field of botany by traveling throughout Central and South America to collect plant specimens during the 1920s. Her father was a Mexican diplomat who had met her mother in Washington, D.C.; when her parents' marriage failed, Mexia spent her childhood years in the United States with her mother and her young adult years in Mexico with her father. As a result, she spoke both English and Spanish.

In 1898 Mexia married a Mexican rancher who died six years later. Four years after his death, still in Mexico, she remarried but fell into a deep depression. She left her husband and moved to the United States, where she joined the Sierra Club and began taking classes related to natural history at the University of California at Berkeley. In 1925 she joined a botanical expedition to Mexico sponsored by nearby Stanford University. Once in Mexico, however, she decided that the other expedition members were not accomplishing much, and she abandoned the group to go off on her own. Her trip was successful, and for the next two years she traveled throughout Mexico to collect more than 1,500 plants, many of them never before classified, for the University of California.

Mexia's reputation as a botanist was now assured. In 1928 she was hired to collect plants in Alaska and from 1929 to 1932 in Brazil and Peru. She made another expedition to South America from 1934 to 1936 and another to Mexico from 1936 to 1938. During her Mexico expedition, she fell ill and discovered that she had lung cancer. She died within months. *See also* LATIN AMERICA.

Further Reading: Tinling, Marion. *Women into the Unknown: A Sourcebook on Women Travelers*. Westport, CT: Greenwood Press, 1989.

Migration

The word "migration" comes from the Latin *migrare*, which means "to travel from one place to another." Women migrate for various reasons. Some go voluntarily, perhaps to find food, work, and better living conditions or to escape religious or political persecution. Some go involuntarily, perhaps as slaves, or accompany families.

Most statistics regarding the number of people who have migrated throughout history are gender-neutral. Approximately 250,000 immigrants traveled to America during the seventeenth century; during the eighteenth century this number was over 1.5 million, and by 1780 there were approximately 3 million colonists on the continent. Some estimates put the number of European immigrants to the United States at 20,000 per year from 1820 to 1830, 60,000 per year from 1831 to 1840, and 260,000 per year from 1851 to 1860. These figures do not include African slaves, who were captured and brought to America in increasingly large numbers. According to some estimates, by 1800 more than 10 million African slaves had been transported to the New World.

Outside of enslavement, immigrants have come to America for many reasons. In the seventeenth century, many Puritans from

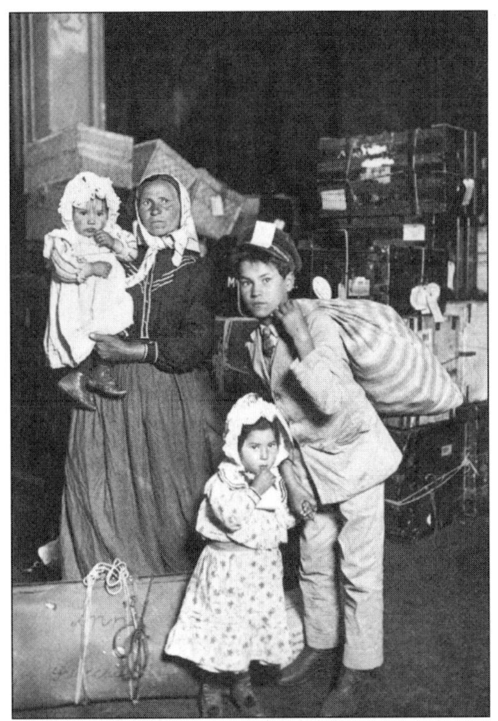

Italian immigrants arrive at Ellis Island in 1905. © *Bettmann/CORBIS.*

England chose to migrate to the New World to escape religious persecution. In the eighteenth and nineteenth centuries, people from countries like Sweden, where land was scarce, were drawn by America's abundance of natural resources. Others were driven from their native countries by tyranny, social unrest, poverty, or famine. A large number of Irish, for example, migrated to the United States between 1845 and 1848 because of a blight on potato crops in Ireland.

Today the greatest number of immigrants to the United States come from Mexico, where poverty drives many people across the U.S.-Mexico border illegally. Some estimates place the number of people entering the United States from Mexico without permission at 2.5 million per year, with approximately 500,000 remaining to become permanent residents. Many of the latter become migrant farmworkers, traveling from place to place to harvest crops in accordance with seasonal demands. Migrant farm families generally follow the same migration pattern because specific crops ripen at the same time every year. For example, on the east coast of the United States, migrant farm workers begin their harvesting season in Florida picking citrus fruit, and end in Maine digging potatoes. Migrant workers exist in other countries as well; for example, many Italians and Portuguese travel north to find seasonal jobs.

People who continually travel in search of work or food are called nomads. Some of the earliest such people were Gypsies. Prior to the 1300s the Gypsies roamed through northwest India, but by the 1300s they had appeared in Europe and the Middle East, by 1500 they were in England, and today they live in the United States and Canada as well. Gypsy bands spend warm months traveling to various towns to work at carnivals, fairs, or similar venues, where they typically sell handmade goods, tell fortunes, run carnival games and rides, or provide entertainment; along the way they sleep in trailers, motor homes, or tents. In the winter they live in large groups, renting houses or apartments together. A few Gypsies maintain fairly settled lives, living within a relatively small area and working at a less migratory job like horseshoeing, but most continue to uphold their tradition of extensive travel, migrating many miles each year. Moreover, Gypsies typically keep to themselves and have therefore managed to retain their own customs and language (Romany, which is derived from the ancient Indian language of Sanskrit).

Refugees

While nomads and immigrants travel by choice, refugees have been forced from their homelands—whether because of a political or religious conflict, a government policy, or a natural disaster—and hope someday to return there. Refugees typically leave a region not as individuals but as representatives of a group. For example, in the fifteenth century most Jewish people were forced to leave Spain, and in the seventeenth century Protestants were forced to leave France. During the nineteenth century, Native Americans were forced to leave their homelands by the United States government, and during World

War I refugees fled invading forces in Belgium, France, Italy, and Romania. The Russian Revolution also created many refugees seeking to leave the new Bolshevik regime, and in the 1910s and 1920s, many Armenians were forced from Syria, Palestine, and Turkey.

The largest number of refugees resulted from World War II. Historians estimate that more than 12 million people in Europe became refugees because of invasions by Nazi forces, and more than 30 million Chinese fled Japanese armies at the beginning of the war. By 1943, according to some estimates, more than 60 million people had been forced from their homes. Some of these returned after the war, but many did not. In fact, the number of European refugees who settled in the United States during this time was so large that new government and charitable agencies were established to deal with them.

Even after World War II, refugees continued to pose a problem for many countries. When the state of Israel was created in 1948, for example, more than a half million Palestinians became refugees, as did many Koreans during the Korean War in the 1950s. During the 1950s, many Chinese refugees settled in Hong Kong, which was controlled by the British, and Taiwan; Cubans began fleeing the dictatorship of Fidel Castro—a practice that continues today. The Vietnam War displaced approximately two million people, not only in Vietnam but also in Laos and Cambodia. The 1980s brought the largest insurgence of refugees from one country when the Soviet invasion of Afghanistan forced more than six million people into Pakistan, Iran, and other countries. The 1990s also saw a large number of refugees, particularly in Uganda, Ethiopia, and other African countries. Out of the roughly 16 million refugees living throughout the world in 1991, approximately four million were African. As of 2000, some estimates place the number of refugees worldwide at 18 million.

Although there are no accurate statistics regarding how many of these refugees are women, sociologists and others who have worked with refugee groups do know that women are in the majority among refugee groups and that their lack of male support increases the stress of their already difficult situation. As Judy Mayotte, an expert on the plight of modern refugees, explains in her book *Disposable People: The Plight of Refugees* (1992):

> Millions of refugee women enter another country without father, husband, or brother—without male support. Generally they come from countries in which the vast majority of women are very dependent on men, especially rural women who form the largest proportion of any refugee population. Seldom do they settle immediately into one place in the land of exile. As they move about, the first home a women can provide for her children is often no more than a blanket or quilt raised on precarious sticks, and food and water are scarce. When the refugees come in large numbers, chaos defies any quest for order. (p. 150)

Mayotte points out that most refugees leave their homes in haste, without taking many of their possessions, and that in many cases they have suffered rape and other indignities at the hands of their persecutors. Deeply traumatized, they must adjust to the fact that often one or more of their family members—usually males—have been left behind. Mayotte explains how this affects women's daily existence:

> With their lives centered around their families [before becoming refugees], caring for and nourishing their husbands and children, they are not prepared for the tearing apart of the very fabric of the family. Young women are not ready to accept widowhood, especially when they are left with five or six young children, yet millions must face the reality that their husbands will never return. The widows do not feel competent to be the chief breadwinner of the family . . . Often nonliterate and unfamiliar with the language and customs of the host country, the women feel with uncommon acuity their isolation, their lack of tools, their inability to provide for their families, and their inferior social status. . . . Fears for family consume the refugee woman. She worries for her husband, whether he is absent in battle or present in despair. . . .

> She must keep his spirits up and be attentive to . . . his potential for violence against those he loves most if they become the symbols of his emasculation. (pp. 150–151)

Of course, women immigrants have also had to deal with some of these difficulties, including language and cultural barriers. Those forced from their homelands by poverty might have had to deal with husbands and/or parents who felt inadequate, while others had to travel alone and afraid. The same has been true for women who migrated within their own countries rather than from one country to another. For example, women migrating west across the North American continent during the mid-nineteenth century endured great hardships in order to reach their destinations, as did poor women who left the American Midwest during the Depression of the 1930s to find work.

Further Reading: Butruille, Susan G. *Women's Voices from the Oregon Trail: The Times that Tried Women's Souls, and a Guide to Women's History along the Oregon Trail*. Boise, ID: Tamarack Books, 1994; Cebula, Richard J. *The Determinants of Human Migration*. Lexington, MA: Lexington Books, 1979; Clark, W. A. V. *Human Migration*. Beverly Hills, CA: Sage Publications, 1986; Esty, Katharine. *The Gypsies: Wanderers in Time*. New York: Meredith Press, 1969; Faragher, John Mack. *Women and Men on the Overland Trail*. New Haven, CT: Yale University Press, 1979; Gamio, Manuel. *Mexican Immigration to the United States: A Study of Human Migration and Adjustment*. New York: Dover Publications, 1971; Gropper, Rena C. *Gypsies in the City: Culture Patterns and Survival*. Princeton, NJ: Darwin Press, 1975; Holmes, Kenneth L. *Covered Wagon Women: Diaries & Letters from the Western Trails*. Lincoln: University of Nebraska Press, 1995; Lewis, G. J. *Human Migration: A Geographical Perspective*. New York: St. Martin's Press, 1982. Mayotte, Judy. *Disposable People: The Plight of Refugees*. Maryknoll, NY: Orbis Books, 1992; McDowell, Bart. *Gypsies, Wanderers of the World*. Foreword by an English Gypsy, Clifford Lee. Washington, DC: National Geographic Society, 1970; McNeill, William H., and Ruth S. Adams, eds. *Human Migration: Patterns and Policies*. Bloomington: Indiana University Press, 1978; Ridge, Martin, ed. *Westward Journeys: Memoirs of Jesse A. Applegate and Lavinia Honeyman Porter who Traveled the Overland Trail*. Chicago: Lakeside Press, 1989.

Military Service

Until 1901 American women who served as nurses, physicians, and support personnel for soldiers and military workers were civilians or contract employees with no military rank. Of the women under contract to the military, the earliest were cooks, seamstresses, and laundresses who traveled from encampment to encampment providing services during the Revolutionary War. Meanwhile some women, like Deborah Sampson, disguised themselves as men so they would not be excluded from the fighting. Other women volunteered their medical skills as nurses or doctors, again as unpaid or contract workers. For example, Margaret Vliet Warne, a New Jersey physician, rode from one Revolutionary War battlefield to another treating soldiers and their families as an unpaid volunteer.

During the War of 1812, the U.S. Navy began placing contract nurses on its ships, and during the Civil War a group of Roman Catholic nuns served on board the Navy's first hospital ship, *Red Rover*. Nurses also served as volunteers on Civil War battlefields, in field hospitals, and as members of the United States Sanitary Commission (USSC), which was established by President Abraham Lincoln to prevent disease among soldiers and to coordinate relief efforts. One member of the USSC, Mary Livermore, became interested in women's issues and sought to find out how many Civil War soldiers were actually women disguised as men. She estimated that there were over 400 such women in the Union Army alone, and that at least 60 of them had been killed or wounded.

In World War I American women could still not fight on the battlefield unless they were in disguise, but they were allowed to join the military as support personnel. When the United States entered the war in April 1917, there were 403 Army nurses on active duty and 170 on reserve. This number increased steadily throughout the war; by June 1918 there were more than 2,000 active and 10,000 reserve Army nurses, and when the war ended in November 1918 there were over 21,000 nurses on active duty. More than 10,000 of these women served overseas.

U.S. Army Nurse in Vietnam in 1969. © *AP/Wide World Photos.*

Women also traveled abroad to serve in the Army as telephone operators, dietitians, and physical and occupational therapists. Women in the Marine Corps primarily performed clerical duties, whereas those in the Navy had a much wider choice of jobs, including electrician, radio and telephone operator, pharmacist, chemist, and torpedo assembler.

World War II provided more job opportunities for American women, although most of these were in the United States. A few women were given the responsibility of flying military planes from their point of manufacture to regions where they would be put in use. However, the majority of military women were, as always, nurses and clerical workers; this remains the situation today. At the same time, beginning in World War II, the tendency to keep women away from the most dangerous war zones has lessened. In World War II more than 200 American Army nurses died overseas in combat zones, and more than 1,000 received decorations for heroism. During the Korean and Vietnam Wars, female nurses were routinely sent to field hospitals in areas of active fighting, and some of them died there.

Many women journalists also went into high-risk areas during these wars. Some of them initially traveled to a war region as someone's girlfriend only to turn to reporting once there. For example, journalist Laura Palmer went to Saigon during the Vietnam War (1964–1975) because she was in love with a pediatrician who was stationed there, but once the romance ended she remained in Vietnam to cover the war for ABC Radio and *Rolling Stone* magazine. However, at one point U.S. General William Westmoreland banned women journalists from being in the field after dark, believing that it was not safe for them. Only protests from women reporters and other media representatives led Westmoreland to end the ban.

Subsequent conflicts have also placed American women in high-risk areas, not only as journalists, nurses, and support personnel but also as soldiers. Women from other countries, most notably Israel, Russia, and Yugoslavia, have served as soldiers, but they typically do not travel outside of their own homelands to do so. *See also* PUBLIC SERVICE; SAMPSON, DEBORAH.

Further Reading: Sherrow, Victoria. *Women and the Military: An Encyclopedia*. Santa Barbara and Denver: ABC-CLIO, 1996; Gendergap. *Women in the Military*, http://www.gendergap.com (May 2000).

Mills, Dorothy (1889–1959)

Born in 1889, British travel writer Lady Dorothy Mills was the first white woman to visit several parts of the world, including Timbuktu (in present-day Mali) in 1923 and parts of West Africa and South America in 1926 and 1930, respectively. Her books include *The Road to Timbuktu* (1924), *Through Liberia* (1926), and *The Country of the Orinoco* (1931). Mills's career began in 1917, two years after her divorce. The daughter of an earl, she was an adventurous woman with the wealth to fund her exotic travels. Her last trip was up the Orinoco River of South America in 1930. She died in 1959. *See also* AFRICA; LATIN AMERICA; TRAVEL WRITERS.

Further Reading: Robinson, Jane. *Wayward Women: A Guide to Women Travellers*. Oxford: Oxford University Press, 1990.

Missionaries

Missionaries have existed since the beginning of the Christian religion, organizing to spread Christianity. However, they were traditionally male until the 1800s, when a few women, such as Sojourner Truth, achieved notoriety through proselytizing. Truth was a former slave who traveled throughout the northeastern United States during the 1830s preaching the gospel; later she became known for her abolitionist activities.

Women's missionary activities remained the province of individuals rather than groups until the early 1860s, when the Women's Missionary movement began in the United States. This movement, which eventually spread to Europe, spawned over 40 societies sponsoring trips to non-Christian countries—usually in Africa and Asia—for single female missionaries.

One of the most prominent of the nineteenth century missionary societies was the China Inland Mission, which in 1878 started sending women, both single and married, to preach Christianity in China. By 1882 there were 95 single and 56 married women participating in Mission activities. Most of the single women worked in groups along with married women and their husbands, while others worked in pairs. A few single women worked alone, including Englishwoman Annie Royle Taylor.

China

Taylor first went to China under the auspices of the China Inland Mission in 1884. In 1892 she decided to preach in Tibet, which was closed to foreigners at that time. The first European woman to enter Tibet, she made a 1,000-mile journey into the country's interior dressed as a Tibetan nun. However, although she sought to reach the capital city of Lhasa, she was unmasked to authorities by a disgruntled servant, arrested, and eventually told to leave the country. She then remained near the Tibetan border to establish her own mission, the Tibetan Pioneer Mission, and asked the China Inland Mission to send her other women missionaries.

Other prominent members of the China Inland Mission include Englishwoman Mildred Cable, Algerian sisters Evangeline and Francesca French, and Canadian Isobel Kuhn. Cable went to China for the China Inland Mission in 1902 to help establish a school and found herself working closely with Evangeline and Francesca French, who had almost been killed in China in 1900 during a massacre of foreign missionaries. Together the three women decided that God meant for them to preach in the Gobi Desert in northwestern China. They took an ox cart to the last city inside the Great Wall of China, a town that had been dubbed the City of the Prodigals for its tendency to attract criminals, and then set out for the settlements that had sprung up along desert trade routes. During their travels they encountered bandits and other hazards, and each winter they had to return to the City of the Prodigals until the bad weather cleared. The three women persisted in their missionary work until 1931, when a new Chinese government restricted them to the city of Tunhwang. After several months they escaped their quarters and fled across the desert, then left China to continue their missionary activities elsewhere.

Isobel Kuhn went to China for the China Inland Mission in 1928. Shortly after her arrival she married fellow missionary John Kuhn, and for 20 years the two worked together among the Lisu people of the western province of Yunnan. In 1950 the area became the site of guerilla activity so they left for Thailand, where they continued their missionary work. Kuhn wrote many articles and books based on her experiences, including *Nests Above the Abyss* (1947), *Green Leaf in Drought-Time: The Story of the Escape of the Last C.I.M. Missionaries from Communist China* (1957), and *Ascent to the Tribes: Pioneering in North Thailand* (1956).

Another book that describes a missionary's experience in China is the autobiographical *My Days of Strength* (1939) by American Anne Walter Fearn. Fearn was a physician who went to China in 1893 as part of a missionary effort run by the Methodist Episcopal Church, South. In 1896, after marrying a fellow missionary, she began her own medical practice in China. She lived most of her life in the city of Shanghai, where she established the Fearn Sanatorium.

Africa

Africa also had many women missionaries—so many, in fact, that when English explorer Mary Kingsley traveled through the continent, many local people incorrectly assumed that she was a missionary herself. However, only one female African missionary became truly famous for her activities, and she was a fictional character who appeared in the movie *The African Queen* (1951).

A more recent example of missionary work is the career of Eunice Bishop, who worked as a missionary in China, Ethiopia, and other African, Asian, and South American countries from 1901 to 1972. The daughter of missionaries in China, she attended college in the United States but returned to China as a young woman to teach in the Methodist Women's Missionary School. She then married a fellow missionary, and together the two continued their missionary work throughout the world. Bishop wrote a book about her experiences entitled *At Home in China and Around the World: The Life Story of Merlin and Eunice Bishop* (1981). *See also* AFRICA; ASIA; CABLE, MILDRED; FEARN, ANNE WALTER; FRENCH, EVANGELINE AND FRANCESCA; KINGSLEY, MARY; KUHN, ISOBEL; TAYLOR, ANNIE ROYLE.

Further Reading: Barr, Pat. *To China with Love: The Lives and Times of Protestant Missionaries in China, 1860–1900*. Garden City, NY: Doubleday, 1973; Kent, Graeme. *Company of Heaven: Early Missionaries in the South Seas*. Wellington: A. H. & A. W. Reed, 1972; Moorhouse, Geoffrey. *The Missionaries*. London: Eyre Methuen, 1973; Mueller, John Theodore. *Great Missionaries to China*. Freeport, NY: Books for Libraries Press, 1972; Romano, Patricia W., ed. *Women's Voices on Africa: A Century of Travel Writings*. New York: M. Wiener Pub., 1992.

Montague, Mary Wortley (1689–1762)

Born in 1689 in England, Lady Mary Wortley Montague (also sometimes spelled Montagu) traveled from England to Turkey in 1716 and wrote about her experiences in a series of letters that were published posthumously as *Letters of the Right Honourable Lady M—y W—y M——e. Written, during her Travels in Europe, Asia and Africa, To Persons of Distinction, Men of Letters, &c. in different Parts of Europe. Which contain, . . . Accounts of the Policy and Manners of the Turks* (1763). Montague undertook the journey because her husband had been named ambassador to Turkey. To reach the Middle East, the couple traveled for almost a year through Holland, Germany, and Austria; 18 months later they returned to England via Greece and North Africa. In 1738 Montague left her husband and moved to Europe, where she lived in various Italian and French towns. She also continued to write letters, which were published in collections after her death in 1762.

Published in 1988, *Embassy to Constantinople: The Travels of Lady Mary Wortley Montague* provides many details about the people and culture of Turkey; her letters offered the English readers of her era a unique glimpse into a world they knew little about. In fact, Montague criticizes previous works by men who misunderstood the life of Turkish women. For example, in one letter she writes:

> Turkish ladies don't commit one sin the less for not being Christians. Now I am a little acquainted with their ways, I cannot forebear admiring either the exemplary discretion or extreme stupidity of all the writers that have given accounts of 'em. 'Tis very easy to see they have more liberty than we have, no woman of what rank soever being permitted to go in the streets without two muslins, one that covers her face all but her eyes and another that hides the whole dress of her head and hangs half-way down her back; and their shapes are wholly concealed by a thing they call a *ferigée*, which no woman of any sort appears without. This has strait sleeves that reach to their fingers' ends and it laps all round 'em, not unlike a riding hood. . . . You may guess how effectually this disguises them, that there is no distinguishing the great lady from her slave, and 'tis impossible for the most jealous husband to know his wife when he meets her. . . . This perpetual masquerade gives them the entire liberty of following their inclinations without danger of discovery. (Morris, pp. 5–6)

Montague explains that this disguise is so complete that not even the women's lovers know who they are. Such frank remarks, about subjects that most writers never addressed, made her letters popular reading.

Further Reading: Grundy, Isobel. *Lady Mary Wortley Montagu*. New York: Clarendon Press, 1999; Halsband, Robert, ed. *The Selected Letters of Lady Mary Wortley Montagu*. New York: St. Martin's Press, 1971; Montague, Mary Wortley. *Embassy to Constantinople: The Travels of Lady Mary Wortley Montague*. Edited and compiled by Christopher Pick; introduction by Dervla Murphy. New York: New Amsterdam, 1988; Morris, Mary. *Maiden Voyages: Writings of Women Travelers*. New York: Vintage Books, 1993.

Moody, Deborah (1586–1659)

Born Deborah Dunch in 1586, Dame Deborah Moody traveled from England to American in 1639 to settle in the Massachusetts Bay Colony. She embarked on this trip because of religious persecution. After her wealthy husband, Sir Henry Moody, died in 1629, she embraced the Anabaptist doctrine that baptism should be administered to believers. This doctrine ran counter to the established church's belief in infant baptism, and its adherents were persecuted.

Left with ample funds for travel, Moody decided to start a new life in America. But once in Massachusetts she discovered that the Puritans who lived there oppressed Anabaptists as well. Consequently she established her own colony, Gravesend, where she and others could experience true freedom of religion. It was the only permanent settlement of its time created by a woman, and it was carefully plotted to provide residents with equal portions of land.

Gravesend was in a region of New York under control of the Dutch; however, the Dutch granted the Gravesend settlers the right to create a self-governing town. But after war broke out between the English and the Dutch in Europe in 1652, tensions rose between these two factions in America. In 1657, Moody angered the Dutch by inviting English Quakers to hold meetings at her home, but Moody did not suffer for her actions, and she continued to promote freedom of worship until her death in 1659. *See also* North America.

Further Reading: Cooper, Victor H. *A Dangerous Woman: New York's First Lady Liberty: The Life and Times of Lady Deborah Moody (1586–1659?): Her Search for Freedom of Religion in Colonial America*. Bowie, MD: Heritage Books, 1995; Crawford, Deborah. *Four Women in a Violent Time: Anne Hutchinson (1591–1643), Mary Dyer (1591?–1660), Lady Deborah Moody (1600–1659), Penelope Stout (1622–1732)*. New York: Crown Publishers, 1970.

Morin, Nea (1905–1986)

Born in 1905, British mountaineer Nea Morin actively promoted the sport in England and France. She participated in several important climbing expeditions, including a successful ascent of Ama Dablam in Nepal. Morin and Fellow mountaineer Janet Adam Smith co-translated *Annapurna, First Conquest of an 8,000-Meter Peak (26,493 Feet)* (1950), a book written by climber Maurice Herzog. She also wrote *A Woman's Reach: Mountaineering Memoirs* (1968), which not only relates her own experiences as a mountaineer but also provides information on other women's mountaineering feats. Morin died in 1986. *See also* MOUNTAINEERING.

Further Reading: Herzog, Maurice. Translated by Nea Morin and Janet Adan Smith. *Annapurna, First Conquest of an 8,000-Meter Peak (26,493 Feet)*. 1950. Reprint. London: Lyons & Burford Publishers, 1997.

Morrell, Jemima (1832–1909)

Born in 1832, Jemima Morrell left behind a journal that detailed the first tour ever guided by Thomas Cook, whose "Cook's tours" revolutionized the field of travel. A member of the Junior United Alpine Club, Morrell went on the three-week tour through Switzerland in 1863, along with 119 other travelers. Her journal was discovered in the rubble of a Thomas Cook office that had been bombed during World War II and was published in 1963 as *Miss Jemima's Swiss Journal: The First Conducted Tour of Switzerland*. According to Morrell, Cook's tour was composed of approximately 140 tourists at the outset. They traveled from London to Paris in one day—from 6 A.M. to 11:30 P.M. Roughly half of them remained in Paris unguided, while the rest left with Cook at 6 A.M. the next morning for Geneva, Switzerland. Once in Geneva the group again split into two. Some—including Morrell—went with Cook to Chamonix while the rest went to the Swiss Alps. Once in Chamonix, Morrell, her brother William, and seven other young men decided to go off on their own, traveling through Switzerland for two weeks before rejoining Cook. The two Morrells later said that they went off on their own because their experiences with Cook were too touristy and they wanted to be among the Swiss people rather than among other Britains. Nonetheless, they were pleased with the travel opportunities that Cook had provided for them. *See also* GRAND TOURS.

Further Reading: Jemima, Miss. *Miss Jemima's Swiss Journal: The First Conducted Tour of Switzerland*. London: Putnam, 1963; Withey, Lynne. *Grand Tours and Cook's Tours: A History of Leisure Travel, 1750–1915*. New York: W. Morrow, 1997.

Morris, Mary (1947–)

Born in 1947, American author Mary Morris has written two travel books, *Nothing to Declare: Memoirs of a Woman Traveling Alone* (1988) and *Wall to Wall: From Beijing to Berlin* (1991). *Wall to Wall* describes a 1986 trip on the Trans-Siberian Express through China, Russia, and eastern Europe, while *Nothing to Declare* describes a trip to Mexico and Central America. *Nothing to Declare* also expresses the discoveries that Morris makes about herself during her journey, explaining the personal significance of travel. Morris presents travel as an inborn need—even an obsession—rather than a choice. She says:

> How do you know if you are a traveler? What are the telltale signs? As with most compulsions, such as being a gambler, a kleptomaniac, or a writer, the obvious proof is that you can't stop. If you are hooked, you are hooked. One sure sign of travelers is their relationship to maps. I cannot say how much of my life I have spent looking at maps, but there is no map I won't stare at and study. I love to measure each detail with my thumb, to see how far I have come, how far I've yet to go. I love maps the way stamp collectors love stamps. Not for their usefulness, but rather for the sheer beauty of the object itself. I love to look at a map, even if it is a map of Mars, and figure out where I am going and how I am going to get there, what route I will take. I imagine what adventures might await me even though I know that the journey is never what we plan for; it's what happens between the lines. . . . I have al-

> ways instinctively found my way and in this I feel blessed, as if somehow I am intended to be a journeywoman, a wanderer of the planet, and, I suppose, of the heart. (pp. 22–23)

Morris is also the author of several novels and the editor of a popular anthology of women's travel writing entitled *Maiden Voyages: Writings of Women Travelers*. Published in 1993, *Maiden Voyages* offers excerpts from the works of more than 50 women travelers from the early eighteenth century to the present. These include Lady Mary Wortley Montague, Mary Wollstonecraft, Frances Trollope, Mary Kingsley, Isabella Bird Bishop, Vita Sackville-West, Freya Stark, and Dervla Murphy. In writing about the essay collection, Morris says that the body of work "charts feminism, over close to three hundred years, through women and their journeys." (pp. xxi) At the same time, she says that the volume does not speak with a multicultural voice, because travel writers as a whole fall into the same ethnic and economic categories. She explains:

> Most of the women in this volume represent what . . . [travel writer Mable Sharman] Crawford refers to as a "woman of independent means and without domestic ties." The early women travelers were women of the upper classes in European society, invariably white and privleged. This trend has not shifted greatly in the past two hundred years as we are left with the legacy of colonialism. Travel literature by both men and women awaits its full range of multicultural voices and perspectives. Yet as feminists the writers gathered here hold surprisingly progressive views considering the times in which they wrote and lived. (p. xxi)

Morris also offers very brief biographical information about each author, although she explains that many of these women left behind few details about their lives apart from their travel literature. *See also* Bishop, Isabella Lucy Bird; Kingsley, Mary; Montague, Mary Wortley; Murphy, Dervla; Sackville-West, "Vita" (Victoria Mary); Stark, Freya Madeline; Trollope, Frances; Wollstone–craft, Mary.

Further Reading: Morris, Mary. *Nothing to Declare: Memoirs of a Woman Traveling Alone*. Boston: Houghton-Mifflin, 1988; ———. *Wall to Wall: From Beijing to Berlin*. New York: Doubleday, 1991; Morris, Mary, and Larry O'Connor. *Maiden Voyages: Writings of Women Travelers.* New York: Vintage Books, 1993.

Mountaineering

Many women adventurers have been drawn to mountaineering, traveling to remote mountains both as tourists seeking an exotic climbing experience and as athletes seeking a physical challenge. One of the first peaks to attract female climbers—and male climbers as well—was Mont Blanc, at 15,771 feet the highest peak in the French Alps. In 1760 Horace-Bénédict de Saussure, a young scientist from Geneva, Switzerland, offered a prize to the first person who scaled Mont Blanc, and for the next 26 years, until the prize was finally won by French physician Michel-Gabriel Paccard and his porter Jacques Balmat, hundreds of men and women flocked to the nearby town of Chamonix to

A woman climbs the Wedge in the Chugach Mountains, Alaska. © *Paul A. Souders/CORBIS.*

participate in or watch the efforts that resulted from the competition. Those who were not capable of making the steep ascent often hired guides to take them on walks up the lower slopes.

The first woman to climb Mont Blanc was Maria Paradis in 1808; In 1838 Henriette d'Angeville made the ascent; she was accompanied by six guides and six porters. After her success, other women ascended the peak. Most of them, including d'Angeville, were severely criticized for their efforts. According to historian Lynne Withey in her book *Grand Tours and Cook's Tours*, d'Angeville's friends called her desire to climb "dangerous and unfeminine," and mountaineer Elizabeth Le Blond, who ascended Mont Blanc in 1881, faced the scorn of an aunt who insisted that Le Blond's mountaineering was "scandalising all London." (pp. 208–209) Strenuous climbing, most people agreed, was suitable only for men.

Neither d'Angeville nor Le Blond was deterred. D'Angeville remained a mountaineer until the age of 69, while Le Blond became a specialist in climbing peaks during winter. She wrote *The High Alps in Winter: Or, Mountaineering in Search of Health* (1883), *My Home in the Alps* (1892), *Adventures on the Roof of the World* (1904), and *Mountaineering in the Land of the Midnight Sun* (1908) to describe her mountaineering experiences throughout the world. Le Blond was also the president of the Ladies' Alpine Club of London, from its formation in 1907 until her death in 1934.

The Ladies Alpine Club was the first all-woman climbing group, and it was established independently of England's Alpine Club (established in 1857) because the male club did not invite women as members. Britain's Pinnacle Club was formed in 1921 with the help of British climber Dorothy Pilley, who scaled mountains in Europe, Asia, the United States, and Canada with her husband, famous mountaineer I. A. Richards. In 1975, the Ladies Alpine Club merged with the Alpine Club; the group currently has more than 100 male and female members.

However, this does not mean that mountaineering in modern times is no longer a male-dominated sport. In a 1993 autobiography, *Beyond the Limits: A Woman's Triumph on Everest*, American Stacey Allison says that only recently have women climbers earned any respect from male mountaineers. She writes: "In the early seventies the majority of male climbers were up-front about their sexism. Women, they said, just didn't have the muscle power to take on the big mountains. That attitude had dominated mountaineering since the sport's beginnings." (p. 88)

Despite such sexism, the number of women climbers has increased steadily since the late 1800s, when a popular book about mountain climbing, *A Lady's Tour round the Monte Rosa: with Visits to the Italian Valleys . . . in a Series of Excursions in the Years 1850–56–58* (1859) by Mrs. H. W. Cole, encouraged women to take up the sport and provided tips on such important matters as how to dress. In addition to Cole, notable British women mountaineers from this period include Lucy Walker, who scaled 98 peaks between 1858 and 1879 and Anna Pigeon, who climbed more than 63 major peaks. With her sister Ellen in 1873, she became the first woman to go completely over the Matterhorn, traveling from the Italian side of the Alps to the Swiss side.

Meanwhile, women's mountaineering was becoming popular in the United States. In fact, one of the foremost female climbers of the late nineteenth and early twentieth centuries was an American: Annie Smith Peck, who became interested in mountaineering at the age of 35 and climbed the Matterhorn in 1895. During the next few years she climbed other mountains in Europe, California, and Mexico, supporting herself by giving lectures on her adventures. In 1908 she became the first person to ascend the south summit of Mount Huascaran in Peru. She wrote about this experience in *A Search for the Apex of America: High Mountain Climbing in Peru and Bolivia, including the Conquest of Huascaran, with Some Observations on the Country and People Below* (1911). Other notable American mountaineers of the late nineteenth and early twentieth centuries include Addie Alexander, who in 1871 became the first woman to climb Longs Peak in Colorado; Fay Fuller, who in 1890 became the first woman to climb Mount Rainier (called Tahoma by the Indians) in Washington State; and Dora Keen, who in 1912 became the first woman to lead an expedition up Alaska's Mount Blackburn and in 1913 the first woman to reach the mountain's peak. Of these, Fuller—who was accompanied on her expedition by four men—received perhaps the most publicity, because her father owned a Tacoma, Washington, newspaper and published her account of the climb just two weeks after the ascent.

Scientists

During the 1920s, another group of climbers was added to those seeking fame and physical challenge: women scientists, who embraced mountaineering in order to study plant and animal life at high altitudes. For example, Dutch mountaineer Jenny Visser-Hooft explored, mapped, and researched the flora and fauna of the Karakorum Glaciers in Pakistan and India. She and her husband, a physician and diplomat, made four expeditions to the region during the 1920s, writing about some of their experiences in *Among the Kara-Korum Glaciers in 1925* (1926). Visser-Hooft also climbed other mountains in Central Asia, as well as in the Alps and the Caucasus Range.

Over the next several decades, advances in technology made possible women's more equal participation in the sport because mountaineering equipment gradually became stronger yet lighter in weight. Although the woman's movement may have increased women's desire to compete with men, women have, from the very beginning, climbed beause they wanted to. The sport is one of independent commitment, not necessarily competition.

Women's "Firsts"

Consequently the last half of the twentieth century has produced the largest number of "firsts" among women climbers. They include Julie Tullis, who in 1985 became the first woman to climb Broad Peak on the Pakistan-China border, one of the tallest mountains in the world; Junko Ishibashi Tabei of Japan, who in 1975 became the first woman to reach the summit of Mount Everest; and Stacy Allison, who in 1988 became the first American woman to reach the summit of Mount Everest. In 1969 Tabei founded the first woman's mountaineering club in Japan, the Joshi-Tohan. She is also the first woman to have ascencded the tallest mountan in each of the world's seven major geographical regions. Also notable among women climbers is American Arlene Blum, who participated in climbing expeditions to Annapurna in Nepal.

Annapurna is part of the Himalayas, a mountain system that features some of the highest mountains in the world. It is a series of three parallel ranges that extend from northwest Pakistan across Kashmir through northern India, southern Tibet, Nepal, Sikkim, and Bhutan, for a total of 1,500 miles. At 29,035 feet, Mount Everest is the highest peak in the region and also the highest in the world. The first all-woman expedition to climb the Himalayas was appropriately named the First Women's Himalayan Expedition. Its members were Monica Jackson, Elizabeth Stark, and Evelyn Camrass. All belonged to the Ladies' Scottish Climbing Club; in 1955 they set out to verify a map of the region and to find a new way through an area of glaciers. After two months they achieved their goals, and Jackson later wrote about their experiences in *Tents in the Clouds: The First Women's Himalayan Expedition* (1956).

The most recent climb to feature women mountaineers also occurred in the Himalayas: American Supy Bullard, who led the first all-woman mountaineering team to climb the Cho Oyu in Nepal, the sixth-tallest peak in the world, without the assistance of local guides or supplemental oxygen. She accomplished this feat in 1999, along with American teammates Caroline Byrd, age 39; Liane Owen, age 40; Cara Liberatore, age 28;

Georgie Stanley, age 31; and Kathryn Miller Hess, age 33. These and more than 25 other women describe their climbing experiences in a book of essays entitled *Leading Out: Women Climbers Reaching for the Top* (1992), edited by Rachel Da Silva.

Several books chronicle failed mountaineering expeditions. For example, *Facing the Extreme: One Woman's Story of True Courage, Death-Defying Survival, and Her Quest for the Summit* (1998) is the story of American Ruth A. Kocour's disastrous experience climbing Mount McKinley in Alaska. She was the only female member of an expedition to reach the summit of the mountain, after 11 of her teammates died during a snowstorm. In the book's preface she writes of how she perceived danger before setting out on her climb: "Even on McKinley, where the weight of the mountain's force had crushed the last breath out of so many other climbers before, I didn't set out with the thought that I was risking my life. I am more afraid of not living than I am of dying." Kocour's attitude is not uncommon among women climbers, who consider a journey to the top of the world worth even the most extreme effort. *See also* ALLISON, STACY; BULLARD, SUPY; COLE, MRS. H. W. (HENRY WARWICK); DUNSHEATH, JOYCE; FULLER, MARGARET; JACKSON, MONICA; KEEN, DORA; LE BLOND, ELIZABETH; PECK, ANNIE SMITH; PIGEON, ANNA; PILLEY, DOROTHY; TABEI, JUNKO; TULLIS, JULIE; VISSER-HOOFT, JENNY.

Further Reading: Angell, Shirley. *The Pinnacle Club: A History of Women Climbing.* Great Britain: The Pinnacle Club, 1988; Birkett, Bill, and Bill Peascod. *Women Climbing: 200 Years of Achievement.* Branson, MO: Mountaineer Books, 1990; Cobb, Sue. *The Edge of Everest: A Woman Challenges the Mountain.* Harrisburg, PA: Stackpole Books, 1989; Da Silva, Rachel. *Leading Out: Women Climbers; Reaching for the Top.* Seattle, WA: Seal Press, 1992; Harper, Stephen. *Lady Killer Peak: A Lone Man's Story of Twelve Women on a Killer Mountain.* London: World Distributors, 1965; Jackson, Monica, Arlene Blum, and Elizabeth Stark. *Tents in the Clouds: The First Women's Himalayan Expedition.* Seattle, WA: Seal Press, 2000. Kocour, Ruth A., as told to Michael Hodgson. *Facing the Extreme: One Woman's Story of True Courage, Death-Defying Survival, and Her Quest for the Summit.* New York: St. Martin's Press, 1998; Robertson, Janet. *The Magnificent Mountain Women: Adventures in the Colorado Rockies.* Lincoln: University of Nebraska Press, 1990; Rowan, Peter, and June Hammond Rowan, eds. *Mountain Summers: Tales of Hiking and Exploration in the White Mountains from 1878 to 1886 as Seen through the Eyes of Women.* Gorham, NH: Gulfside Press, 1995; Smith, Cyndi. *Off the Beaten Track: Women Adventurers and Mountaineers in Western Canada.* Jasper, Alberta: Coyote Books, 1989; Williams, Cicely. *Women on the Rope: The Feminine Share in Mountain Adventure.* London: Allen & Unwin, 1973; Withey, Lynne. *Grand Tours and Cook's Tours: A History of Leisure Travel, 1750–1915.* New York: W. Morrow, 1997.

Murie, "Mardy" (Margaret) (1901–)

Born in 1901, American conservationist Margaret Murie, commonly known as "Mardy," traveled throughout Alaska with her husband Olaus, a wildlife biologist, and later wrote about her experiences in *Island Between* (1977) and *Two in the Far North* (1978). An environmental activist, she also worked to established and subsequently protect the Arctic National Wildlife Refuge and to pass the Wilderness Act of 1964. In 1977 she convinced the U.S. Congress to pass the Alaska Lands Bill, preserving Alaskan wilderness areas. Murie currently lives in Jackson Hole, Wyoming, where she moved with her late husband in 1927 to study elk. The couple had previously studied caribou during a 1924 dogsled expedition along Alaska's Koyukuk River. In 1998 Murie received the Presidential Medal of Freedom to honor her achievements in environmentalism. *See also* ECO-TOURISM.

Further Reading: Murie, Margaret E. *Island Between.* Fairbanks: University of Alaska Press, 1977; ———. *Two in the Far North.* Seattle, WA: Alaska Northwest Books, 1997.

Murphy, Dervla (1931–)

Born in 1931, Irish travel writer Dervla Murphy has written many books about her adventures in foreign countries. As a young woman she was an avid bicyclist but could not travel long distances because she had to care for her elderly parents. The year after their deaths, in 1962, she decided to take a cycling trip across Europe to India, staying in inns and hostels along the way. She dressed in masculine clothes and carried a gun to protect herself on the road, and when the terrain was difficult she rode the train or hitchhiked a ride on a truck, stowing her bicycle in the back. She made many friends during her journey but also experienced illness and injury.

When Murphy reached India, she visited a camp for Tibetans who had fled when their country was taken over by Chinese. Moved by their need, she decided to volunteer her services to relief workers there and ended up staying for four months. She then returned home to publicize the Tibetans' plight. She also wrote her first book, *Full Tilt: Ireland to India with a Bicycle* (1965). In 1965 she made another trip to visit displaced Tibetans, this time at a camp in Nepal. This experience inspired two more books, *Tibetan Foothold* (1966) and *The Waiting Land: A Spell in Nepal* (1967).

One of Murphy's most popular books, *The Waiting Land,* describes her 1965 trip to Nepal and includes her own black-and-white photographs. Written as a series of diary entries, it offers details about scenery, weather, and people, as well as comments regarding the political situation in Nepal. As an example of the last, Murphy says:

> Already the bazaar price of kerosene has gone up to two pounds per gallon and there is a possibility that all internal . . . flights may be stopped because Nepal is dependent on India for her petrol supplies [and India and Pakistan are at war]. We have just spent two hours listening with great difficulty to snatches of English Language news giving very dissimilar interpretations of the situation from Pakistani, Indian, Nepalese, Chinese, American and British viewpoints. Peking is being even more lurid than usual and describing America as "a vicious wolf" and "the most rabid oppressor mankind has ever known," America is being no less puerile on a slightly more sophisticated level and everyone else is also reacting according to form. The whole thing leaves one sunk in depression. . . . [And] the country I pity most is poor little Nepal, who is now shaking in her shoes . . . between China plus Pakistan and India plus America. (p. 131)

Murphy expresses even more concern for political refugees in Nepal; in 1965 approximately 500 Tibetans sought sanctuary in Nepal after being persecuted by the Chinese. In the epilogue to *The Waiting Land*, Murphy comments on the exiled Tibetans' situation, saying:

> It should be obvious that our first duty is to help the refugees to help themselves, rather than merely to feed, clothe and nurse them. Immediately after their arrival in a strange land they obviously do need considerable material support, as well as sensible guidance on how best to adjust to their new environment. Yet the essence of a refugee tragedy is not disease, hunger or cold. It is the loss of the emotional security of belonging to a stable community in a certain country, and therefore, in exile, the refugee's most valuable possession is his self-respect—which should be used by us as a tool to help him re-establish his identity as a responsible individual. (p. 208)

The author concludes *The Waiting Land* by returning to the issue of Nepal, expressing concern for the country's ability to maintain its cultural traditions in the face of progress. She says:

> There . . . are many institutions that do need reforming; but to reform them in the image and likeness of the West would be a subtle genocide, for there is much, too, that should be cherished . . . However, it is of no avail to think or write thus. The West has arrived in Nepal, bubbling over with good intentions . . . and soon our insensitivity to simple elegance, to the proud work of individual craftsmen, and to all the fine strands that go to make up a traditional culture will have spread material ugliness and moral uncertainty like plagues through the land. (p. 212)

After writing *The Waiting Land*, Murphy traveled through Ethiopia on foot and by mule; the book that resulted from this trip was *In Ethiopia with a Mule* (1968). She then took a break from travel to have a daughter, Rachel, but when the girl was five she returned to her adventures, taking Rachel along. Together the two went to India, Northern Ireland, Peru, the United States, and Madagascar. Murphy's subsequent books include *Eight Feet in the Andes* (1983); whose title refers to her feet, her daughter's feet and a pack mule's feet making a 1,300-mile journey through the Peruvian Andes, *Muddling through in Madagascar* (1989); and an autobiography entitled *Wheels within Wheels* (1979).

In *Muddling through in Madagascar,* she often comments on the differences between the Madagascar and Western cultures. For example, she says that in Madagascar: "If an old man is heavily laden, any young man catching up with him insists on carrying his load for some distance, though they may be total strangers. And young people ask permission before overtaking their elders on the [path]. Is it a measure of the uncouthness of the modern West that we marvelled so to observe these courtesies?" (pp. 325–326). Murphy's most recent work is *The Ukimwi Road: From Kenya to Zimbabwe* (1995), which describes a sub-Saharan bicycle ride. *See also* BICYCLES.

Further Reading: Murphy, Dervla. *Eight Feet in the Andes*. London: J. Murray, 1983; ———. *Full Tilt: Ireland to India with a Bicycle*. Woodstock, NY: Overlook Press, 1986; ———. *Muddling through in Madagascar*. NY: Overlook Press, 1989; ———. *Tales from Two Cities: Travel of Another Sort*. London: J. Murray, 1987; ———. *The Ukimwi Road: From Kenya to Zimbabwe*. Woodstock, NY: Overlook Press, 1995; ———. *The Waiting Land*. London: John Murray, 1967.

Murray, Amelia (1795–1884)

Born in 1795, Englishwoman Amelia Murray was an amateur botanist who toured North America and Cuba between 1854 and 1855. Because Murray had connections to British nobility, she traveled in style and was given the honor of inaugurating Canada's Grand Trunk Railway. Upon her return to England, she wrote about her experiences in *Letters from the United States, Cuba and Canada* (1856). Her book describes not only plants but also people, places, and customs; it caused great controversy because of its support of slavery. Most people in England were abolitionists, and Murray's defense of slave-owning caused her to lose her job as maid of honor to Queen Victoria. Moreover, although the book included information on how to purchase copies of its illustrations—sketches hand drawn by Murray—no one responded to the offer. Murray never wrote another book, but spent her remaining years advocating improvements in women's education. She died in 1884.

Further Reading: Robinson, Jane. *Wayward Women: A Guide to Women Travellers*. Oxford: Oxford University Press, 1990.

Murray, Margaret Alice (1863–1963)

Born in 1863, British archaeologist Margaret Alice Murray was one of the foremost experts in Egyptology. She became interested in the subject while in college, but because of barriers against women participating in any study program involving field work, she earned her degree in linguistics instead, specializing in Egyptian hieroglyphics. Her expertise in this field earned her an invitation to accompany a noted male archaeologist, Sir Flinders Petrie, to Egypt in the late 1890s. There she helped excavate ruins at Abydos, proving herself a competent archaeologist. Her work on Egyptian archaeology, *The Splendor That Was Egypt* (1931), is still considered a classic.

In addition, because of the gender inequality inherent in her profession, Murray was involved in the early feminist movement. She applied her views to some of her work, writing *The Witch-Cult in Western Europe* (1921) to express her opinion that the persecution of witches was actually a male attempt to repress women in general. Murray died in 1963. *See also* ARCHAEOLOGISTS AND ANTHROPOLOGISTS.

Further Reading: Murray, Margaret Alice. *Egyptian Temples*. 1931. Reprint. New York: AMS Press, 1977; ———. *The Splendour that was Egypt*. 1931. Reprint. London: Sidgwick and Jackson, 1972.

N

Nature. *See* ADVENTURE TRAVEL; ECO-TOURISM.

Nichols, Ruth (1901–)

Born in 1901, American socialite Ruth Nichols was a pioneering aviator who set a world distance record in 1931, flying nonstop from Oakland, California, to New York City. Nichols was recovering from a spinal injury at the time. Three months earlier, she had crashed a plane while attempting to fly nonstop from St. John, New Brunswick, to Paris. Nichols also promoted the use of aviation to bring medical supplies to civilians during World War II. Her charity group, Relief Wings, was under the direction of the Civil Air Patrol. *See also* AIR TRAVEL.

Further Reading: The Ninety-Nines: International Organization of Women Pilots. "Ruth Nichols." http://www.ninety-nines.org (May 2000).

Nightingale, Florence (1820–1910)

Born in 1820, English nurse Florence Nightingale traveled extensively as part of her volunteer and nursing duties, visiting sick people throughout England. In 1854 she accompanied a group of volunteers to Turkey to provide medical care for soldiers wounded during the Crimean War. Even before this trip, however, she had been overseas, having traveled with her wealthy family as a girl. She had trained as a nurse against her parents' wishes, becoming superintendent of the Institution for the Care of Sick Gentlewomen in London; once established in Turkey, she was appointed head of the nurses in the military hospitals of Scutari. By enforcing sanitary

Florence Nightingale. © *Bettmann/CORBIS.*

regulations, she reduced the death rate at her hospitals from 45 percent to 2 percent. She also bought hospital supplies with her own money, and even after nearly dying of Crimean fever she refused to leave her patients. She became famous for her efforts, and, after returning to England, she was a prominent activist who crusaded for better hospital conditions in England. In addition, she established a nursing school with money awarded her by the British government. She died in 1910.

Further Reading: Hume, Ruth Fox. *Florence Nightingale*. New York: Random House, 1960.

Niles, Blair (1880–1959)

Born in 1880, American explorer Blair Niles—also known as Mary Beebe or Mary Blair Rice—was the first white woman to visit Devil's Island, a penal colony in French Guiana. She also led expeditions, primarily to hunt birds, in South America and wrote many books about her adventures. In 1925 she helped found the Society of Women Geographers.

Niles's first travel writings appeared under the name Mary Blair Beebe; her husband, Charles Beebe, collected bird specimens for the New York Zoological Society, and she began traveling with him shortly after their marriage in 1903. Together the Beebes went to Mexico and South America before embarking on a 20-country tour to study pheasants. Their journey lasted from 1909 to 1911, during which their marriage deteriorated. In 1913 they divorced. Almost immediately, Mary Beebe married architect Robert Niles and began writing as Blair Niles. Her new husband enjoyed travel, and the couple went on a long tour to South America. Niles wrote several articles and books about this trip, including *Casual Wanderings in Ecuador* (1923), *Columbia, Land of Miracles* (1924), and *A Journey in Time: Peruvian Pageant* (1937), the last of which brought her several awards. In her later years, she wrote novels and biographies; she died in 1959.

Further Reading: Niles, Blair. *Casual Wanderings in Ecuador*. New York and London: Century, 1923; ———. *A Journey in Time: Peruvian Pageant*. Indianapolis, IN, and New York: Bobbs-Merrill, 1937.

North, Marianne (1830–1890)

Born in 1830, English artist Marianne North traveled widely in search of subjects for her oil paintings, which featured plants from around the world. She came from a prominent family, and after her parents died in 1870, she used her inheritance to travel, first through Europe and then to the eastern United States, Jamaica, and Haiti. From 1872 to 1874 she visited South America, and from 1876 to 1878 Japan, Singapore, and parts of South Asia. En route to the Far East she crossed the United States and toured California. During the late 1870s and early 1880s, she also visited several Pacific and Atlantic islands, as well as Australia and South Africa, and returned to some of the sites of her earlier travels. Her last journey was to Chile in 1884.

In all of her travels, North's main focus was the foliage. Whenever she heard about an unusual plant, she went to its habitat to sketch it. In 1882 a gallery opened in England's Royal Botanical Gardens to display some of her work. When she died in 1890 she left numerous paintings and enough autobiographical writings to produce two books: *Recollections of a Happy Life: Being the Autobiography of Marianne North* (1892) and *Some Further Recollections of a Happy Life* (1893). Both were edited by North's sister, Mrs. John Addington Symonds.

Further Reading: Brenan, J.P.M. *A Vision of Eden: The Life and Work of Marianne North*. New York: Holt, Rinehart and Winston, 1980.

North America

North America has produced many notable women travelers and has long been an important travel destination for women as well. The first nonindigenous women to travel through the region were colonists from Europe. They established communities along the east coast of what would become the United States, beginning in the 1600s, and then spread west in phases. The first phase occurred after the American Revolution, when pioneers began to cross the Appalachian Mountains and move into the Ohio and Mississippi River valleys. The second phase took place during the mid-nineteenth century, when settlers started crossing the Mississippi River in greater and greater numbers. After gold was discovered in California in 1848, there was a rush to the west coast, with many areas in between being bypassed. By the end of the nineteenth century, however, most of the country had been settled.

Although most women coming to the region were colonists, some were explorers who sought challenging travel adventures. For example, Englishwoman Frances Hornby Barkley embarked on a trading expedition from Belgium to the Pacific Northwest with her husband in 1786, and together they explored much of western Canada. Marie-Anne Lagemodière accompanied her fur-trader husband on an 1806 expedition from Quebec across Canada to a western outpost on the Red River, while Isabel Gun sailed to Canada from Scotland disguised as a man to work for the Hudson's Bay Company. Another Canadian explorer was Agnes Dean Cameron, who traveled to the Arctic and wrote about her adventures in *The New North: An Account of a Woman's Journey through Canada to the Arctic* (1909).

In contrast, many of the upper-class European women who visited North America in the early nineteenth century were tourists who sought not exploration but enlightenment. Many of them also wanted to share their travel experiences with their peers back home. For example, Englishwomen Frances Trollope, Emmeline Stuart Wortley, Isabella Bird Bishop, Frances Wright, and Amelia Murray and Irishwoman Anna Brownell Jameson all wrote travel books about their adventures in North America. A few American women also became well known for their travel writings during this period. Perhaps the most notorious of these women was Anne Royall, whose *Sketches of History, Life and Manners in the United States, by a Traveller* (1826), *The Black Book: or, A Continuation of Travels in the United States* (1828–1829; three volumes), *Mrs. Royall's Pennsylvania: or, Travels Continued in the United States* (1829; two volumes), and *Mrs. Royall's Southern Tour: or, Second Series of Black Books* (1830–1831; three volumes) were highly critical of certain aspects of the places she visited. An outspoken critic of people and organizations she disliked, Royall made so many enemies by publishing her books that eventually she feared for her life.

A few women settlers also wrote about their experiences in North America. These include Eliza Farnham, whose *Life in Prairie Land* (1846) describes her pioneer life in Illinois, and Willa Cather, who drew on her experiences as an immigrant in Nebraska for numerous novels. Homesteaders such as Elinore Pruitt Stewart, who claimed a 160–acre plot of land in Burtfork, Wyoming, in the early 1900s, also wrote about their experiences; in Stewart's case her intent was partly to encourage other women to become homesteaders.

Many women published their letters and other writings during the California Gold Rush, hoping either to encourage other women to move west or to warn them not to do so. For example, Louise Clappe's letters detailing her experiences in California from 1851 to 1852 were published in *The Pioneer: Or, California Monthly Magazine* in 1854–1855 and appeared in book form as *The Shirley Letters* in 1933. A letter from Virginia Reed, who was one of 47 survivors from the Donner Party, was published in the *Illinois Journal* in December 1847; it was first to bring the news of the major wagon train disaster to Easterners.

Of course, most women pioneers in North America during the nineteenth and early twentieth centuries were not authors and many were Native Americans and African American slaves who had neither the education nor the opportunity to tell their stories. The Chinese women who came to California to work during the Gold Rush were also without voice. The names and experiences of many travelers during this period who were members of minority ethnic groups are lost to history. Among the few that are known are African American Clara Brown, a former slave who financed and escorted a wagon train of 16 African Americans to Colorado in the 1860s, and Polly Bemis, who was enslaved in China and brought to California during the Gold Rush to work in a brothel. Bemis is representative of hundreds of other women who traveled not by choice but by force or obligation through North America during the 1800s.

In the late nineteenth century, an increasing number of women traveled for enjoyment. Tourists were aided by improved roads and vehicles, and they were encouraged to travel by the railroads, which heavily promoted the idea of tourism to increase profits. One women who was involved in such promotional efforts was Carrie Adell Strahorn, who, with her husband, traveled throughout the country to write advertising pamphlets for a railroad company during the latter part of the century.

During the Great Depression (1929–1940), severe poverty coupled with a drought in the Midwest forced thousands of farmers off their land. This, in turn, increased migration as people struggled to find places where they could earn a living. The crisis did not end until World War II, which created new demands for labor and goods. Once the war was over, tourism again increased. Although the train remained popular, the automobile quickly became the vehicle of choice, with families embarking on new highways to visit America's growing system of national parks and other tourist destinations. The automobile continues to be the most widely used means of long-distance transport within the United States, although as air transportation becomes more affordable many people are choosing to fly instead of drive across the continent. *See also* BARKLEY, FRANCES; BISHOP, ISABELLA LUCY BIRD; BROWN, CLARA; CAMERON, AGNES DEAN; CARR, EMILY; CATHER, WILLA; CLAPPE, LOUISE; FARNHAM, ELIZA; GOLD RUSH; GUN, ISABEL; JAMESON, ANNA BROWNELL; LAGEMODIÉRE, MARIE-ANNE; MURRAY, AMELIA; REED, VIRGINIA; ROYALL, ANNE NEWPORT; STEWART, ELINORE PRUITT; STRAHORN, CARRIE ADELL; TROLLOPE, FRANCES; WORTLEY, EMMELINE STUART AND VICTORIA; WRIGHT, FRANCES.

Further Reading: Bird, Isabella L. *A Lady's Life in the Rocky Mountains*. 1879. Reprint. Norman: University of Oklahoma Press, 1960; Davis, Gwenn. *Personal Writings by Women to 1900: A Bibliography of American and British Writers*. Norman: University of Oklahoma Press, 1989; Farnham, Eliza W. *Life in Prairie Land*. New York: Harper & Brothers, 1846; Stefoff, Rebecca. *Women Pioneers*. New York: Facts On File, 1995; Stewart, Elinore Pruitt. *Letters of a Woman Homesteader*. 1914. Reprint, Boston: Houghton Mifflin, 1982; Tinling, Marion, compiler and ed. *With Women's Eyes: Visitors to the New World, 1775–1918*. Hamden, CT: Archon Books, 1993; Trollope, Frances. *Domestic Manners of the Americans*. 1832. Reprint with introduction by Donald Smalley. New York: Alfred A. Knopf, 1949.

O

Oakley, Annie (1860–1926)

Born in 1860 as Phoebe Anne Oakley Mozee, American markswoman and trick rider Annie Oakley is representative of women entertainers who traveled as part of their profession. In fact, she is one of the most famous traveling entertainers in history. Throughout much of her life she toured the United States in vaudeville, circus, and Wild West shows. As a girl she won a shooting contest that launched her career. When word of her shooting skills spread, she was invited to perform overseas for royalty. Her tricks included shooting a hole in the center of a playing card held a great distance away and shooting a cigarette out of someone's mouth from 30 paces away. Oakley died in 1926.

Further Reading: Alderman, Clifford Lindsey. *Annie Oakley and the World of her Time*. New York: Macmillan, 1979; Cooper, Courtney Ryley. *Annie Oakley, Woman at Arms, A Biography*. New York: Duffield, 1927; Women of the West. http://www.over-land.com/westpers2.html (May 2000).

Studio portrait of Phoebe Anne Oakley Mozee (Annie Oakley), probably between 1885 and 1901. *Denver Public Library, Western History Collection.*

Oakley, Barbara

During the 1980s, American author Barbara Oakley lived on Russian trawlers to work as a translator. She was part of a cooperative venture between the United States and Russia, which meant that she primarily sailed off the coast of the Pacific Northwest. Oakley wrote a book about her experiences entitled

Hair of the Dog: Tales from Aboard a Russian Trawler (1996).

Further Reading: Oakley, Barbara. *Hair of the Dog: Tales from Aboard a Russian Trawler*. Seattle: Washington State University Press, 1996.

O'Brien, Kate (1897–1974)

Born in 1897, Irish author Kate O'Brien wrote two travel books, *Farewell Spain* (1937) and *My Ireland* (1962). *Farewell Spain* was one of the most controversial travel books of its time. Upon publication, the book was banned in Spain, and O'Brien was told she could not visit the country again. This ban was not lifted until 1957.

Farewell Spain includes political comments about the changes that had taken place in that country between O'Brien's 1922 visit and her 1935 visit. For example, she criticizes the military regime that was established after the Spanish Civil War (1936–1939), saying:

> We all know the dreary story of autumn 1934 in Spain and all the woes that followed from it . . . [including] a war . . . openly aimed at the murder of every democratic principle, and for the setting up of [General Franco's] little self as yet another Mussolini [then dictator of Italy]—such a war strikes not merely for the death of Spain, but at every decent dream or effort for humanity everywhere. It kills not only the slow, creeping growths, the sensitive plants of social and economic justice which the last hundred years have gardened so painfully and at so great a price, but it sterilises the whole future. (pp. 220–221)

Farewell Spain is not solely political commentary, however. It also offers descriptions of the sights and people that O'Brien encountered while traveling through the country and addresses the subject of tourism in general. O'Brien suggests that someday the world will be so homogeneous that people will no longer travel to learn new things. She explains:

> There will be no point . . . in going out . . . [if] air-travel, radio and television will have made all possible novelties into boring fireside matters-of-fact. The world will be flat and narrow, with the Golden Horn a stone's throw from the Golden Gate and nothing unknown beyond any hill. Antarctica, where no one lives, will be a weekend joyride, and our descendants, should any records survive to catch their eyes, will marvel at our naïve interest in our neighbors, smiling to discover that once an Arab differed somewhat in his habits from a Dutchman, and a Tibetan from a Scot. (p. 4)

At the end of *Farewell Spain*, O'Brien expresses the hope that Spain will somehow retain its individuality.

O'Brien also wrote several novels, including *Without My Cloak* (1931), *The Ante-Room* (1934), *Mary Lavelle* (1936), and *The Land of Spices* (1941), as well as plays and an autobiography, *Presentation Parlour* (1963). O'Brien died in 1974.

Further Reading: Dalsimer, Adele. *Kate O'Brien*. Boston: Twayne Publishers, 1990; O'Brien, Kate. *Farewell Spain*. 1937. Reprint. London: Virago Press, 1985; ———. *The Land of Spices*. Garden City, NY, Doubleday, Doran, 1941. ———. *My Ireland*. New York: Hastings House, 1962.

O'Malley, Grace (ca. 1530–1603)

Born in approximately 1530, Grace O'Malley was a notorious Irish pirate who traveled the seas around the British Isles attacking English ships. She was the daughter of Dudara "Black Oak" O'Malley, an Irish chieftain who owned a fleet of herring ships and engaged in a variety of trading activities—some legitimate, some illegitimate. When she turned 15, O'Malley was forced to marry the leader of another powerful clan. Several years later, after she had given birth to several children, her husband was killed while trying to regain a fortress stolen by an enemy clan. She then took up his cause, leading his men into battle and regaining the fortress herself. Nonetheless, Irish law would not allow her to become clan leader, so she returned to her birthplace and took over the ships of her newly deceased father. She immediately turned to piracy, raiding merchant fleets along Scotland, England, and Europe. She particularly targeted English ships, because she perceived the English as an enemy of the Irish. Later, however, she became a friend of the English Court, using her ships to support England in a war with its enemies on the Continent. O'Malley's contributions proved so valuable that Queen Elizabeth I knighted one of O'Malley's many children as a way to honor her. (Women could not be knighted.) This child, Tibbott-ne-Long ("Toby of the Ships"), had been born at sea after O'Malley remarried; her new husband, Richard-in-Iron, was an Irish chieftain who owned a large fleet of trading vessels and was also given to plundering rival ships. Even the rigors of childbirth did not prevent O'Malley from performing her piratical duties. The day after her son was born, she helped her men fight off an attack from a Turkish ship, capturing the crew and subsequently hanging them. O'Malley continued to sail the seas until her retirement in 1601; she died in 1603. *See also* PIRATES.

Further Reading: Chambers, Anne. *Granuaile: Life and Times of Grace O'Malley c. 1530–1603*. Portmarnock, Ireland: Wolfhound Press, 1979; De Pauw, Linda Grant. *Seafaring Women*. Boston: Houghton Mifflin, 1982; Druett, Joan. *She Captains: Heroines and Hellions of the Sea*. New York: Simon & Schuster, 2000.

Organizations and Associations

A variety of organizations and associations are devoted to promoting travel and exploration. Most were established by men for men, although today these groups offer membership to women. A few, however, were established by women explorers and adventurers who wanted to support and encourage their peers.

The oldest group dedicated to women's exploration activities is the Society of Women Geographers. Founded in 1925 by noted Americans explorers Harriet Chalmers Adams, Amelia Earhart, Marguerite Baker Harrison, Blair Niles, and Grace Seton Thompson, from the outset the society has encouraged women in all forms of geographical exploration and research, interpreting "geographer" to include such allied disciplines as anthropology, archaeology, biology, ecology, geology, and oceanography. Its members have included American hunter Delia Akeley, American mountaineer Annie Smith Peck, American anthropologist Margaret Mead, and British archaeologist Mary Leakey. The first person to win a gold medal from the society was American pilot Amelia Earhart for outstanding achievement.

Today the society offers active and associate memberships. Active members are women involved with research activities in the field who have furthered human knowledge of a subject by publishing written or visual representations of their work. Associate members are women who have traveled extensively in the pursuit of knowledge but have not produced a record of their findings. According to society literature, high achievement is the main criterion for membership, which is by nomination only. Since 1953, the group has awarded more than 100 fellowships to support women graduate students in geography or related sciences at Rutgers University, the University of California at Los Angeles (UCLA), and the University of California at Berkeley.

Another prominent women-only group is the Ninety-Nines, an international organization of women pilots that is dedicated to the advancement of women's aviation. It was founded in 1929 by 99 licensed women pilots, one of them Amelia Earhart, after the women participated in the first National Women's Air Derby together. Today the Ninety-Nines sponsors women's flying events and scholarship programs.

Some of the oldest organizations established by males also offer scholarships and financial support to women. For example, the National Geographic Society, which was established in 1888 in Washington, D.C., currently offers extensive support. A nonprofit scientific and educational organization dedicated to furthering knowledge related to geographical exploration, the society has funded expeditions and field research projects for American undersea explorer Sylvia Earle, who is preparing to launch a five-year series of expeditions designed to explore the deepest parts of 12 U.S. National Marine Sanctuaries. As leader of the National Geographic

Society's Sustainable Seas Expeditions (SSE), she will pilot a one-person submarine into each sanctuary and broadcast information about her discoveries on the Internet. Other women who have been supported by the National Geographic Society include primate researchers Jane Goodall (British) and Dian Fossey (American) and Australian explorer Robyn Davidson. In exchange for funding, they wrote articles for the *National Geographic* magazine; some of their work appears in an anthology of *Geographic* articles, *From the Field: The Best of National Geographic* (1997), which also discusses the early years of the Society.

Another geographical society that offers support to women interested in geographical exploration is the Association of American Geographers. Founded in Philadelphia in 1904 but currently located in Washington, D.C., this group funds scientific research, scholarly studies, and educational programs related to geography and works to develop better materials for teaching geography in classrooms. This group encountered little resistance when it decided to include women in its membership, but another geographic society, the Royal Geographical Society of London, did not readily admit women as members. Founded in 1830, the Royal Geographical Society is the largest geography organization in Europe, having merged with the Institute of British Geographers in 1995, and today it has approximately 13,000 elected members of both genders. However, its early history saw many heated debates regarding whether or not women should be invited to join. In fact, the group rescinded its first invitations to women, which went to 22 potential members including explorer May French Sheldon, after several male members threatened to leave the group if women were allowed to join.

Another group that resisted the inclusion of women is the international Explorers Club, headquartered in New York City, which grew out of the males-only Arctic Club established in the late 1800s by Henry Collins Walsh. Walsh was a journalist and explorer who wanted to join together the survivors of an 1894 Arctic expedition led by Frederick A. Cook, a physician and ethnologist who claimed to be the first person to reach both the North Pole and the summit of Mount McKinley. The Arctic Club became the Explorers Club in 1905. Unitl 1981, these and subsequent members rejected women's efforts to join the organization, but today the Explorers Club has no membership restrictions based on gender. Its members are male and female field scientists and explorers from over 60 countries, working in such physical, natural, and biological sciences as anthropology, archaeology, astronomy, biology, ecology, entomology, mountaineering, oceanography, paleontology, polar exploration, and zoology. A nonprofit international organization, its mission is to advance scientific exploration on land, in the air, in the ocean, and in space. To this end, the Explorers Club provides grants and assistance to people planning expeditions of scientific exploration, and it encourages exploration among the general public via educational and travel programs.

In addition to sponsoring expeditions through financial support, today's Explorers Club allows expeditions to petition for the privilege of carrying its flag. For example, Spanish explorers Vicente Pledel and Marian Ocana carried the club flag while following an ancient nomadic caravan route, the Route of the Empires, in the Middle East and Africa; the couple embarked on their two-year expedition in June 1999. The Explorers Club also offers guided tours for nonscientists, who can join the club as associate members.

In addition to organizations that offer financial support to women explorers and provide sightseeing opportunities for women travelers, many groups teach outdoor skills to women in order to strengthen their confidence in participating in these activities. One such group is Outward Bound (www.outwardbound.com), founded in Wales in 1941 by educator Kurt Hahn. Outward Bound provides people with physically challenging, organized expeditions that include alpine mountaineering, canoeing, sailing, rock climbing, sea kayaking, dogsledding, whitewater rafting, backpacking, and similar activities as a way to build self-esteem and encourage introspection. *See also* ADAMS, HARRIET CHALMERS; AKELEY, DELIA; DAVIDSON, ROBYN; EARHART, AMELIA; EARLE, SYLVIA; FOSSEY, DIAN; *FROM THE FIELD;* GOODALL, JANE; HARRISON, MARGUERITE BAKER; LEAKEY, MARY; MEAD, MARGARET; NILES, BLAIR; PECK, ANNIE SMITH; SETON THOMPSON, GRACE GALLATIN.

Further Reading: The American Association of Geographers website, www.aag.org (May 2000); The Explorers Club website, www.explorers.org (May 2000); The Ninety-Nines website, www.ninety-nines.org (May 2000); The Royal Geographic Society, www.rgs.org (May 2000).

O'Shaughnessy, Edith (1870–1939)

Born in 1870, American travel writer Edith O'Shaughnessy was the wife of a diplomat who held a series of posts in Europe and South America between 1904 and 1915. While he was stationed in Mexico from 1911 to 1914, O'Shaughnessy wrote her first travel book, *A Diplomat's Wife in Mexico* (1916), and a sequel, *Diplomatic Days* (1917). Each of these works was a collection of letters that O'Shaughnessy wrote to her mother. O'Shaughnessy went on to write several other books related to her travels, including *My Lorraine Journal* (1918), *Alsace in Rust and Gold* (1920), and *Other Ways and Other Flesh* (1929), the last of which is set in Leichtenstein. She also wrote novels and short stories. O'Shaughnessy died in 1939.

Further Reading: O'Shaughnessy, Edith. *Diplomatic Days*. New York: Harper & Brothers, 1917; ———. *A Diplomat's Wife in Mexico*. 1916. Reprint. New York: Arno Press, 1970.

P

Parker, Mary Ann

Mary Ann Parker was the first woman to write about a sea voyage around the world. Her book *A Voyage round the World, in the Gorgon Man of War: Captain John Parker* (1795), written after she was widowed, describes her life as the wife of a sea captain on a journey from England to Australia and New Zealand, then around Cape Horn and the Cape of Good Hope. The trip took approximately nine months, with an additional three months in Australia for Parker's husband to collect plant and animal specimens. *See also* SEA TRAVEL.

Further Reading: Coleman, Deidre, ed. *Maiden Voyages and Infant Colonies: Two Women's Travel Narratives of the 1790s*. London and New York: Leicester University Press, 1999.

Parkhurst, Charley (ca. 1808–1881)

Charley Parkhurst was a stagecoach driver in California during the mid-1800s. Born around 1808 in New Hampshire as Charlotte Parkhurst, she apparently escaped from an orphanage by disguising herself as a boy, a deception she maintained for the rest of her life. Parkhurst traveled west during the Gold Rush and worked for a stagecoach line until her death in 1881. It was not until then that her acquaintances discovered the truth about her gender. *See also* DISGUISES; GOLD RUSH.

Further Reading: Øydegaard, Floyd D. P. "She Was a Man." http://www.squick.sptddog.com/sotp/parkhurst.html; Women of the West, http://www.over-land.com/westpers2.html (May 2000).

Parrish, Maud (1878–1976)

Born in 1878, Maud Parrish claimed have gone almost everywhere, insisting she had traveled around the world a total of 16 times. This was not an impossible boast given that she lived to be 98 and was on the road for much of her life. Her memoir, *Nine Pounds of Luggage* (1939), describes her early travel experiences. In it she writes about running away from a bad marriage when she was very young, being brought back by her parents, and getting a divorce. But even then, Parrish could not settle down. She says: "Wanderlust can be the most glorious thing in the world sometimes, but when it gnaws and pricks at your innards, especially in spring, with your hands and feet tied, it's awful. So I left. Without telling a soul." (Morris, p. 174) She went to Alaska, where gold had recently been discovered and became a dancehall girl. Later, she went to Beijing, China, where she ran a gambling house. Parrish died in 1976.

Further Reading: Morris, Mary. *Maiden Voyages: Writings of Women Travelers*. New York: Vintage, 1993; Parrish, Maud. *Nine Pounds of Luggage*. New York: J. B. Lippincott, 1939.

Patten, Mary (1837–1860)

Born in 1837, in Boston, Massachusetts, Mary Patten was perhaps the first woman to captain a nineteenth-century sailing vessel west around Cape Horn, becoming famous for her effort. She was the wife of a sea captain and had sailed with her husband for three years when he suddenly fell ill during a sailing race from the east to the west coast of the United States. Patten took over the helm despite being pregnant. She remained in charge for 52 days until her husband was well enough to resume command. Before then, however, she was violently attacked by a crewman unhappy that she had been given the helm and not he. Patten recovered from the attack, and the man was sent to prison when the ship reached California.

Patten's ship came in second in the three-vessel race, but it was considered an amazing feat given the temporary captain's gender. She received awards and a cash prize in honor of her accomplishment. When her husband died shortly after the adventure ended, however, Patten grew despondent and fell ill herself. She died in 1860, leaving behind a three-year-old child. *See also* SEA TRAVEL.

Further Reading: De Pauw, Linda Grant. *Seafaring Women*. Boston: Houghton Mifflin, 1982.

Peck, Annie Smith (1850–1935)

Born in 1850, American mountaineer Annie Smith Peck was one of the foremost female climbers of the late nineteenth and early twentieth centuries. She was born in Providence, Rhode Island and originally worked both as teacher and professor at Purdue University and Smith College. While attending the American School of Classical Studies in Athens, Greece, she started mountain climbing, and in 1895 she reached the summit of the Matterhorn in the Alps. During this ascent, she wore a pair of pants, called knicker–bockers, rather than a skirt; this earned her a great deal of publicity in newspapers back home.

During the next few years Peck climbed other mountains in Europe, California, and Mexico. She lectured on the Matterhorn climb and others including peaks in Mexico. She was a founding member of the American Alpine Club in 1902. Around this time she became determined to be the first person—not just the first woman—to climb a mountain that had never been climbed before. She searched for a likely prospect and at first chose Illampu in Bolivia. But after leading three unsuccessful climbing expeditions there (in 1903, 1904, and 1906), she decided to try Mount Huascaran in Peru. After multiple attempts, in 1908 Smith succeeded in ascending the south (lower) summit and climbed two of its three summits. In 1911, at the age of 61, she became the first woman to climb another South American mountain, Mount Coropuna in Peru, and in subsequent years she also toured South America by air. She made her last climb, of

Annie S. Peck in her mountain climbing outfit in 1934. © *Bettmann/CORBIS.*

Mount Madison in New Hampshire, at the age of 82.

Throughout her life, Peck wrote articles and books about her climbing experiences. Her works include *A Search for the Apex of America: High Mountain Climbing in Peru and Bolivia, including the Conquest of Huascaran, with Some Observations on the Country and People Below* (1911) and *Flying over South America: Twenty Thousand Miles by Air* (1932). Peck died in New York in 1935. *See also* MOUNTAINEERING.

Further Reading: Robinson, Jane. *Wayward Women: A Guide to Women Travellers*. Oxford: Oxford University Press, 1990; Tinling, Marion. *Women into the Unknown: A Sourcebook on Women Travelers*. Westport, CT: Greenwood Press, 1989.

Perham, Margery Freda (1895–1982)

Born in 1895, English author Dame Margery Freda Perham frequently traveled to Africa between 1922 and 1932, and later she wrote books that not only reported on her trips throughout that continent but expressed her views on current African politics. Her works include *African Apprenticeship: An Autobiographical Journey in Southern Africa* (1929), *East African Journey: Kenya and Tanganyika, 1929–30* (1976), and *West African Passage: A Journey through Nigeria, Chad, and the Cameroons, 1931–32* (1983), as well as a book on the South Pacific, *Pacific Prelude: A Journey to Samoa and Australasia, 1929* (1988). In addition, a collection of her correspondence with fellow African traveler Elspeth Huxley, *Race and Politics in Kenya: A Correspondence between Elspeth Huxley and Margery Perham*, was published in 1944.

Perham first went to Africa in 1922 to visit her sister, whose husband had decided to settle in British Somaliland after serving there during the Boer War. In 1929 Perham made her second trip to Africa on a research fellowship from Oxford University. A history scholar, she had been assigned the task of studying indigenous cultures, not only in Africa but also in other parts of the world. However, it was Africa alone that held her interest for the rest of her life. She became active in African political discussions and received many awards for her involvement in important issues. She also helped establish Oxford's Institute of Colonial Studies. She died in 1982. *See also* AFRICA; HUXLEY, ELSPETH JOSCELINE.

Further Reading: Huxley, Elspeth Joscelin Grant. *Race and Politics in Kenya: A Correspondence between Elspeth Huxley and Margery Perham*, 1944. Reprint, with an introduction by Lord Lugard. Westport, CT: Greenwood Press, 1975; ———. *East African Journey: Kenya and Tanganyika, 1929–30*. London: Faber and Faber, 1976; ———. *West African Passage: A Journey through Nigeria, Chad, and the Cameroons, 1931–1932*. London and Boston: Peter Owen, 1982; Perham, Margery. *African Apprenticeship: An Autobiographical Journey in Southern Africa, 1929*.1929. Reprint. New York, Africana Publishing, 1974; Perham, Margery, and J. Simmons, eds. *African Discovery: An Anthology of Exploration*. Evanston, IL: Northwestern University Press, 1963; Smith, Alison, and Mary Bull, eds. *Margery Perham and British Rule in Africa*. London: F. Cass, 1991.

Pfeiffer, Ida Reyer (1797–1858)

Born in 1797, Austrian travel writer Ida Reyer Pfeiffer helped popularize world travel during the nineteenth century. She made her first trip in 1842, after her children were grown and she had separated from her much older husband. Pfeiffer spent nine months visiting the Middle East, ostensibly on a pilgrimage because she knew it would please friends and family to think she was traveling for religious purposes. Unused to solo travel, she was nervous about her safety and made out a last will and testament before leaving home. When she returned she published her first book, *Visit to the Holy Land, Egypt, and Italy* (1843). Its success funded her next trip, a six-month sojourn in Iceland that resulted in *A Visit to Iceland and the Scandinavian North* (1852). In 1847 she embarked on a 19-month world tour, and this too resulted in a book: *A Lady's Voyage Round the World* (1850). Its publication turned Pfeiffer into a celebrity, and her next book was even more popular. Published in 1855, its full title was *A Lady's Second Journey Round the World: From London to the Cape of Good Hope, Borneo, Java, Sumatra, Celebees, Ceram, The Moluccas, etc., California, Panama, Peru, Ecuador, and the United States.*

A Lady's Second Journey documents Pfeiffer's second world tour, during which she went from Vienna to London in March 1851 and then spent the next four years, from 1851 to 1855, traveling around the Cape of Good Hope to Borneo, Java, Sumatra, and several other islands before heading across the Pacific Ocean to California, Panama, Peru, Ecuador, the eastern United States, and finally across the Atlantic Ocean back to London. The work begins with a description of Pfeiffer's first stop, London, where she witnessed the opening of Prince Albert and Queen Victoria's Great Exhibition. From England, Pfeiffer sailed on the ship *Allandale* for the Cape of Good Hope, then spent four weeks in Cape Town, South Africa. She had intended to embark from there through the African interior but cancelled this portion of her trip after learning about the region's poor roads and risk of illness. Instead she decided to sail for Australia by way of Singapore, a 54-day voyage. She had visited Singapore during her first trip around the world, and in the interim a lighthouse had been built, as well as a cottage where she stayed for a few days. While there she decided not to go to Australia, but instead to travel to Sarawak on the west coast of Borneo. She was detained there for several days by bad weather, after which she went inland by river and saw a tribe of indigenous people whose men showed her some human heads that had been taken as war trophies.

In her book Pfeiffer discusses headhunting and other customs among these people and others in Borneo, which she thoroughly explored. She also offers detailed descriptions of her experiences in Java, Sumatra, and other islands in what was then known as Dutch India. As part of these descriptions, she discusses Dutch missionaries and the Dutch officials who controlled the area, criticizing them for oppressing the indigenous people.

After touring several islands Pfeiffer had difficulty deciding where to go next. She finally embarked on a ship for San Francisco, about which she found much to dislike. It was the time of California's Gold Rush, and Pfeiffer says:

> San Francisco is unanimously declared the City of Wonders. . . . There are, indeed, only two forces capable of effecting such wonders—gold and despotism. The former has been the lever in this case; for the thirst of gold, which is the greatest of despots, has drawn people hither from all corners of the earth, and dwellings of wood and stone have arisen for them as if by magic. But what are all these simple works compared with the antique cities of Hindostan, the ruins of which even still attest their magnificence, and which are stated to have been built in a no less incredibly short time. . . . Truly it is only those who place all happiness in money who could submit, for the sake of gain, to live in such a place, and perhaps forget at last that there are such things as trees, or a green carpet lovelier than that which covers the gold-laden gaming tables. (pp. 293–294)

However, at the end of her book Pfeiffer apologizes for such criticisms, saying:

> Should I any where have spoken too strongly with respect to the manners and customs of countries through which I have passed, or have taken up erroneous views concerning them, I can only beg for the indulgent consideration of my readers, and repeat what I said in the first book of travels I ever published, that I am by no means to be counted among the fortunately gifted of my sex, but that I am a most simple and unpretending person, and can claim as a writer no merit whatever beyond that of describing truly and without exaggeration what I have seen and experienced. (pp. 499–500)

After leaving San Francisco, Pfeiffer went east to the city of Sacramento and then on to several Gold Rush towns before returning to San Francisco to take a ship to South America. After touring several cities on that continent's west coast, she crossed the Isthmus of Panama by rail and by carriage, then took a steamer north to New Orleans. There she witnessed a public slave auction, and in her book she condemns the practice of slavery. She also discusses the women's rights movement, particularly as she describes her trip northeast from New Orleans to Boston via St. Louis, Chicago, Milwaukee, Montreal, and New York. In describing New York, she mentions "that uneasy longing for . . . emancipation that characterizes American women" (p. 468) and expresses some reservations regarding the necessity for the women's rights movement. She then returns to her travel narrative to briefly describe her journey back across the Atlantic to London, with a detour to visit friends in the Azores and Portugal before returning to Vienna.

Pfeiffer next traveled to Madagascar in 1857. This trip proved to be her most difficult, because, after a period of political unrest, the government in that country detained Pfeiffer and several other foreigners, then forced them to march from the interior of the island to the coast, where a ship awaited to take them home. Pfeiffer fell ill while imprisoned and never recovered. She remained in poor health during her return trip and died in Austria in 1858.

Further Reading: Pfeiffer, Ida. *A Journey to Iceland and Travels in Sweden and Norway*. Translated by Charlotte Fenimore Cooper. New York: G. P. Putnam, 1852; ———. *A Lady's Second Journey round the World: From London to the Cape of Good Hope, Borneo, Java, Sumatra, Celebes, Ceram, the Moluccas, etc., California, Panama, Peru, Ecuador, and the United States*. London: Longman, Brown, Green, and Longmans, 1855; ———. *A Lady's Voyage round the World: A Selected Translation from the German of Ida Pfeiffer*. 1851. Reprint, with an introduction by Maria Aitken. London: Century, 1988; ———. *Visit to the Holy Land, Egypt, and Italy*. Translated by H. W. Dulcken. London: Ingram, Cooke, 1853.

Philip, Leila (1962–)

Born in 1962, American author Leila Philip is representative of modern travelers who visit a place specifically to learn a skill. During the 1980s she spent two years in Miyama, Japan, studying unique pottery techniques, and she later wrote about her experiences in *The Road Through Miyama* (1989). In her book, Philip describes Miyama's long association with the craft; Korean potters settled the small village in the seventeenth century after being taken captive by the Japanese. Philip's book also provides insights about Japanese rural life and offers many details about pottery techniques both past and present. Philip left Miyama in 1985 to attend college in the eastern United States. She currently lives in New York.

Further Reading: Philip, Leila. *The Road Through Miyama*. Reprint. New York: Vintage Books, 1991.

Photographers and Artists

Certain types of photographers and artists typically travel in search of subjects. For example, wildlife photographers go to remote regions to take pictures of animals in their natural habitat; photojournalists visit places of human crisis to capture images of war, political turmoil, natural disasters, or other important events; and impressionists travel to scenic regions to paint out-of-doors.

Some of the best-known women travelers have been photojournalists. Therese Bonney, for example, was a prolific American photographer whose images of World War II were among the best known of the era. She moved from the United States to France in 1919, at the age of 25, and when her adopted country became embroiled in the war she was determined to show the conflict's horrors to the rest of the world. Her photographs were published in many popular magazines, and she produced two books of her photographs, *War Comes to the People* (1940) and *Europe's Children* (1943). Her work also appeared in museums both in the United States and Europe. During the 1940s she was decorated twice by the military for her bravery.

Similarly, American photographer Toni Frissell traveled throughout Europe capturing images of American and allied military men and women as well as innocent victims during World War II. She was a fashion photographer for the magazines *Vogue* and *Harper's Bazaar* before volunteering in 1941, at the age of 34, to be a war photographer for the American Red Cross. Her work was often used to encourage the American public to support the war effort, and she is particularly noted for her photographs of women and African Americans in uniform. These images helped counter prejudices against minorities.

Yet another prominent American photojournalist was Margaret Bourke-White, who covered news stories throughout the world. Her photographs and articles appeared in major magazines such as *Life* and *Look*, and she wrote several books, including *Eyes on Russia* (1931), *Say, This Is the U.S.A.* (1941), *Shooting the Russian War* (1942), and *They Called It "Purple Heart Valley": A Combat Chronicle of the War in Italy* (1944). Bourke-White's first photographs were of factories in the eastern United States, work that led to an assignment photographing factories and factory equipment in Germany and Russia in 1930. From 1934 to 1937, when she was in her early thirties, she photographed scenes of American poverty in the Midwest and South; in 1937 these photographs appeared in a book, *You Have Seen Their Faces*. In 1938 Bourke-White returned to Europe to photograph Czechoslovakian villages, which resulted in a book, *North of the Danube* (1939). In World War II, she worked for the U.S. Air Force photographing Allied troops throughout Europe, sometimes shooting pictures in dangerous combat conditions; when the war ended she photographed bodies at a German concentration camp, Buchenwald. She subsequently went to India to photograph the dead in a religious conflict there, and in 1950 she took pictures of diamond miners in South Africa. The following year, she developed a serious illness that prevented her from continuing her work.

To photojournalists like Bourke-White, photography is the primary reason for travel. But photography is also used by people who travel for other reasons. Tourists, naturalists, and explorers have all used the medium as a way to document their travel experiences. For example, Englishwoman Isabella Bird Bishop carried photographic equipment on some of her journeys, even though the camera equipment of her day was heavy and cumbersome. Englishwoman Rosita Forbes, who in 1920 became the first white woman to enter the sacred city of Kufara in the Sahara Desert (in present-day Libya), also documented her journey with photographs. American Fanny Bullock Workman included many black-and-white photographs in her book, *Through Town and Jungle: Fourteen Thousand Miles A-Wheel Among the Temples and People of the Indian Plain* (1904), as did Irishwoman Dervla Murphy in *The Waiting Land: A Spell in Nepal* (1967). In 1917, American Osa Johnson became the first woman explorer to make movies of her adventures; now many modern tourists carry video cameras on their sightseeing expeditions. Austrian-born naturalist Joy Adamson produced still photographs of the lions she studied on an African game preserve, and filmmakers were eventually drawn to photograph her and her lions as well.

In the case of artists, some have traveled solely because of their art while others have seen the chance to draw and paint new subjects as a pleasant by-product of travel. Among the former is early-twentieth-century Canadian artist Emily Carr, who traveled through British Columbia specifically to paint Native American villages, people, and landscapes. Similarly, Ethel Tweedie traveled from England to Russia and China in 1925–1926 to find subjects for her watercolors. In the latter category is Englishwoman Marianne North, who used an inheritance to travel through Europe, the United States, Jamaica, Haiti, South America, Japan, Singapore, parts of South Asia, several Pacific and Atlantic islands, Australia, and South Africa during the late nineteenth century. She enjoyed travel for its own sake, but also produced oil paintings based on her travels featuring plants from around the world. *See also* Bishop, Isabella Lucy Bird; Johnson, Osa; Tweedie, Ethel Brilliana; Workman, Fanny Bullock.

Further Reading: Ayer, Eleanor H. *Margaret Bourke-White: Photographing the World*. New York: Dillon Press, 1992; Bourke-White, Margaret. *Shooting the Russian War*. New York: Simon & Schuster, 1942; Brown, Theodore M. *Margaret Bourke-White, Photojournalist*. Ithaca, NY: Andrew Dickson White Museum of Art, Cornell University, 1972; Mitchell, Margaretta. *Recollections: Ten Women of Photography*. New York:

Viking Press, 1979; Sullivan, Constance, ed. *Women Photographers*. New York: Abrams, 1990.

Pigeon, Anna (1833–1917)

Born in 1833, Anna Pigeon was among Great Britain's most accomplished mountaineers. Usually accompanied by her sister Ellen, she climbed over 63 major peaks, including Mount Blanc on the French–Italian border and the Matterhorn on the Swiss–Italian border. Pigeon and her sister ascended the Matterhorn twice; on their second climb in 1873, they became the first women to go completely over the mountain, traveling from the Italian side to the Swiss side. Pigeon continued to be an active mountaineer into her 70s. She died in 1917. *See also* MOUNTAINEERING.

Further Reading: Robinson, Jane. *Wayward Women: A Guide to Women Travellers*. Oxford: Oxford University Press, 1990.

Pilgrims

Pilgrims are people who travel to holy places to do penance, seek divine help, or fulfill an obligation of their religion. Such journeys, called pilgrimages, were particularly popular during the Middle Ages, when women like Margery Kempe traveled to holy sites in the Middle East as part of their Roman Catholic faith. Kempe made her pilgrimage around 1413, when she left England with the specific goal of visiting as many sites mentioned in the New Testament of the Bible as possible.

Two even earlier pilgrims, Saint Helena (mother of Emperor Constantine the Great) and Abbess Etheria, made a pilgrimage from Rome to Jerusalem in approximately 324 C.E. and established two churches, one on the spot where she believed Christ had been crucified and the other on the spot where she believed he had been born. Etheria was a European abbess who some historians believe went on a pilgrimage to the Holy Lands of the Middle East sometime between 381 and 384 C.E. and recorded her impressions in a manuscript that still exists. In discussing this work, historian Jane Robinson says that it makes Etheria the first travel writer, as well as the honorary "patron saint" of women travelers. (p. 159)

A much later female pilgrim was Lady Evelyn Cobbold, who was the first Englishwoman to make a true religious pilgrimage to Mecca. Cobbold was a convert from Christianity to Islam; the teachings of Islam require the faithful to travel to the holy city of Mecca at least once. Cobbold undertook this pilgrimage in the 1930s, writing about her experiences in a 1934 book entitled *Pilgrimage to Mecca*.

In the early 1990s travel writer Bettina Selby set out to retrace the route of the ancient pilgrims who traveled through France and across northern Spain to the shrine of St. James the Apostle. She wrote about her experiences in *Pilgrim's Road: A Journey to Santiago de Compostela* (1994).

The word pilgrim is also used to refer to a specific group of people who left England to escape religious persecution in the 1600s. These people sailed to America on a ship called the *Mayflower*, settling in New England in 1620. *See also* COBBOLD, EVELYN; ETHERIA; HELENA, SAINT; KEMPE, MARGERY; NORTH AMERICA.

Further Reading: Lambdin, Laura C., and Robert T. Lambdin. *Chaucer's Pilgrims: An Historical Guide to the Pilgrims in The Canterbury Tales*. Westport, CT: Greenwood Press, 1996; Robinson, Jane. *Wayward Women: A Guide to Women Travellers*. Oxford: Oxford University Press, 1990; Selby, Bettina. *Pilgrim's Road: A Journey to Santiago de Compostela*. Boston: Little, Brown, 1994.

Pilley, Dorothy (1894–1986)

Born in 1894, British mountaineer Dorothy Pilley helped found Britain's Pinnacle Club, one of the first all-woman climbing clubs, in 1921. Pilley scaled mountains in Europe, Asia, the United States, and Canada, often with her husband and fellow mountaineer I. A. Richards. Together the two were the first to climb the north face of Dent Blanche in Switzerland, a feat they accomplished in 1928. Although Pilley enjoyed climbing with her husband, she would undoubtedly have remained in the sport even if he had not been interested in it. She first began climbing as a girl, and as a young woman she went on mountaineering vacations in the Alps several times a year. After her marriage in 1921 she tackled other mountain ranges as well, although historian Jane Robinson reports that Pilley "always considered the Alps her spiritual home." (*Wayward Women,* p. 73). Pilley died in 1986. *See also* MOUNTAINEERING.

Further Reading: Pilley, Dorothy. *Climbing Days*. 1965. Reprint. London: Hogarth Press, 1989.

Piozzi, Hester Lynch (1741–1821)

Born in 1741, Englishwoman Hester Lynch Piozzi is representative of women who became travelers relatively late in life. In 1784, after marrying, raising several children, and being widowed, she shocked her friends by getting married again, this time to an Italian musician, and moving to Italy. She then began traveling through continental Europe, keeping detailed journals about her adventures. In 1789 she published them in book form as *Observations and Reflections Made in the Course of a Journey Through France, Italy, and Germany*. By this time she had already had two other books published, *Anecdotes of the Late Samuel Johnson* (1786) and *Letters to and from the Late Samuel Johnson* (1788), which chronicle her relationship with noted essayist Dr. Samuel Johnson. Piozzi died in 1821.

Further Reading: Hamalian, Leo, ed. *Ladies on the Loose: Women Travellers of the 18th and 19th Centuries*. New York: Dodd, Mead, 1981.

Pirates

Pirates are thieves of the sea who travel great distances to attack innocent vessels and steal their cargo. They have existed since the beginning of sailing, and, although most pirates were male, a few were female. These women had to be just as ruthless as their male counterparts and, in fact, some were known for being more brutal than the men.

One of the earliest female pirates to become famous for her activities was Jane De Belleville, a French pirate who sailed up and down the coast of Normandy during the mid-fourteenth century attacking only ships from her own country. She was angry because the French government had killed her husband after—unjustly, in her mind—deeming him to be an English spy. For the same reason, she contributed the services of her three pirate ships to the English in 1345 during the Hundred Years' War between the English and French.

During the sixteenth century Irish pirate Grace O'Malley gained fame for attacking English ships. Later, however, she became a friend of the English Court, using her fleet to support England in a war with its European enemies. Another woman pirate during this period was Lady Killigrew, who sometimes sailed the coast of England with her husband Sir John in search of treasure but also spent a great deal of time on land hiding their loot. Because Sir John was a nobleman, he agreed to attack only foreign ships that were not

Mary Read, a West Indies pirate. © *Bettmann/CORBIS*.

under England's protection; in return, English authorities left him alone. But Lady Killigrew did not share her husband's loyalty to his country, and in 1582 she and some of their men boarded a protected ship, killed its crew, and stole its goods. Shortly thereafter she was arrested, tried, and sentenced to death. Later her life was spared in deference to her gender, but she remained in prison.

Women pirates became even more ruthless between the mid-seventeenth century and the early nineteenth century, in a period that historians have designated the Golden Age of Piracy. During this time, pirates called buccaneers used the Caribbean as a base from which they attacked trading vessels in the Atlantic. These pirates had a rule that women were not to be brought on board ship unless they were captured at sea, and even then they were taken to land as soon as practical. Nonetheless, the Golden Age had several famous women pirates. Of these, perhaps the most notorious were Anne Bonney and Mary Read, who crewed with pirate captain John Rackham, also called Calico Jack, from 1718 to 1720.

Bonney was the illegitimate daughter of a wealthy English attorney from South Carolina when she eloped with a sailor and found herself in the port town of New Providence in the Bahamas. There she fell in love with Calico Jack and, at his urging, disguised herself as a man to join his crew. The two managed to conceal her gender until Anne became pregnant. She abandoned the baby shortly after giving birth, whereupon she again dressed as a man. By this time Calico Jack had replaced his crew, so once more the other pirates on board did not know Bonney's true gender.

Mary Read was another pirate on Calico Jack's ship who disguised herself as a man. Read first shipped out from her native England as a child, dressed as a cabin boy to secure a position on a British warship. After returning from sea, she continued to pretend she was male, becoming first a foot soldier and then a mounted cavalryman during a war between the English and the French. Later she abandoned her disguise to marry and run a tavern, but she soon missed her life of adventure and signed on as a crewman on a Dutch ship bound for the West Indies. This time her vessel was captured by English pirates, and they asked her to join them, thinking she was a man. Read sailed with these pirates until she joined the crew of Calico Jack.

While in disguise, Read became close friends with Anne Bonney, and eventually the two women dropped their deception and became widely known as women pirates. Both were the equal to any male pirate in skills and ruthlessness. In 1720, while anchored off Jamaica, their ship was captured by a privateer who turned Read and Bonney over to the Jamaican court. The two women were convicted of piracy and sentenced to hang, as was Calico Jack. However, when officials found out that both women were pregnant, their lives were spared. Bonney was eventually paroled, but Read died in prison of a fever, possibly due to complications of pregnancy.

Two other eighteenth-century women pirates were Maria Lindsey Cobham and Rachel Wall. Cobham operated out of England, sailing the Atlantic with her husband to attack ships and steal their cargo. After her husband retired from piracy, Cobham took one trip without him, during which she captured a trading vessel and killed the entire crew by poisoning her prisoners' food. Back on land, she grew depressed over her husband's refusal to return to piracy and killed herself. Wall operated out of America, sailing along the eastern seaboard with her husband, who attacked ships using Wall as a lure. After a storm, Wall's band of pirates would pretend to be ordinary seamen whose ship had been damaged; Wall was their distraught passenger. When a trading vessel came alongside to offer help, the pirates would attack and kill everyone on board. On one occasion, however, the pirates' ship really was damaged in a storm, and Wall's husband drowned. Without him, she abandoned piracy for a life of crime on land.

Also operating out of America was Fanny Campbell, the only woman to captain a war ship—if only for a brief period—during the American Revolution. Prior to the war, she had been a pirate commanding a fleet with her pirate boyfriend; together they attacked British ships along the east coast of North America. Once the war broke out, she allied her ships with England.

Another pirate who turned to military activities was Hsi Kai Ching Yih of China, also known as Madame Ching, who sailed during the early nineteenth century. She began her life on board ship as the captive of a pirate captain but soon became co-captain of his large fleet. The two worked together until he died in 1807, from which point Ching commanded the fleet alone. She was so successful that eventually the Chinese government convinced her to trade piracy for the command of official war ships, paying her well for her efforts. She led part of the emperor's fleet for several years before retiring to live in a palace given to her by the government. However, she continued to be an outlaw for the rest of her life, secretly directing smuggling activities from her palace.

Meanwhile, male pirates were responsible for sending large numbers of women on voyages against their will, raiding port towns and transporting captured women to sell as slaves. This practice was particularly common in the seventeenth century among the pirates of the Barbary Coast, a region in the Mediterranean along North Africa and Morocco. These pirates would travel as far away as Ireland to obtain women, although Italy was their most frequent target, and bring them to North African slave auctions. They would also capture women at sea; several wives and daughters of sea captains were taken, although some were released when relatives paid ransoms for their return. *See also* CAMPBELL, FANNY; CHING YIH, HSI KAI; O'MALLEY, GRACE; WALL, RACHEL.

Further Reading: Druett, Joan. *She Captains*. New York: Simon & Schuster, 2000; Klausmann, Ulrike, Marion Mainzerin, and Gabriel Kuhn. Translated by Tyler Austin and Nicholas Levis. *Women Pirates and the Politics of the Jolly Roger*. Montreal, Quebec: Black Rose Books, 1997; Stanley, Jo, ed. *Bold in her Breeches: Women Pirates Across the Ages*. London and San Francisco: Pandora, 1996.

Poles, North and South

The first women travelers to approach the poles were the wives of sea captains on whaling vessels. One such woman was American Mary Brewster, who in 1849 became the first woman to journey to the Arctic Ocean on a whaling ship. However, male explorers reached the poles first, as early as 1909 for the North Pole and 1910 for the South Pole.

The first woman to walk on Antarctica was Norwegian Caroline Mikkelsen in 1935, and in the late nineteenth and early twentieth centuries, a few women traveled to the Arctic, although both poles remained primarily male preserves. Victorian British traveler Elizabeth Taylor became the first woman to reach the Mackenzie River where it crosses into the Arctic in 1892, and Canadian Agnes Dean Cameron visited the Arctic in 1908 so she could write a book about the region, *The New North* (1909). But the woman who by far did the most extensive exploration of the North Pole was American Louise Boyd, who made seven expeditions to the Arctic during the 1920s and 1930s to collect scientific information. Boyd also provided photographs of the region to the U.S. military during World War II, and in 1955 she became the first woman to fly over the North Pole.

The airplane did much to change the nature of polar exploration. In the early years of travel to the Arctic or Antarctic, explorers had to go by ship and/or on foot and by dogsled. With the advent of air travel, people and supplies could be dropped off near the poles. However, flying in the region is not without risk. Ice crystals forming in gas lines and on airplane parts impair operation, and severe storms can make it difficult for pilots to see. Therefore, although surveying polar plant and animal life by air is possible, most research is conducted at scientific research facilities called polar bases.

At both the North and South Poles, these bases can be very difficult to reach because of severe storms, winds, ice, snow, and cold; when someone falls ill at a polar base it can take a long time for the person to get treatment. One incident that highlights this problem occurred in October 1999, when

Anne Dalvera, on transantarctic expedition, South Pole, 1993. © *Susan Giller.*

physician Jerri Nielsen had to be flown from the Amundsen-Scott South Pole Station after being diagnosed with breast cancer. Nielsen was one of 10 women and 31 men, all 41 of them scientists, who lived at the American base as part of a research program run by the National Science Foundation. She discovered that she had cancer five months before being evacuated, but although medicine could be airdropped to her, she could not leave until the climate made a landing possible, and even then it was difficult. In addition to serious illnesses like cancer, polar scientists also have to deal with depression, tension, disorientation, and insomnia, all caused by their bleak environment.

Nielsen was the sole physician at the base and ran its hospital, but the fact that she and nine other residents were women was unusual. Woman have not been a large part of polar science, a point that journalist Sara Wheeler emphasizes in her book *Terra Incognita: Travels Through Antarctica* (1996), which she wrote after staying at an Antarctic research facility for several weeks. She reports that, in 1995, only 61 of the 244 people to spend the winter at Antarctic scientific research facilities were women. Moreover, women were not even selected to participate in Antarctic research programs until 1957, when the Soviets allowed one woman to join an Antarctic expedition. No American woman participated in South Pole explorations until 1969, although a few male polar explorers and researchers took their wives with them on their journeys. For example, in 1946–1947 Jennie Darlington accompanied her husband on an expedition to the Antarctic and later wrote about her experiences in *My Antarctic Honeymoon* (1956). Also on this expedition was Edith Ronne, whose husband Finn Ronne was its leader. Because these two American women actively helped their husbands with scientific studies, some historians consider them to be the first women researchers to take part in a polar exploration, even though their participation was by virtue of marriage. Moreover, they were the first two women to spend an entire winter in the Antarctic.

During the 1980s and 1990s, women began to explore the Antarctic without companions. For example, Ann Bancroft who, with three other women, participated in the 1992–1993 American Woman's Antarctic Expedition, the first all-woman expedition to the South Pole. During the trip, Bancroft and her companions traveled on skis and participated in scientific tests that examined the effects of a polar environment on the human body and mind. Bancroft was also the only women out of six participants in the 1986 Steger International Polar Expedition, a 900-mile expedition that was the first to use satellite equipment to verify that its members really did reach the geographical South Pole. In addition, she is the founder of the American Women's Expedition Educational Foundation, which is dedicated to disseminating information about Antarctic exploration. Another notable Antarctic explorer is American Shirley Metz, the first woman to ski to the South Pole. She accomplished this feat in 1988 when she was 39 years old, braving violent winds and subzero temperatures to reach her goal.

As for modern exploration in the North Pole, among the most notable women to travel to the region were Scottish mountaineer Myrtle Lillias Simpson, American adventurers Helen Thayer and Pam Flowers, and British filmmaker Caroline Hamilton. In 1964 Simpson became the first woman to ski across Greenland's polar icecap, and in 1968 she traveled closest to the geographical North Pole of any previous woman explorer. Thayer also chose skis as her mode of Arctic travel; in 1988 she became the first woman to ski to the magnetic North Pole. Accompanied only by her dog, she spent 27 days traveling 345 miles across a polar icecap to reach the site. Flowers, however, traveled through the Arctic by dogsled, having gained experience competing in Alaska's Iditarod dogsledding competition in 1983. She led the first successful women's expedition to the magnetic North Pole in 1987, and from February 1993 to January 1994 she traveled 2,500 miles alone, also via dogsled, to become the first woman to traverse the Arctic. Her journey was also the longest solo dogsled journey by a woman. For her efforts, Flowers received a gold medal from the Society of Women Geographers in 1996.

The following year, Caroline Hamilton led the first all-woman expedition to the geographical North Pole. The expedition had five 20-woman relay teams, and each team walked and/or skied to a predesignated point where an airplane picked them up and left the next team. Each team traveled for approximately two weeks, although two guides—Matty McNair from the United States and Denise Martin of Canada—went the entire 625 miles from Ward Hunt Island in northern Canada to the North Pole. Hamilton was part of the final team that reached the North Pole, after a 13-hour day of traveling across ice. *See also* Boyd, Louise Arner; Cameron, Agnes Dean; Hamilton, Caroline; Iditerod Trail Sled Dog Race; Simpson, Myrtle Lillias; Taylor, Elizabeth; Thayer, Helen; Wheeler, Sara.

Further Reading: Chipman, Elizabeth. *Women on the Ice: A History of Women in the Far South.* Carlton, Australia: Melbourne University Press, 1986; Rothblum, Esther D., Jacqueline S. Weinstock, and Jessica Morris, eds, *Women in the Antarctic*. New York: Harrington Park Press, 1998; Thayer, Helen. *Polar Dream: The Heroic Saga of the First Solo Journey by a Woman and Her Dog to the Pole.* New York: Simon & Schuster, 1993; Wenzel, Dorothy. *Ann Bancroft: On Top of the World.* Minneapolis, MN: Dillon Press, 1990.

Poole, Sophia (1804–1891)

Born in England in 1804, Sophia Poole became famous in that country in the 1840s for her letter collections about Egyptian life. She and her two small children lived in Egypt from 1842 to 1849, immersing themselves in Egyptian culture, after her brother, Edward Lane, had suggested she would enjoy living there. Consequently Poole was able to offer the first descriptions of many aspects of Egyptian life as seen from a foreigner's perspective. Poole's works include *The Englishwoman in Egypt: Letters from Cairo, Written during a Residence there in 1842, 3, and 4* (1844) and *The Englishwoman in Egypt: Letters from Cairo, Written During a Residence There in 1845–46* (1846). Poole died in 1891. *See also* Africa.

Further Reading: Robinson, Jane. *Wayward Women: A Guide to Women Travellers*. Oxford: Oxford University Press, 1990.

Pregnancy. *See* CHALLENGES FOR THE MODERN TRAVELER.

Public Service

Duties related to public service have prompted many women to travel, either by choice or necessity. For example, nurses sometimes travel great distances to reach areas where the are needed. Volunteer relief workers, such as those in the Red Cross, Salvation Army, and Peace Corps, also accept travel assignments to help people in distant lands. Meanwhile, heads of state and ambassadors routinely travel to foreign countries to perform diplomatic functions that will benefit their own people.

One common reason for nurses to travel is to minister to the poor in third-world countries. Of these women, among the most notable are Swiss Maria Haseneder and American Ida Sophia Scudder. Haseneder was a nurse and Seventh-Day Adventist missionary who traveled through Africa, primarily in Ethiopia and Rwanda-Urundi, from 1928 to 1941, to nurse the sick. She wrote about her experiences, *A White Nurse in Africa* (1951). Scudder traveled as a medical missionary to India in 1900, remaining there until her death in 1960; she was in charge of the first medical school in India to train women as physicians rather than nurses. Still other nurses have gone abroad in search of medical knowledge. These include Englishwoman Kate Marsden, who traveled from her homeland to Siberia in 1890 to try to find a cure for leprosy.

An even more common reason for nurses to go abroad has been war. Many women have traveled great distances to help soldiers in foreign wars. One of the best known is Englishwoman Florence Nightingale, who went with a group of volunteers to Turkey in 1854 to minister to wounded British soldiers during the Crimean War. Many other nurses have served with equal dedication, but their names are unrecorded. These include hundreds of American women, both nurses and physicians, who went to France during World War I to care for wounded soldiers and civilians. Several organizations, including the American Fund for French Wounded, the American Committee for Devastated France, and the Medical Women's National Association (later the American Medical Women's Association) sent women medical personnel abroad during this period. In addition, Great Britain, Canada, and the United States each sent all-female ambulance units to serve in France. In May 1918 one American ambulance driver, Ethel Drake, wrote an article for the *New York Times* in an attempt to encourage other women to take up ambulance driving.

Prior to 1901, American women who served as medical personnel in wartime were civilian volunteers or contract workers; it was not until World War I that they could join the active military and not until World War II that they could move up in rank. Meanwhile, women continued to disguise themselves as men in order to become soldiers, while others contributed to the war effort by working for various relief organizations. One of the most prominent of these is the Red Cross. This organization began in Switzerland in 1863 as the International Red Cross; American nurse Clara Barton created the American Association of the Red Cross (later renamed the American Red Cross) in 1881. During World War I, the group shipped hundreds of nurses to France; approximately 2,500 women traveled overseas to serve under the auspices of the American Red Cross from 1914 to 1919. The organization also sent volunteers to regions where people had suffered from natural disasters. The earliest relief efforts of this type were related to forest fires in Michigan (1881), the Ohio and Mississippi River floods (1884), the Johnstown Flood (1889), a Russian famine (1892), and a hurricane on the Georgia Sea Islands (1893).

Two other organizations that sent women abroad during World War I were the Salvation Army and the Young Men's Christian

Association (YMCA). Approximately 132 Salvation Army women provided food, clothing, and religious comfort to American soldiers in France, while the YMCA sent more than 4,000 women abroad to establish "Y huts," or comfort stations, where soldiers could go to socialize and obtain free food and supplies like toothbrushes and soap. One YMCA worker, Gertrude Ely, had her car shipped to France in 1917 so she could drive from one military encampment to the next distributing supplies, hot soup, and cookies. The YMCA is also notable in that it is believed to be the only World War I relief group to send African American women to France; there were 19 such women who established Y huts for African American soldiers. A companion group to the YMCA, the Young Women's Christian Association (YWCA) also offered aid to World War I soldiers but concentrated its efforts on military bases within the United States.

During World War II and later conflicts, these and other relief organizations continued to support soldiers in the field; however, once the U.S. military allowed women within its ranks, women increasingly chose military service instead of public service as a way to contribute their skills during wartime. In peacetime, many have chosen to work with volunteer agencies such as the Peace Corps to provide aid to third-world countries. The Peace Corps is an American organization that works internationally, with approxiately 7,000 people currently serving in the field. Established by the United States government in 1961 through an executive order of President John F. Kennedy, the group first sent American volunteers to poor villages in Ghana, Tanzania, Colombia, the Philippines, Chile, and St. Lucia to help residents improve their living conditions; today 78 countries are being served. Peace Corps workers typically spend two years in a foreign country with no pay except for a modest monthly living allowance. The largest number of workers serve in the fields of education and health, with most of the remainder providing assistance related to the environment and business concerns.

Some of the Peace Corps' programs are specifically targeted to women. For example, the organization is currently holding seminars in Panama on women and family rights, and it is offering grants to women in this region who want to start their own businesses. In Uganda, the Peace Corps is sponsoring a program called "Women Build" in cooperation with an organization called Habitat for Humanity International. As part of this program, women learn to build houses that are then donated to the needy.

Notable women who have served in the Peace Corps include Donna Shalala, Priscilla Wrubel, and Loret Miller Ruppe. Shalala, who was appointed the U.S. Secretary of Health and Human Services in 1993, served in the Peace Corps from 1962 to 1964 as a teacher in rural Iran. Wrubel served as a Peace Corps volunteer in Liberia in from 1963 to 1965, during which she witnessed rain forest destruction. This experience led to her cofounding of The Nature Company, a store that uses its products to promote environmental awareness, in 1973. Ruppe was the director of the Peace Corps from 1981 to 1989, the longest directorship in the organization's history.

Women like Shalala, Wrubel, and Ruppe travel extensively to offer aid outside their own country. In contrast, ambassadors and heads of state travel primarily to benefit the people in their own countries by promoting various public agendas. Countries with diplomatic ties typically exchange ambassadors. In 1949 the United States appointed its first woman ambassador: Eugenie Anderson, who was sent to Denmark. Her career, however, did not receive the same attention as that of former child actress Shirley Temple Black. Black's appointment as ambassador to Czechoslovakia in 1989 was a media event; she held the ambassadorship until 1992. Another well-known U.S. ambassador was Jean Kirkpatrick, who did not serve abroad but was appointed ambassador to the United Nations in 1981. Eleanor Roosevelt, widow of President Franklin D. Roosevelt, also worked at the United Nations during the 1940s as chairperson to the Commission on

Human Rights. *See also* BARTON, CLARA; MARSDEN, KATE; NIGHTINGALE, FLORENCE; YOUNG WOMEN'S CHRISTIAN ASSOCIATION (YWCA).

Further Reading: Day, Frances Martin, and Phyllis Spence, eds. *Women Overseas: Memories of the Canadian Red Cross Corps*. Vancouver, British Columbia: Ronsdale Press, 1998; Hutchinson, John. *Champions of Charity: War and the Rise of the Red Cross.* Boulder, CO: Westview Press, 1997; Kennedy, Geraldine. *From the Center of the Earth: Stories out of the Peace Corps.* Santa Monica, CA: Clover Park Press, 1991; Morin, Ann Miller. *Her Excellency: An Oral History of American Women Ambassadors*. New York: Twayne Publishers, 1995; Peace Corps. "Celebrating Women's History Month. http://www.peacecorps.gov/essays/women/index.html (June 2000).

Q

Queens

Many monarchs have traveled to view their lands. Still more have encouraged their subjects to travel in order to acquire more land for the Crown because places discovered and settled by their subjects were easier to claim and defend. In modern times, many rulers have encouraged tourism as a way to strengthen their country's economy. Four queens are particularly associated with travel and exploration: Queen Hatshepsut of Egypt, Queen Isabella of Spain, and English queens Elizabeth I and Victoria. Queen Hatshepsut (?–1483 B.C.E.) declared herself Pharaoh of Egypt in 1503 B.C.E., succeeding to the throne after her husband's death. She then launched several trading expeditions, including one that established a new route down the Nile River and overland to the Red Sea. After her death, scenes of these trading expeditions were painted on the walls of her tomb. Queen Isabella (1451–1504) also launched a voyage related to trade, financing Christopher Columbus's 1492 sea expedition to find a new trade route to the Orient; he discovered North America instead. Queen Elizabeth I (1533–1603) encouraged explorers like Sir Francis Drake to travel to new lands; during her 45-year reign, England greatly increased its colonial holdings. During the 64-year reign of Queen Victoria (1819–1901), the British government supported colonists' efforts to settle these lands in order to strengthen Britain's claim to them. *See also* AFRICA; GREAT BRITAIN; NORTH AMERICA.

Further Reading: Frasier, Antonia, ed. *The Lives of the Kings and Queens of England*. Berkeley: University of California Press, 1998; MacUrdy, Grace Harriet. *Hellenistic Queens*. Westport, CT: Greenwood Publishing, 1975; Parsons, John Carmi. *Medieval Queenship*. New Yotk: St. Martin's Press, 1994; Tyldesley, Joyce. *Hatshepsut: The Female Pharaoh*. New York: Viking Press, 1996.

Quimby, Harriet (1875–1912)

Born in 1875, American photojournalist and aviator Harriet Quimby was the first woman in the United States to become a licensed pilot. She earned her license in 1911, the same year she learned to fly. Also in 1911 she became the first woman to make a night flight, flying over Staten Island, New York, in front of 20,000 spectators. The following year she became the first woman to fly a plane across the English Channel. Most of her flight took place in heavy fog, so she had to use a compass to guide her Bleriot monoplane.

In addition to her aviation skills, Quimby was a highly regarded photojournalist. From 1903 to 1912 she worked for *Leslie's Illustrated Weekly* in New York, providing the magazine with both articles and pictures. Her story subjects ranged from political issues to household tips to theater reviews. Several of her articles were travel pieces on places like Cuba, Egypt, the Middle East, Scandinavia,

Harriet Quimby starts her plane manually, by turning the propeller. Undated photograph, circa 1910s. © *Bettmann/CORBIS.*

France, the Panama Canal, and U.S. cities like New York and San Francisco. More than 150 photos accompanied these articles. Quimby wrote over 300 articles during her career under her own name and several pseudonyms, earning enough money to support herself and her parents. Eventually she was made an editor for the magazine.

Quimby continued her photojournalism work while touring the United States and Mexico in 1912 to perform in aviation shows. In July of that year, she was flying near Quincy, Massachusetts, over a crowd of hundreds of spectators to promote an upcoming airshow when disaster struck. Her open-air plane suddenly tipped forward, and she and a passenger—the manager of the airshow—fell out. They died immediately upon striking the ground as Quimby's plane crashed into the nearby sea. Aviation experts continue to debate the cause of the accident.

Further Reading: Hall, Ed Y. *Harriet Quimby: America's First Lady of the Air: The Story of Harriet Quimby, America's First Licensed Woman Pilot and the First Woman Pilot to Fly the English Channel.* Spartanburg, SC: Honoribus Press, 1990; The Harriet Quimby Research Conference, http://www. harrietquimby.org (June 2000); Holden, Henry M. *Her Mentor Was an Albatross: The Autobiography of Pioneer Pilot Harriet Quimby.* Mt. Freedom, NJ: Black Hawk Publishing, 1993.

R

Ramsey, Alice Heyler (1886–1983)

Born in 1887, American motorist Alice Heyler Ramsey was the first woman to drive across the United States. She accomplished this feat in 1909 with three women passengers; everyone in the group was a member of the Women's Automobile Club of New York, and Ramsey was its president.

Ramsey's cross-country route took her over 3,800 miles from New York City to San Francisco, California. Some accounts report that her trip took 41 days, others 59. In either case, Ramsey made it across the United States more quickly than any of the men who had previously attempted the same drive.

Ramsey also had many adventures on her trip, which was sponsored by the Maxwell-Briscoe Car Company. For example, her group encountered Native Americans on horseback and a sheriff's posse that briefly detained them on suspicion of committing murder in a nearby town. *See also* TRANSPORTATION, GROUND.

Further Reading: Hyatt, Patricia Rusch. *Coast to Coast with Alice*. Carolrhoda Books, 1995; Mathison, Richard R. *Three Cars in Every Garage*. New York: Doubleday, 1968.

Rape. *See* CHALLENGES FOR THE MODERN WOMAN TRAVELER.

Rau, Santha Rama (1923–)

Born in India in 1923, Santha Rama Rau is representative of modern women who grow up traveling because their parents are involved in international politics. Rau is also unique in that she has written more extensively about her experiences than most of her non-European counterparts. Many of her autobiographical works were first published in magazines, then collected in books like *Gifts of Passage: An Informal Autobiography*.

Published in 1961, *Gifts of Passage* describes Rau's experiences traveling and living in India, Africa, China, Indonesia, Ceylon, Spain, Afghanistan, Russia, Japan, England, and the United States. The daughter of an Indian diplomat, Rau attended school in England and vacationed in Europe as a girl. As a teenager she lived with her family in South Africa, where her father had been sent as High Commissioner for India. In South Africa she faced a great deal of prejudice because of her dark skin. Rau attended college in the United States but again lived with her father when he became India's first ambassador to Japan. Meanwhile her mother was working in India as the head of the largest women's organization in the country. After a year in Japan, Rau spent two years traveling through Asia. By this time she had developed "an incurable addiction to travel" and toured parts of the United States and Europe, writing magazine articles based on her observations. She then married a fellow journalist equally addicted to travel, and together the two went to Russia, China, and many other places—with a baby in tow. About her child she says:

> "He has traveled with us on all our extensive journeys ever since and so far we have had no cause to regret it; still the situation has occasioned innumerable outraged arguments with friends, relatives, educators, stray busybodies about our cavalier attitude towards his 'security.' But this gratuitous outside concern has been virtually our only problem and we, on the whole, have been far more conscious of the rewards of having him with us and enjoying his growing awareness of a very big world." (pp. 122–123).

See also CHILDREN, TRAVELING WITH.

Further Reading: Rau, Santha Rama. *Gifts of Passage: An Informal Autobiography*. New York: Harper & Brothers, 1961.

Reed, Virginia (1834–1921)

Born in 1834, American pioneer Virginia Reed was one of 47 survivors of the Donner Party, a wagon train attempting to travel across America to California in 1846–1847. She is one of the most famous of the survivors because of a letter that she wrote about her experience. Written to a cousin in Illinois, it was published in the *Illinois Journal* in December 1847 and brought the first news of the Donner Party disaster to Easterners.

Virginia was 12 years old when she accompanied her mother and stepfather, James Reed, on the journey to California. Her family set out from Springfield, Illinois, along with the families of George Donner and George's brother, Jacob. They met up with a larger group in Independence, Missouri. Once joined, this wagon train intended to travel to California using the directions in a book, *The Emigrant's Guide to Oregon and California* (1845). The book's author, Lansford W. Hastings, had described a shortcut to the West that required travelers to temporarily leave the conventional route near western Wyoming.

At the critical juncture, some members of the group insisted on remaining on the traditional, well-traveled trail, having heard from soldiers that Hastings's new route was not as easy as it sounded. Others wanted to take the shortcut. After some argument, the group split into two. Twenty wagons, including those of the Donners and the Reeds, decided to follow the Hastings directions, and George Donner was voted its leader; there were 89 members of the Donner Party in all.

The short cut did indeed prove difficult—so difficult, in fact, that the Donner Party's wagons were too large and heavy to manage it. Consequently the group chose an alternate path, but soon the way was blocked with thick brush. Unwilling to turn back, the pioneers—both men and women—chopped their way through the foliage but made slow progress. By the time they reached the eastern edge of the Great Salt Lake Desert, they were short on supplies, and their water ran out before they crossed to the other side and found a stream. Nonetheless, they were all alive at this point, and they soon found themselves back on the main route with the shortcut behind them. In actually, they had traveled 150 miles more than those who had remained on the main road, and they were far behind their former associates. Even worse, the delay meant that unless they hurried, they would reach the Sierra Nevada mountain range in the winter, when the roads were impassable.

As they pressed on, the Donner Party's anxiety over their situation increased, and James Reed got into a fight with another man and stabbed him to death. The rest of the group then banished Reed from the wagon train, unwilling to hang him in front of his wife and children, and he rode off by himself. Meanwhile, two men who had ridden ahead to get help returned with extra food. While this seemed to be fortunate, in actuality it led to the tragedy that followed. Buoyed by the additional supplies, the group decided to rest for a few days before crossing the nearby Sierra Nevada range. By the time they finally got on their way, a snowstorm had begun. Still, the Donner Party did not turn back. They kept heading into the mountains until at last their wagons became so bogged down in snowdrifts that they could go no further.

There were now 81 people in the Donner Party, 56 of them women or children. They built cabins and hoped their families would survive the winter. But this expectation was unrealistic given their poor provisions, and in December 1846, 17 members of the party, five of them women, decided to walk out of the mountains. Most of them died, and those who survived ate their bodies. Meanwhile, the people who had remained in their cabins began dying of starvation, and there, too, the survivors ate the dead, although Virginia Reed always insisted that her own family had never resorted to cannibalism.

In February 1847 a rescue expedition finally arrived to help the Donner Party. Ironically, it had been organized by James Reed, who had reached Sutter's Fort in California safely after leaving the wagon train. The rescuers found 47 members of the Donner Party still alive. Their plight was already famous in California, where many people had contributed money to fund their relief effort. In 1891, Reed again wrote about the experience in an article for *Century Magazine*. By this time she was married and settled in California. She died in 1921.

Further Reading: Lavender, David Sievert. *Snowbound: The Tragic Story of the Donner Party*. New York: Holiday House, 1996; Stewart, George. *Ordeal by Hunger: The Story of the Donner Party*. Boston: Houghton Mifflin, 1960.

Refugees. *See* MIGRATION.

Riddles, Libby (1957–)

In 1985, American Libby Riddles became the first woman to win the Iditarod, a long-distance dogsledding race in Alaska. Despite being caught in a major windstorm along the route, she arrived at the finish line five hours ahead of the second-place winner. She initially took up dogsledding in 1978, in a small race where she placed first, and first competed in the Iditarod in 1980, placing eighteenth. Competing again in 1981, she placed twentieth. She then began breeding and training her own dogs, along with partner Joe Garnie; Garnie placed third in the 1984 Iditarod and second in the 1986 competition. Riddles has written a book about the Iditarod, *Race Across Alaska: First Woman to Win the Iditarod Tells Her Story* (1988). *See also* IDITAROD TRAIL DOG SLED RACE.

Further Reading: Riddles, Libby, and Tim Jones. *Race Across Alaska: First Woman to Win the Iditarod Tells Her Story*. Harrisburg, PA: Stackpole Books, 1988.

Rijnhart, Susie Carson (1868–1908)

Born in 1868, Canadian physician Susie Carson Rijnhart is best known for her attempt to enter the Tibetan city of Lhasa, which was closed to foreigners at the time. She and her husband traveled through China and Tibet as missionaries from 1894 to 1898, attending to people's medical needs along the way, and in 1897 she gave birth to a son. The following year, she became determined to preach Christianity in Lhasa. But before she could reach it, tragedy struck. Her son suddenly became ill and died, and she and her husband were robbed before becoming lost in difficult terrain. Rijnhart's husband left her to get help but never returned. Rijnhart then set out on her own. She walked for miles, enduring danger from the elements, other robbers, and lack of food. Finally she reached a mission, where she learned that her husband had been killed, possibly by the same bandits who had attacked the couple earlier.

Rijnhart returned alone to Canada and wrote about her experiences in *With the Tibetans in Tent and Temple: The Narrative of Four Years' Residence on the Tibetan Border, and of a Journey into the Far Interior* (1901). She remained committed to missionary work, and in 1902 she went back to China, where she met and married a fellow missionary and became pregnant. She became ill during the pregnancy; in 1907, she returned to Canada, and she gave birth to a son in 1908. Three weeks later she died. *See also* ASIA; MISSIONARIES.

Further Reading: Rijnhart, Susie Carson. *With the Tibetans in Tent and Temple: The Narrative of Four Years' Residence on the Tibetan Border, and of a Journey into the Far Interior*. Chicago and New York: F. H. Revell, 1901.

Rogers, Susan Fox

Susan Fox Rogers is a prominent author in the field of feminist travel literature. She makes her living writing travel guidebooks and editing collections of travel writing exclusively for and about women. Her works include *Solo: On Her Own Adventure* (1996), which includes 23 essays describing women's solo travel experiences, and *Another Wilderness: New Outdoor Writing by Women* (1994), which includes 20 essays by women enjoying wilderness experiences. *See also* GUIDEBOOKS, TRAVEL; TRAVEL WRITERS.

Further Reading: Rogers, Susan Fox, ed. *Another Wilderness: New Outdoor Writing by Women.* Seattle, WA: Seal Press, 1994; ———. *Solo: On Her Own Adventure*. Seattle, WA: Seal Press, 1996; ———. *Two in the Wild: Tales of Adventure from Friends, Mothers, and Daughters.* New York: Vintage Books, 1999.

Roosevelt, Eleanor (1884–1962)

Born in 1884, Eleanor Roosevelt traveled extensively as the wife of President Franklin D. Roosevelt and, after his death in 1945, as a delegate to the United Nations and chair of the United Nations Commission on Human Rights. Throughout much of her life she was also involved in humanitarian activities and public speaking. In all of her official capacities she met with world leaders and visited sites of human crisis, circling the globe several times to do so. During World War II she spent time with troops on U.S. military bases. The niece of President Theodore Roosevelt, Eleanor Roosevelt was born in New York City but educated in England. She married Franklin Roosevelt in 1905 and helped with his early political campaigns. After he was crippled by polio, she began traveling in his stead and became more active in politics herself. She describes her life in her autobiographies, *This Is My Story* (1937), *This I Remember,* and *On My Own,* later collected into one volume. Roosevelt died in 1962.

Eleanor Roosevelt steps out of a plane in Miami in 1934. © *Bettmann/CORBIS.*

Further Reading: Black, Allida M. *Casting Her Own Shadow: Eleanor Roosevelt and the Shaping of Postwar Liberalism*. New York: Columbia University Press, 1996; Cook, Blanche Wiesen. *Eleanor Roosevelt*. New York: Viking, 1992; Hoff-Wilson, Joan, and Marjorie Lightman, eds. *Without Precedent: The Life and Career of Eleanor Roosevelt*. Bloomington: University Press, 1984; Roosevelt, Eleanor. *The Autobiography of Eleanor Roosevelt.* Reprint. Da Capo Press, 2000.

Rough Guides

Published by Rough Guides Ltd. of London, England, Rough Guides are a series of travel guidebooks, each of which is devoted to a particular country, city, or subject. In the last category are three books related to women's travel, *Half the Earth: Women's Experiences of Travel Worldwide* (1986), *Women Travel: A Rough Guide Special* (1990, with an updated version in 1999) and *More Women Travel: Adventures and Advice from More than 60 Countries* (1995). Co-edited by Natania Jansz and Miranda Davies, these books offer travel essays and advice from women travelers throughout the world. They also provide reading suggestions and other valuable resources for women's travel. In writing about their approach to the material, Jansz and Davis say:

> We knew that there was a rich seam of knowledge and experience that women shared with each other when they crossed

> paths abroad—in the private ways that women have always found to pass on the information that matters—telling of how we would be perceived and treated, the pleasures and pitfalls we might face. It was time to bring this to a wider audience. Our formula was simple. We advertised for travellers returning from abroad to write to us about their experiences, not in the authoritative and detached style of travel journalism, but woman to woman. Clearly we had touched a chord. Articles, advice, and contacts listings arrived in sackfulls. . . . (1995, pp. xi–xii)

Today many women travelers recommend the Rough Guides books to others embarking on journeys abroad. *See also* GUIDEBOOKS, TRAVEL.

Further Reading: Jansz, Natania, and Miranda Davies, eds. *Half the Earth: Women's Experiences of Travel Worldwide*. London: Rough Guides, 1986; *More Women Travel: Adventures and Advice for More than 60 Countries*. London: Rough Guides, 1995; ———. *Women Travel: A Rough Guide Special*. Updated ed. London: Rough Guides, 1999.

Round-the-World Travelers. *See* CIRCUMNAVIGATORS AND ROUND-THE-WORLD TRAVELERS.

Royall, Anne Newport (1769–1854)

Born in 1769, American author Anne Newport Royall wrote about her travels through the United States between 1817 and 1831 in *Sketches of History, Life and Manners in the United States, by a Traveller* (1826), *The Black Book: Or, A Continuation of Travels in the United States* (1828–1829; three volumes), *Mrs. Royall's Pennsylvania: Or, Travels Continued in the United States* (1829; two volumes), *Mrs. Royall's Southern Tour: Or, Second Series of Black Book* (1830–1831; three volumes), and *Letters from Alabama on Various Subjects: To Which Is Added, an Appendix Containing Remarks on Sundry Members of the 20th and 21st Congress* (1830). Royall supported herself in part by selling her books in advance via reader subscriptions. In addition, as her work became better known, she was often allowed to stay in hotels for free because her books praised specific innkeepers who did her favors and condemned those who did not. Royall was an outspoken critic of people and organizations she disliked. In fact, her second work, *The Black Book*, was so named because it reported on the "black deeds" of people she encountered during her travels; she believed these people were evil. Its publication earned her many enemies, as did its sequel, and by 1830 she decided it was no longer safe for her to travel. Searching for another way to support herself, she established a newspaper in Washington, D.C., in 1831, but it failed. In 1836 she tried again with another newspaper, but it too did poorly. When Royall died in 1854 she was a pauper. *See also* NORTH AMERICA.

Further Reading: Porter, Sarah Harvey. *The Life and Times of Anne Royall*. New York: Arno Press, 1972; Royall, Anne. *The Black Book: Or, A Continuation of Travels in the United States*. Washington, DC: Printed for the author, 1828–29; ———. *Mrs. Royall's Southern Tour: Or, Second Series of the Black Book*. Washington, DC: 1830–31.

S

Sacchi, Louise (?–1997)

Between 1955 and 1980, American aviator, navigator, and mechanic Louise Sacchi made more than 340 ocean crossings, more than any other private pilot. She first began flying in 1939 and three years later became the first female navigation instructor for England's Royal Air Force at their British Flying Training School in Texas. In 1944 she studied airplane mechanics, and by 1947 she was managing a seaplane base in New Jersey. Less than 10 years later she had begun transporting people and cargo across the Atlantic, and in 1965 she started the Sacchi Air Ferry Enterprises (SAFE) overseas ferry company, becoming the first female international air ferry pilot. She most frequently flew to Europe, the Philippines, Africa, England, and the United States. Sacchi also competed in air races, and in 1971 she set a record for the fastest New York-to-London flight in a single-engine land plane. She received numerous awards for her contributions to aviation and in 1979 published a book about her experiences, *Ocean Flying.* Sacchi died in 1997.

Further Reading: Sacchi, Louise. *Ocean Flying.* New York: McGraw-Hill, 1979.

Sacajawea (ca. 1787–?)

Born in approximately 1787, Sacajawea (also spelled Sacagawea) was a Native American woman of the Shoshone tribe who became a vital member of the Lewis and Clark expedition of 1804–1805. As a girl she had been captured by a rival tribe and sold into slavery, eventually becoming the property of a Canadian trapper named Touissant Charbonneau. When he was hired by explorers Meriwether Lewis and William Clark to act as their interpreter and guide on their journey, Sacajawea accompanied him.

Funded by the United States government, the Lewis and Clark expedition, had a twofold goal. First, the men were to explore land that had recently been purchased from France, including parts of what are now North Dakota, Montana, Idaho, Washington, and Oregon. Second, the explorers were to look for the legendary Northwest Passage, a direct water route from the Mississippi River to the Pacific Ocean.

Sacajawea's relatives lived in Idaho, and while there the expedition encountered her brother, Chief Cameahwait. This meeting proved critical to the mission's success. Cameahwait provided the group with necessary provisions, fresh horses, and information regarding the best way to cross the Rocky Mountains, and Sacajawea helped guide Lewis and Clark along the route. In 1805 the expedition reached the Pacific Ocean at the mouth of the Columbia River; by then Lewis and Clark had confirmed that there was no Northwest Passage.

Sacajawea and Charbonneau accompanied Lewis and Clark back to North Dakota,

Sacajawea. *Denver Public Library, Western History Collection.*

then went their own way. However, Clark later rewarded them with land in Missouri. Historians are unclear what happened to Sacajawea. Some say that she and Charbonneau remained together, farming their land, until Sacajawea died in 1812. Others believe that Sacajawea returned to her tribe and was sent with her relatives to an Indian reservation, where she died in 1884.

Further Reading: Clark, Ella E., and Margot Edmonds. *Sacagawea of the Lewis and Clark Expedition*. Berkeley: University of California Press, 1979; Howard, Harold P. *Sacajawea*. Norman: University of Oklahoma Press, 1971; Moulton, Gary E., ed. *Atlas of the Lewis & Clark Expedition*. Lincoln: University of Nebraska Press, 1983.

Sackville-West, "Vita" (Victoria Mary) (1892–1962)

Born in 1892, English author Victoria Mary "Vita" Sackville-West wrote many novels, short stories, poems, and gardening columns, as well as one important work of travel literature, *Passenger to Teheran* (1926), which describes a 1925 journey through the Middle East and Russia. In it she writes about such things as the relaxed attitude of the Persians toward property, saying that while she is sitting in someone's garden:

> No one will come up and say that I am trespassing. . . . All are equally free to come and enjoy. Indeed there is nothing to steal, except the blossom from the peach trees, and no damage to do that has not already been done by time and nature. The same is true of the whole country. There are no evidences of law anywhere, no sign-posts or milestones to show the way; . . . you may travel along any of those three roads for hundreds of miles in any direction, without meeting anyone or anything to control you; even the rule of the road is nominal, and you pass by as best you can. If you prefer to leave the track and take to the open, then you are free to do so. One remembers—sometimes with irritation, sometimes with longing, according to the fortunes of the journey—the close organisation of European countries. (Morris, pp. 142–143)

Outside of *Passenger to Teheran*, Sackville-West is best known for her writings about her sexuality. Although married and the mother of two children, she was a lesbian and had affairs with women. Sackville-West died in 1962. *See also* ASIA.

Further Reading: Morris, Mary. *Maiden Voyages: Writings of Women Travelers*. New York: Vintage, 1993; Sackville-West, Vita. *Passenger to Teheran*. 1926. Reprint. New York: Moyer Bell, 1990.

Safari. *See* AFRICA.

Sampson, Deborah (1760–1827)

Born in 1760, American soldier Deborah Sampson disguised herself as a man so she could fight in the Revolutionary War. She traveled throughout the eastern United States as part of the 4th Massachusetts Regiment from 1782 to 1783, when her gender was discovered during an illness. Sampson was also possibly the first woman to travel across the country as a professional lecturer. After the war she went from town to town, wearing her Continental Army uniform, to describe her experiences in battle. Sampson died in 1827.

Further Reading: Mann, Herman. *The Female Review: Life of Deborah Sampson; The Female Soldier in the War of Revolution*. 1866. Reprint. New York: Arno Press, 1972.

Schele, Linda (1942–1998)

Born in 1942, American artist Linda Schele is representative of women whose travel experiences change the course of their lives. While on a vacation to Mexico with her husband in 1970, Schele visited some ancient Mayan ruins and became fascinated with them. From that point on, she dedicated her life to studying the Mayan people, their culture, and their art. During the mid-1970s she began a major project to decipher Mayan hieroglyphics, which became the subject of her doctoral dissertation in Latin American studies in 1982. In subsequent years she conducted workshops on hieroglyphics and continued her research into Mayan culture. Schele, who died in 1998, wrote many books on the Mayans. They include *The Code of Kings: The Language of Seven Sacred Maya Temples and Tombs* (co-authored with Peter Mathews, 1998), *The Blood of Kings: Dynasty and Ritual in Maya Art* (co-authored with Mary Ellen Miller, 1986), and *The Mirror, the Rabbit, and the Bundle: "Accession" Expressions from the Classic Maya Inscriptions* (1982).

Further Reading: Schele, Linda, and Peter Matthews. *The Code of Kings: The Language of Seven Sacred Maya Temples and Tombs*. New York: Scribner, 1998; Schele, Linda, and Jeffrey H. Miller. *The Mirror, the Rabbit, and the Bundle: "Accession" Expressions from the Classic Maya Inscriptions*. Washington, DC: Dumbarton Oaks, Trustees for Harvard University, 1982; Schele, Linda, and Mary Ellen Miller. *The Blood of Kings: Dynasty and Ritual in Maya Art*. New York: G. Braziller; Fort Worth, TX: Kimbell Art Museum, 1986.

Scidmore, Eliza Ruhamah (1856–1928)

Born in 1856, American travel writer and photographer Eliza Ruhamah Scidmore was one of the first correspondents for *National Geographic* magazine. Beginning in the 1890s, she wrote about her experiences in various parts of Asia, observing the aftereffects of a Japanese tsunami in 1896 and worshipers at a holy stretch of India's Ganges River in 1907. Scidmore's writing was of such high quality that Charles McCarry, the editor of an anthology of *National Geographic* articles, *From the Field,* calls her the best author the magazine ever had. He further says:

> Her dispatches from Japan, from Manchuria, from one forbidden Asian vastness after another, fascinated me. She was a meticulous reporter, a fluently confident writer. A ripple of amusement ran across the crystalline surface of her prose; the essential mystery of life on Earth stirred in its depths. Her style was the expression of her personality, which was that of an independent, educated, late-Victorian woman of principle whose visible passions were those of the mind. . . . Reading one of her stories was like reading a letter from my Aunt Carolyn, a teacher of geography who, every summer, set off for some interesting foreign destination aboard a tramp steamer. (pp. 4–5)

Scidmore also accompanied her articles with her own photographs. According to McCarry, her photography equaled her writing. He says: "She was as good at the one craft as at the other, as though the two hemispheres of her brain, like the sexes in the world she desired to live in, worked in effortless harmony as absolute equals." (p. 5) Scidmore wrote for *Geographic* for much of her life. She died in 1928. *See also* From the Field.

Further Reading: McCarry, Charles, ed. *From the Field: The Best of National Geographic*. Washington, DC: National Geographic Society, 1997; Scidmore, Eliza Ruhamah. *Java: The Garden of the East*. 1897. Reprint. Singapore and New York: Oxford University Press, 1984; ———. *Winter India*. New York: Century, 1903.

Scientists

Many women travelers in modern times have been scientists. In the nineteenth century, the majority were specimen collectors who used the need to gather plants, animals, and/or insects as an excuse to travel or as a means to support their travels. Beginning in 1869, American botanist Jeanne Carr supported herself in part by collecting plant specimens in what is now California's Yosemite National Park. She was the first white woman to explore the park from Tuolumne Canyon to Hetch-Hetchy Valley, and she climbed several of the region's mountains in order to gather plants for later study. Beginning in 1890, English amateur entomologist Margaret Fountaine visited every continent and most major countries, including Syria, Turkey, Algeria, Spain, Africa, the Caribbean, India, New Zealand, Fiji, South America, the United States, and the Far East, under the guise of collecting butterflies, when in fact she primarily enjoyed the thrill and romance of travel. Between 1905 and 1930, American hunter, explorer, and adventurer Delia Denning Akeley collected animal specimens in Africa and sold them to museums. Between 1925 and 1938, American botanist Ynes Mexia collected thousands of plant specimens in South America for universities. During the 1920s and 1930s, Scots botanist Isobel Wylie Hutchison traveled to Greenland, Alaska, Canada, and the Aleutian Islands to collect specimens of Arctic plants for the British Museum. In the 1990s, American artist and biologist Natalie Prosser Goodall, who is a member of the Society of Women Geographers, went on several expeditions to search for fossils of marine mammals, finding more than 2,000 of them; she is currently planning to establish her own museum.

Some scientists travel to study animals in their native habitats. For example, American ichthyologist and scuba diver Eugenie Clark has traveled throughout the world researching shark behavior. She has worked on 24 television specials about marine life in the United States, Bermuda, England, Egypt, Israel, Japan, and Mexico, and conducted 71 dives off Grand Cayman, Bermuda, the Bahamas, California, and Japan to study deep sea fish. Similarly, in 1979 U.S. Fish and Wildlife Service biologist Diane Boyd traveled to Montana's Glacier National Park to study Canadian wolves that had crossed the U.S. border to form new packs. She continues to study Montana wolf packs there today, recording their movements and evaluating their behavior, and is involved in a project to restore wolves to their native wilderness areas. Her work is similar to that of Englishwoman Jane Goodall, American Dian Fossey, and German Birute Galdikas, who have studied chimpanzees, mountain gorillas, and orangutans, respectively.

Other scientists travel to study entire ecosystems rather than just one animal within a region. American meteorologist Peggy Dillon is interested in weather rather than forests. She was the first woman to work at the Mt. Washington Observatory, a global weather observation site in New Hampshire that is reputed to experience the worst weather in the world. She has also worked at a geology field camp in the Transantarctic Mountains. Many other scientists venture into the field for research purposes, either living away from home for long stretches or routinely traveling between the field site and the laboratory. *See also* Akeley, Delia Denning; Caufield, Catherine; Fossey, Dian; Fountaine, Margaret; Galdikas, Birute; Goodall, Jane; Mexia, Ynes.

Further Reading: Caufield, Catherine. *The Emperor of the United States of America & Other Magnificent British Eccentrics*. New York: St. Martin's Press, 1981; ———. *In the Rainforest*. Chicago: University of Chicago Press, 1986; ———. *Multiple Exposures: Chronicles of the Radiation Age*. Chicago: University of Chicago Press, 1990; Kaufman, Polly Welts. *National Parks and the Woman's Voice*. Albuquerque: University of New Mexico Press, 1996.

Scott, Blanche

Blanche Scott was the first woman to drive an automobile (an Overland) across the United States alone, traveling from New York to San Francisco in 1910. That same year she was also the first American woman to fly an airplane; however, because her solo flight took place while she was taxiing rather than as part of a planned flight, and because the plane only rose a few feet off the ground, she is not considered by many aviation historians to be the first woman to fly. Instead that honor has gone to Bessica Raiche, who flew a plane a few feet in New York only two weeks after Scott. *See also* AIR TRAVEL; TRANSPORTATION, GROUND.

Further Reading: Mathison, Richard R. *Three Cars in Every Garage*. New York: Doubleday, 1968.

Sea Travel

Women have traveled by boat ever since the first humans built them in prehistoric times. But whereas men have sought adventure over the seas, until modern times women typically traveled by ship only to get from one place to another in the easiest possible way. The exceptions were unconventional women like the Vikings and pirates who sailed vast distances in search of treasure.

In conventional European and American society, women were not allowed on board certain types of vessels unless they were wives of the ship's officers. Officers' wives were permitted on ships even in times of war. As Joan Druett reports in her book *She Captains: Heroines and Hellions of the Sea*:

> It was not just the wives of sea captains who were accorded the privilege [of being on a Navy vessel], for the wives of the Boatswain, Gunner, and Carpenter often sailed, too, along with the wives of the Sailmaker, Cooper, Cook, Steward, and Purser. At any given moment on any wooden man-of-war, in fact, it was almost guaranteed that there would be a company of respectable matrons on board, apparently quite undismayed by the prospect of going into battle. (p. 185)

Druett explains that the reason wives were allowed on board had to do with the length of service required of a ship's officer. She says:

> "The ship, in effect, was their lifetime home, for they were 'in constant employ.' . . . When the officers changed ships, their wives and families went with them, just like moving from one workplace to another today. These wives and children were supposed to be landed when the ship was commissioned for a voyage, but often they were not, simply because there was nowhere else for them to go. While it was against regulations for petty officers' families to sail, it happened all the time. All that was necessary was for the captain to turn a blind eye." (2000, pp. 185–186)

Many of these wives gave birth on board ship, sometimes even during battles, and they also offered help in times of crisis. Many records tell of wives carrying powder to the cannons or dressing wounds.

The wives of captains who were not in the military could also accompany their husbands on long voyages. For example, American Mary Brewster's husband was captain of a whaling ship, and because she went to sea with him, in 1849 she became the first European woman to travel to the Arctic Ocean by ship. Americans Susan Hathorn and Mary Lawrence, and Englishwoman Annie Slade were also the wives of nineteenth-century captains, and all accompanied their husbands on long sea voyages. Slade was also the daughter of a sea captain and had grown up on board ship. Born in 1865, her father had trained her in seamanship out of necessity when he was shorthanded, and Slade in turn taught her husband her navigation skills. In fact, he acquired his captainship directly as a result of her training. While at sea, Slade bore several children, and they remained on board ship with their parents just as Slade had. This was the case as well for Englishwoman Alice Rowe Snow. Born in 1869, the daughter of an English sea captain, Snow kept a detailed diary of her experiences. Her work was published in 1944 as *Log of a Sea Captain's Daughter: With Adventures on Robinson Crusoe's Island by Alice Rowe Snow, who sailed with her father, Captain Joshua N. Rowe, on the Bark Russell during a Voyage of Four Years and a Half.* American Joanna Carver Colcord, who grew up on board ship, abandoned the sea entirely as soon as she was old enough to set her own course in life.

During the nineteenth century, so many wives and daughters of whalers and traders were at sea that most ships had a gamming chair on board. This device was either a chair or a barrel with a seat, hoisted by ropes, that was used to convey a woman from a rowboat up to the deck of a sailing ship. Captain's wives from whaling vessels called their visits with one another "gamming" because they had to rely on gamming chairs if they wanted to go aboard one another's ships. But, as with the wives on navy vessels, women on board private sailing ships did more than visit with one another. In times of trouble, they were often called upon to serve as part of the ship's crew. Hanna Burgess, for example, took the helm for a month when her husband, the ship's captain, fell ill and died while on a voyage in 1856, thus becoming one of the first women to command a nineteenth-century sailing vessel. As soon as she got the ship to port, another male captain was found to take her place despite her qualifications for the job. Burgess had been accompanying her husband on all his voyages since their marriage in 1852, and she was well versed in sailing and navigation skills. Nonetheless, after her command she was relegated to life on land. American Mary Patten, the first woman to captain a nineteenth-century sailing vessel around Cape Horn, also took command only because her husband was ill and was never given enough credit for her skills. In fact, while she was in command she was attacked by a crew member incensed that a woman was at the helm. Most captain's wives, however, never got the chance to command a ship; they were forever passengers, considered inferior by the men around them.

Women in Disguise

Women who wanted to take a more active role on board sometimes disguised themselves as men in order to be part of the crew. For example, "Tom" Bowling and "William" Brown served for years on board British warships without their true gender being discovered, and Ann Jane Thornton served as a cabin boy and cook on board sailing ships from 1832 to 1835 before being discovered. Rose de Freycinet, wife of French geographer Louis-Claude de Saulces de Freycinet, sneaked on board her husband's ship, *L'Uranie,* disguised as a boy and accompanied him on an expedition dedicated to science and exploration ordered by Louis XVIII. Although her husband was the ship's captain, at the time the French did not allow any women on board ship; however, she dropped her disguise once she was safely at sea. Her voyage was uneventful for the most part until February 1820, when *L'Uranie* struck a rock near the Falkland Islands and went aground. It took weeks for the stranded crew to find a ship willing to transport them home.

In all, de Freycinet's voyage lasted three years. During that time her husband's scientific research required their vessel to anchor frequently near land, sometimes in populated areas. In contrast, another seagoing wife, American Abby Jane Morrell, spent most of her journey far out to sea. Morrell's husband, Benjamin, had embarked on an expedition to chart whaling and sealing grounds and was also studying tropical lagoons far from civilization. He had not wanted to take his wife along, believing that she would be better off on land, but the young woman insisted. After some argument, Benjamin reluctantly agreed. But in the end Abby Jane was not completely happy to have gotten her way—and Rose de Freycinet, too, had some small measure of regret regarding her situation. As Druett explains: "Abby Jane, for her part, would have liked to spend a lot more time on land, 'to see land, men, shipping, churches, &c.; things I had been accustomed to all my life.' Rose, by contrast, found to her dismay that the lengthy spells on shore that were her lot involved all kinds of unexpected hazards, including dirt and boredom." (2000, p. 220) Druett adds that Rose was never happier than when she was at sea.

Against Their Will

But some women sailors were not pleased to be going on a voyage at all: those transported against their will as slaves or convicts. For example, between the seventeenth and nineteenth centuries approximately 15 million African slaves were transported to the Americas by ship; in the eighteen century alone, approximately 6 million slaves were transported. The usual route for these sailors was to go to America from Africa by way of the West Indies, where there was also a demand for plantation workers. Conditions on board were usually brutal, and many slaves died before ever reaching their destination.

Convicts, too, often endured brutal conditions. For example, in 1787 during an eight-month voyage from England to Botany Bay, Australia, where the British had established a penal colony, 48 of 1,350 convicts died of various illnesses. One of the convicts transported to Botany Bay was Mary Bryant, who soon found herself on another, more difficult voyage. In 1791, she and a group of fellow convicts escaped from the colony by stealing an open rowboat and heading north along the coastlines of Australia and New Guinea for 69 days, eventually reaching the island of Timor in the East Indies. Within a short time they were recaptured and sent back to England.

While Bryant traveled unwillingly, and sea captains' wives traveled only because of their husbands' jobs, modern women typically sail for enjoyment. Some of them have shunned male companionship entirely, choosing to sail solo as a way to challenge themselves physically and mentally. Of these, seven women have succeeded in circumnavigating the globe alone: Naomi James of Ireland (1977–1978); Krystyna Chojowska-Liskiewicz of Poland (1977–1978); Brigitte Oudry of France (1977–1978); Kay Cottee of Australia (1988); Isabelle Autissier of France (1990–1991); Samantha Brewster of England (1995); and Karen Thorndike of the United States (1997–1998). Another American woman, Tania Aebi, believed that she was the first American woman—as well as the youngest woman, at age 18—to solo circumnavigate the globe by sea. However, a year after completing her trip and declaring her success, she discovered that she could not claim the title because she had a friend on board for 80 miles in the South Pacific.

The first woman to sail solo across the Atlantic Ocean was Englishwoman Ann Davison in 1953. However, she did not make her journey nonstop. That same year, Nicolette Milnes Walker earned the honor of being the first nonstop solo Atlantic sailor traveling from Great Britain to the United States in 1953. She wrote a book about her experience—*When I Put Out to Sea* (1972). She has also written articles about sailing and seamanship, as well as a book entitled *Introduction to Dinghy Sailing* (1981). Since then, many more women have embarked on challenging sailing adventures, and some have competed in sailing races previously reserved for men, including the America's Cup. Other women have chosen less difficult sailing experiences, choosing relaxation over challenge.

Many books are devoted to helping women find the type of sailing experiences they desire. One of the best is *Sailing: A Woman's Guide* by Doris Colgate (1999), who is the president of the National Women's Sailing Association and a sailing instructor; it provides how-to information as well as tips from sailors of all types. Another is *The Cruising Woman's Advisor: How to Prepare for the Voyaging Life* by Diana B. Jessie (1997), which describes the sailing life-style in depth. For women interested in more modest boating experiences, there is *Sea Kayaking: A Woman's Guide* by Shelley Johnson, Molly Mulhern Gross, and Doug Hayward (1998), which offers both instruction and travel tips regarding kayaking for women of all ages. Modern travel books that offer first-person accounts of women's experiences at sea include Susan Tyler Hitchcock's *Coming About: A Family Passage at Sea* (1998), which describes the author's experiences on a nine-month Caribbean sailing trip with her husband, eight-year-old son, and six-year-old daughter in a 34-foot sloop, and Lydia Bird's *Sonnet: One Woman's Voyage from Maryland to Greece* (1997), which describes the author's solo 5,000-mile voyage across the

Atlantic Ocean and through the Mediterranean Sea in a 42-foot sailboat. *See also* AEBI, TANIA; AUSTRALIA; BRYANT, MARY; CIRCUMNAVIGATORS AND ROUND-THE-WORLD TRAVELERS; COLCORD, JOANNA CARVER; DAVISON, (MARGARET) ANN; HATHORN, SUSAN; JAMES, NAOMI; LAWRENCE, MARY; PATTEN, MARY.

Further Reading: Brewster, Mary. *"She Was a Sister Sailor": The Whaling Journals of Mary Brewster, 1845–1851*. Mystic, CT: Mystic Seaport Museum, 1992; Creighton, Margaret S., and Lisa Norling, eds. *Iron Men, Wooden Women: Gender and Seafaring in the Atlantic World, 1700–1920*. Baltimore, MD: Johns Hopkins University Press, 1996; De Pauw, Linda Grant. *Baptism of Fire*. Pasadena, MD: The MINERVA Center, 1993; ———. *Seafaring Women*. Boston: Houghton Mifflin, 1982; Druett, Joan. *Hen Frigates: Passion and Peril, Nineteenth-Century Women at Sea*. New York: Touchstone, 1998; ———. *She Captains: Heroines and Hellions of the Sea*. New York: Simon & Schuster, 2000; Druett, Joan, and Mary Anne Wallace. *The Sailing Circle: 19th-century Seafaring Women from New York*. With an introduction by Lisa Norling. East Setauket, NY: Three Village Historical Society; Cold Spring Harbor, NY: Cold Spring Harbor Whaling Museum, 1995; Gunter, Helen Clifford. *Navy WAVE: Memories of World War II*. Fort Bragg, CA: Cypress House Press, 1992; Huntington, Anna Seaton. *Making Waves: The Inside Story of Managing and Motivating the First Women's Team to Compete for the America's Cup*. Arlington, TX: Summit Publishing Group, 1996; Iorns, Jann. *Sailing Away: New Zealand Women in Sailing, 1920–1990*. Auckland, New Zealand: New Women's Press, 1991; Johnson, LouAnne. *Making Waves: A Woman in This Man's Navy*. New York: St. Martin's Press, 1986; Milnes Walker, Nicolette. *When I Put Out to Sea*. London: Collins, 1972.

Seniors

Today tour groups, travel agencies, and other organizations are devoted to providing quality travel experiences for senior tourists. In addition, books such as *Season of Adventure: Traveling Tales and Outdoor Journeys of Women Over 50* (1996) by Jean Gould, which offers a collection of travel essays by women age 50 and over, encourage older women to travel by showing them that age is not a barrier to any adventure.

One might assume that this was not always the case. However, many of the best known women travelers of the late nineteenth and early twentieth centuries were far from young. For example, Englishwoman Lucy Evelyn Cheesman made eight expeditions to islands of the South Pacific between the ages of 42 and 74, and American Louise Arner Boyd was 68 when she became the first woman to fly over the North Pole. Frenchwoman Alexandra David-Neel embarked on her difficult journey to Tibet at the age of 54, while Englishwoman Mary Kingsley did not even begin traveling in Africa until the age of 31. Englishwoman Freya Stark's first travel book was published when she was 41, and she continued to travel and write about her experiences into her late seventies. American Ynes Mexia traveled throughout South America from the age of 45 until her death at age 58. Englishwoman Isabella Bird Bishop made her last expedition, to Morocco, at the age of 66. Dutchwoman "Alexine" Tinne took her mother Harriet with her when she explored Africa between 1856 and 1863, when Harriet died at the age of 65.

Age alone did not confine any of these women to their homes, and some of the oldest women travelers even engaged in strenuous activities. For example, American mountaineer Annie Smith Peck became the first woman to climb Mount Coropuna in Peru at the age of 61, and she made her last climb, of Mount Madison in New Hampshire, at the age of 82. Older travelers have benefited from the fact that they do not have young children to hinder their travels and generally have more money to spend. In addition, as the age of the general population has increased, older women are finding more social support for their desire to travel. As Faith Conlon writes in the introduction to a collection of travel essays, *Gifts of the Wild*:

> We are radically redefining age. Recently, on a hike through the North Cascades with eleven other people, I discovered that the youthful, strong women I befriended were all over fifty. In the nineteen-fifties, we thought that women were old by their late forties. If they had been Anasazi women

> [an ancient people in the American Southwest], they would have been wise elders—had they lived to turn forty. (p. viii)

In addition to essay collections, many other books offer tips to older travelers. One is *Unbelievably Good Deals and Great Adventures That You Absolutely Can't Get Unless You're Over 50* by Joan Rattner Heilman, which is frequently updated to provide information about discounts and special travel opportunities for seniors. Tips related to senior travel can also be found online at such sites as Age of Reason (http://www.ageofreason.com), which offers travel-related articles and links. *See also* BISHOP, ISABELLA LUCY BIRD; BOYD, LOUISE ARNER; CHEESMAN, LUCY EVELYN; DAVID-NEEL, ALEXANDRA; *GIFTS OF THE WILD;* KINGSLEY, MARY; MEXIA, YNES; PECK, ANNIE SMITH; STARK, FREYA MADELINE; TINNE, "ALEXINE" (ALEXANDRINE).

Further Reading: Conlon, Faith, Ingrid Emerick, and Jennie Cook, eds. *Gifts of the Wild: A Woman's Book of Adventure.* Seattle, WA: Seal Press, 1998; Gould, Jean, ed. *Season of Adventure: Traveling Tales and Outdoor Journeys of Women Over 50.* Seattle, WA: Seal Press, 1996; Heilman, Joan Rattner. *Unbelievably Good Deals and Great Adventures That You Absolutely Can't Get Unless You're Over 50.* Lincolnwood, IL: Contemporary Books, 1999.

Seton Thompson, Grace Gallatin (1872–1959)

Born in 1872, American feminist Grace Gallatin Seton Thompson traveled throughout world to study women of various cultures. In 1919 she founded a camping and hiking group for girls, initially called the Girl Pioneers but later the Camp Fire Girls. Seton Thompson first took up camping herself after marrying naturalist Ernest Thompson in 1896. By this time she had worked for several years as a journalist, and she soon wrote a book based on her outdoor experiences with her husband, *A Woman Tenderfoot* (1900). Other books followed, including *Nimrod's Wife* (1907), *A Woman Tenderfoot in Egypt* (1923), *Chinese Lanterns* (1924), *"Yes, Lady Saheb": A Woman's Adventurings with Mysterious India* (1925), and *Log of the "Look See": A Half-Year in the Wilds of Matto Grosso and the Paraguayan Forest; over the Andes to Peru* (1932).

Seton Thompson's international travel experiences began during World War I, when she went to France to drive transport trucks; this eventually led to a job directing a women's motor unit. After the war, in 1921, she went to Egypt to observe a newly founded women's political group there. This launched her investigation of women's issues in other countries, including China and India. She also went on several hunting trips in both the United States and India. In the late 1920s she visited South America with a scientific expedition studying flora and fauna; Seton Thompson not only wrote about the group's activities but also helped label the many specimens its scientists collected. She then left the expedition to hunt and to attend a women's conference in Chile. In her later years, Seton Thompson attended several other such conferences. She also traveled throughout the Orient, again studying women's place in various cultures. Seton Thompson died in 1959.

Further Reading: Seton Thompson, Grace Gallatin. *Log of the "Look-see": A Half-year in the Wilds of Matto Grosso and the Paraguayan Forest, over the Andes to Peru.* London: Hurst & Blackett, Ltd., 1932; ———. *Nimrod's Wife.* New York: Doubleday, Page, 1907; ———. *A woman Tenderfoot.* New York: Doubleday, Page and Co., 1900; ———. *A Woman Tenderfoot in Egypt.* New York: Dodd, Mead and Company, 1923. ———. *Yea, Lady Saheb": A Woman's Adventurings with Mysterious India.* New York; London: Harper & Brothers, 1925.

Shaw, Flora (1852–1929)

Born in 1852, Irish writer Flora Louisa Shaw was one of the most prominent newspaper correspondents of the late nineteenth century. Beginning in 1886, she traveled abroad and submitted articles based on her experiences to such British publications as the *Manchester Guardian* and the *Pall Mall Gazette*. Gradually her writings became more political, as she became interested in how British colonialism was affecting policies in such places as South Africa, Australia, and New Zealand. Her travels also took her to Morocco, Egypt, North America, and Canada.

One series of articles in 1898 reported on the Klondike Gold Rush in Alaska. In 1901 she retired from journalism but continued to travel, although in her last years she settled in England. Shaw died in 1929. *See also* GOLD RUSH.

Further Reading: Robinson, Jane. *Wayward Women: A Guide to Women Travellers*. Oxford: Oxford University Press, 1990.

Sheldon, May French (1847–1936)

Born in 1847, American scientist and explorer May (also known as Mary) French Sheldon was one of the first women to lead her own African safari, an 1891 trek with more than 100 porters and servants into the interior of East Africa from the coastal city of Mombasa, Kenya. She was also one of the first explorers to concentrate on studying the women and children in the area being visited. In addition, the journey made Sheldon one of the first white people to have seen Lake Chala, which rests inside a volcanic cone, and one of few to travel partway up the slope of Mount Kilimanjaro. Her adventures were cut short when she was slightly injured in a fall. When she returned to England, she wrote a book about her travels, *Sultan to Sultan: Adventures among the Masai and Other Tribes of East Africa* (1892); it became a best-seller in both England and the United States. In addition, Sheldon was one of only 22 women invited to become a Fellow of the Royal Geographical Society, although the invitation was withdrawn when the group decided it would not allow women after all.

In 1903 Sheldon returned to Africa to explore the Belgian Congo, and in 1905 she went to Liberia. When World War I broke out in 1914, she supported the Belgian Red Cross by lecturing about this region and about her other travels. Although Sheldon was American, she lived in England for most of her life and traveled extensively throughout Europe and on world tours; in all, she made four trips around the world. She also supported the women's rights movement. In addition to her travel books, she wrote a popular novel entitled *Herbert Severance* (1889). Sheldon died in 1936. *See also* AFRICA.

Further Reading: Sheldon, Mary French. *Sultan to Sultan: Adventures among the Masai and Other Tribes of East Africa*. 1892. Reprint. Freeport, NY: Books for Libraries Press, 1972.

Shields, Mary

In 1974 Mary Shields became one of two women to be the first women to enter the Iditarod Trail Dog Sled Race, a major cross-country dogsledding competition in Alaska. She was also the first women to finish the competition (that some year), placing twenty-third. At the time, Shields had only recently moved to Alaska, and she was new to dogsledding. She was introduced to the sport when she borrowed a sled and dog team to transport firewood to her new home; within a short time she had her own team of eight dogs, the smallest team to enter the Iditarod that year. Today Shields is a professional husky breeder in Fairbanks, Alaska. She also offers tours of her breeding operation and hosts nature walks. She has written books about dogs and dog training as well as a book about her sledding experiences, *Sled Dog Trails* (1984). *See also* IDITAROD TRAIL DOG SLED RACE.

Further Reading: Shields, Mary. *Sled Dog Trails*. Anchorage, AK: Alaska Northwest Publishing, 1984; ———. *Small Wonders: Year-round-Alaska*. Fairbanks, AK: Pyrola Publishing, 1987.

Ships. *See* SEA TRAVEL.

Simpson, Myrtle Lillias (1931–)

Born in 1931, Scots mountaineer Myrtle Lillias Simpson was the first woman to ski across Greenland's polar icecap, an experience she describes in her 1967 book *White Horizons*. In 1968 she also traveled the closest to the North Pole of any previous female explorer; she discusses this adventure in her 1970 book *Due North*. Simpson was in the Arctic because her husband, a scientist, was conducting research there. She has also traveled in other regions and has climbed peaks in New Zealand and Peru. She discusses some of these other travels in her book *Home*

Is a Tent (1964). *See also* MOUNTAINEERING; POLES, NORTH AND SOUTH.

Further Reading: Simpson, Myrtle. *Due North*. London: Gollancz, 1970; ———. *Home is a Tent*. London: Gollancz, 1964. ———. *White Horizons*. London: Gollancz, 1967.

Smeeton, Beryl (1905–1979)

Born in 1905, Canadian Beryl Smeeton was the wife of travel author Miles Smeeton and is featured in all of his books. The couple traveled the world in a 46-foot sailboat. Before their marriage, however, Beryl Smeeton was already an experienced traveler. One of her most challenging adventures was a solo horseback trip through the southern mountains of Patagonia in South America, which she made in the late 1930s. She also traveled to Tibet with Swiss explorer Ella Maillart. Smeeton's only published work, *Winter Shoes in Springtime* (1961), describes her travels in Hong Kong, China, and Burma. She died in 1979. *See also* MAILLART, ELLA.

Further Reading: Smeeton, Beryl. *Winter Shoes in Springtime*. With an introduction by Ella Maillart. London: R. Hart-Davis, 1961.

Smith, Gwendolen Dorrien (1883–1969)

Born in 1883, Englishwoman Gwendolen Dorrien Smith is perhaps best known for being the companion of Lady C. C. Vyvyan in a 1926 trip across Canada, which Vyvyan wrote about 35 years later in *Arctic Adventure* (1961; later retitled *The Ladies, the Gwich'in, and the Rat*). However, she was also a traveler in her own right. During the early twentieth century she visited the Balkans, North Africa, Morocco, and South America; after World War II she visited Australia and New Zealand with her sister. To finance her travels, she painted watercolors of the sights she saw and was particularly noted for her water and boat scenes. She also collected botanical specimens during her travels, and on at least one occasion offered them for sale. Smith never married, preferring travel to a settled home life. She died in 1969. *See also* NORTH AMERICA; VYVYAN, C. C.

Further Reading: MacLaren, I. S., and Lisa N. LaFramboise, eds. *The Ladies, the Gwich'in, and the Rat: Travels on the Athabasca, Mackenzie, Rat, Porcupine, and Yukon Rivers in 1926 by C. C. Vyvyan*. Edmonton: The University of Alberta Press, 1998.

Somerset, Susan (?–1936)

In the late 1880s, Susan Margaret McKinnon Somerset, the Duchess of St. Maur, embarked on a seven-month hunting and fishing expedition across Canada with her husband Algernon. She went from Liverpool, England, to Newfoundland, then took a train to Calgary, Alberta, and Vancouver, British Columbia, hunting along the train route as well as on Vancouver Island. When she returned to England, Somerset published a book about her travels, *Impressions of a Tenderfoot during a Journey in Search of Sport in the Far West* (1890). She died in 1936.

Further Reading: St. Maur, Mrs. Algernon. *Impressions of a Tenderfoot during a Journey in Search of Sport in the Far West*. London: John Murray, 1890.

Spies

Women working as spies have sometimes traveled great distances to gather and deliver information, particularly in times of war. During the American Revolution, for example, patriot Emily Geiger rode 50 miles through enemy territory to deliver a message to an American general. Ann Trotter Bailey and Sarah Bradlee Fulton, the latter of whom is often credited with inspiring the Boston Tea Party, went back and forth through British territory to deliver messages to the Americans. During the Civil War, Belle Boyd and Nancy Hart both carried messages for the Confederate Army. Both spent time in prison for their activities, although Hart escaped after stealing a guard's weapon and shooting him. Meanwhile Pauline Cushman spied for the Union Army while on tour as an actress.

During World War I, American journalist Marguerite Baker Harrison worked as a spy while traveling throughout Europe as a reporter. However, she did not achieve much fame for her spying activities. Far more notorious were World War I spies Edith Clavel, who was also a British nurse, and Dutchwoman Mata Hari because both were executed for their activities, the former by the Germans in 1915 and the latter by the French in 1917. However, contrary to popular belief neither traveled much in her work; they passed information via couriers.

Perhaps the most harrowing travel-related experience for a woman spy occurred during World War II, when Virginia Hall narrowly escaped capture. An American working in France, in the winter of 1941 she fled from the Nazis on foot over the Pyrenees into Spain. Her journey was made even more difficult by the fact that she had a wooden leg, having lost her own in a hunting accident. Later she returned to France to resume spying, but because the Nazis were still looking for her she disguised herself and learned to walk without a limp. After the war she worked in the United States, first for the Office of Strategic Services (OSS) and later for the Central Intelligence Agency (CIA). In modern times, women spies continue to be employed for various clandestine activities, but none have received widespread attention for their efforts.

Further Reading: McIntosh, Elizabeth P. *Sisterhood of Spies*. Annapolis, MD: Naval Institute Press, 1998.

Spinsters Abroad

Published in 1989, *Spinsters Abroad: Victorian Lady Explorers* by Dea Birkett is an important reference work on the experiences of unmarried women explorers during the nineteenth century. It offers in-depth discussions related to why women traveled in the Victorian era as well as biographical information about such travelers as Gertrude Bell, Isabella Bird Bishop, Elizabeth Bisland, Alexandra David-Neel, May French Sheldon, and Fanny Workman. More than 50 women are discussed, and short biographies of 21 of these travelers are at the back of the book. *See also* Bell, Gertrude; Bishop, Isabella Lucy Bird; Bisland, Elizabeth; David-Neel, Alexandra; Sheldon, May French; Workman, Fanny Bullock.

Further Reading: Birkett, Dea. *Spinsters Abroad: Victorian Lady Explorers*. Oxford and New York: B. Blackwell, 1989.

Spouses

Among the earliest women travelers were those who migrated to new worlds with their families or those whose spouses traveled as part of their employment. Of the latter, most were wives of sea captains and other ship's officers. Sometimes these women traveled quite extensively and were participants in major geographical discoveries. For example, in 1769, Englishwoman Frances Hornby Barkley, whose husband was a fur trader and ship's captain, was among the first European women to visit western Canada, and several locations in the region bear portions of her name.

Until the mid-twentieth century, it was unusual for a woman to travel alone or accompanied by a man who was not her husband, brother, or father. Married women, however, were sometimes able to travel alone without causing too much consternation, particularly if they were older women. Therefore a husband was seen as an asset for women wanting to travel—particularly if the male offered his financial support to the traveling woman's endeavors.

One prominent explorer who took advantage of her husband in this way was Frenchwoman Alexandra David-Neel, the first European woman to visit Tibet's capital, Lhasa. She married railroad engineer Philippe Francois Neel in 1904, but after only a few days she decided she couldn't live with him. They separated but remained friends, and he financially supported all of her travels until his death in 1941.

Another woman who traveled without her spouse but with his money was American Molly Brown, famous for surviving the sinking of the ocean liner *Titanic* in April 1912. In 1893 her husband, miner James Joseph "J. J." Brown, discovered one of the largest and purest veins of gold in the United States, and Molly Brown set out to transform herself from a simple country woman to a cultured, well-educated lady of society. She studied languages, art, music, and manners, bought lavish home furnishings and clothes, and went on European and world tours. At first her husband accompanied her on her travels, but he felt uncomfortable among the upper-class socialites and soon decided to remain home. Consequently Molly traveled without him, and in 1909 the couple legally separated. The terms of their separation provided her with enough money to continue her extensive travels through Europe and the United States.

But many other women had husbands who were true travel partners. For example, in 1877 Isobel Gill and her husband, British astronomer David Gill, went together to Ascension Island in the South Atlantic Ocean so that he could get a better view of the stars. During the 1910s and 1920s, American photographer Osa Johnson and her husband worked together to discover indigenous tribes in the South Pacific and film them; they also filmed African wildlife together in the 1920s and 1930s. Also in the early twentieth century, American Delia Denning Akeley went on African safaris with her husband, hunter and explorer Carl F. Akeley; after they divorced, she continued traveling in Africa alone. Englishwomen Florence Baker and Isabel Burton were also wives of well-known explorers, but although they traveled extensively, unlike Akeley, they never set out on their own.

Not only explorers but also scientists, journalists, missionaries, colonists, emigrants, and many other types of women traveled with spouses. Until the twentieth century, however, their journeys did not receive the same attention as did their husbands. For example, after Baker and her husband discovered the source of the Nile River in Africa in 1864, he received much acclaim back in their native England while her contributions to the expedition were ignored. This kind of dismissal of women's travel experiences could occur even within the traveler's own family. Although Englishwoman Lucy Atkinson traveled with her husband through Russia from 1848 to 1853, he never mentioned her in his writings about their exploits; it was as though she had not been along on the trip. Atkinson wrote her own book, *Recollections of Tartar Steppes and Their Inhabitants* (1863) in part because she wanted to reinsert herself into the story of their adventures.

In contrast, after Englishwoman Marika Hanbury-Tenison traveled with her husband to study indigenous cultures in South America during the 1970s, her contributions were recognized not only by her family but also by the public. In fact, her books were more popular than her husband's, although this was in large part because his works were written for a scholarly audience. Interestingly, even in this case, with two equal partners who both wrote about their travel experiences, the man's views were considered more "serious" than the woman's. *See also* AKELEY, DELIA DENNING; ATKINSON, LUCY; BAKER, FLORENCE; BARKLEY, FRANCES; BROWN, MOLLY; BURTON, ISABEL; DAVID-NEEL, ALEXANDRA; GILL, ISOBEL; HANBURY-TENISON, MARIKA; JOHNSON, OSA.

Stanhope, Hester (1776–1839)

Born in 1776, Lady Hester Stanhope was a wealthy Englishwoman who became the first white woman to enter the holy city of Palmyra in present-day Syria in 1813. An exotic figure, she began traveling in 1810, when she decided that she wanted to go to France to meet Emperor Napoleon. At the time, England was not issuing passports to France because of a war between the two countries. Consequently Stanhope sailed to Turkey, believing that the ambassador there would give her a passport. Once in Turkey, however, her plans were stymied, and she decided instead to explore the Middle East. She was shipwrecked off Greece in 1812 and lost most of her clothes, whereupon she began dressing in the clothing of Turkish men. Her fellow Britains considered this practice scandalous, particularly because Stanhope found the clothing so comfortable that she refused ever to return to her traditional dress. Meanwhile, Middle Eastern men found her fascinating, and she claimed that when she entered Palmyra she was not only celebrated, but also named "Queen of the Desert."

Some historians, however, doubt this claim. According to Jane Robinson in *Wayward Women: A Guide to Women Travellers*:

> Opinion has long been divided over Lady Hester Stanhope. Some say she was a great explorer, a traveller *extraordinaire* whose life was spent in a heroic search for new lands and philosophies; others argue that she was a privileged, passionate, and frustrated woman whose journeys were not so much crusades as retreats from an increasingly uncomfortable reality of depression and mental illness. (p. 57)

In either case, Stanhope eventually settled in a Syrian monastery, Dar Djoun. She lived there for 25 years and gradually fell into poverty and insanity. She died in 1839. *See also* AFRICA; ASIA; CLOTHING; DISGUISES.

Further Reading: Hughes, Jean Gordon. *Queen of the Desert: The Story of Lady Hester Stanhope*. London: Macmillan, 1967; Robinson, Jane. *Wayward Women: A Guide to Women Travellers*. Oxford: Oxford University Press, 1990.

Stark, Freya Madeline (1893–1993)

Born in 1893, travel writer Dame Freya Madeline Stark produced an extremely large volume of work. Most of her articles and books are about her trips to the Middle East, although she visited many other parts of the world as well. Her works include *The Valleys of the Assassins and Other Persian Travels* (1934), *Baghdad Sketches* (1937), *Letters from Syria* (1940), *A Winter in Arabia* (1940), *Ionia: A Quest* (1954), *Riding to the Tigris* (1959), and *The Minaret of Djam: An Excursion in Afghanistan* (1970), as well as several autobiographical works, letter collections, and essay collections.

Of these, one of her most popular works was *A Winter in Arabia*. The book describes a visit she made to Arabia in 1937–1938, during which she stayed in the town of Hureidha and directed an archaeological dig. While spending part of her time searching for signs of an ancient city, Stark participated in the daily lives of the Arabian people, and she wrote about their customs and rituals. But even as she interacted with them, she hoped this contact would not alter their way of life. She explains:

> We are in a proud country still new to Europeans, the first foreigners to live in its outlying districts for any length of time; and the hope that I cherish is that we may leave it uncorrupted, its charm of independence intact. I think there is no way to do this and to keep alive the Arab's happiness in his own virtues except to live his life in certain measure. One may differ in material ways; one may sit on chairs and use forks and gramophones; but on no account dare one put before these people, so easily beguiled, a set of values different from their own. Discontent with their standards is the first step in the degradation of the East. Surrounded by our mechanical glamour, the virtues wrung out of the hardness of their lives easily come to appear poor and useless in their eyes; their spirit loses its dignity in this world, its belief in the next. That this unhappy change may come here as elsewhere is only too probable; but it will be no small winter's achievement if it does not come through *us*. (1987, pp. 44-45)

A Winter in Arabia also tells of Stark's difficulties during her stay in Arabia, including a bout with illness. In addition, it reports on her experiences joining a camel caravan to retrace an ancient trading route, the first European woman to do so.

Stark had long been fascinated with travel, having been raised in France, England, and Italy, by wealthy parents who enjoyed travel themselves. During World War I, Stark went to Italy to volunteer as a nurse, and after the war she decided to settle on the Italian Riviera. However, she soon missed travel and began studying Arabic so she could visit the Middle East. In 1927 she went to Syria for a few months to complete her study of the language and introduce herself to Middle Eastern culture. In 1929 she went to Baghdad and from there embarked on a tour of Persia (present-day Iran) looking for archaeological artifacts and learning the Persian language. During the 1930s she made several more tours of the Middle East, and during World War II she provided information about the region to the British government. In 1943 she toured the United States, lecturing about her travel experiences. She also visited India in the late 1940s, Turkey during the early 1950s, and Afghanistan in the early 1960s. Meanwhile she continued to write and was honored for her work until her death in 1993.

Further Reading: Geniesse, Jane Fletcher. *A Passionate Nomad: The Life of Freya Stark*. New York: Random House, 1999; Moorhead, Caroline. *Freya Stark*. New York: Viking, 1985; Stark, Freya. *Alexander's Path: A Travel Memoir*. Woodstock, New York: Overlook Press, 1987; ———. *Baghdad Sketches*. 1937. Reprint. Evanston, IL: Marlboro Press/Northwestern, 1996; ———. *Bridge of the Levant, 1940–43*. Reprint. Salisbury, England: M. Russell, 1977; ———. *The Broken Road, 1947–52*. Reprint. Salisbury, England: M. Russell, 1981; ———. *A Winter in Arabia*. 1940. Reprint. Woodstock and New York: Overlook Press, 1987.

Stark, Mariana (ca. 1762–1838)

Born in approximately 1762, Englishwoman Mariana Stark wrote one of the most important travel guidebooks of the nineteenth century, *Travels on the Continent Written for the Use and Particular Information of Travellers* (1820), which was reissued as *Information and Directions for Travellers in the Continent* in 1924. It provided detailed information for tourists heading for Europe, and later editions included information on Russia and Scandinavian countries as well. Stark made several trips to Europe to keep her guidebooks current, telling her readers not only about the best places to visit, eat, and stay, but also about how to pack and budget for a trip. She died in 1838 while traveling in Italy.

Further Reading: Robinson, Jane. *Wayward Women: A Guide to Women Travellers*. Oxford: Oxford University Press, 1990.

Stewart, Elinore Pruitt (1876–1933)

Born in 1876, Elinore Pruitt Stewart was a pioneer of the American West who became famous for her book *Letters of a Woman Homesteader* (1914) about her experiences as a homesteader; this work encouraged other women to travel west to become homesteaders themselves. Stewart came from a poor family in Oklahoma and helped raise five siblings. As a teen she worked for a railroad crew, primarily doing laundry, and in approximately 1902 she married an Oklahoma homesteader. Accounts differ on whether he died or whether she merely left him within five years, whereupon she traveled to Denver, Colorado, with her two-year-old daughter to take a housekeeping job. Shortly thereafter she saw an advertisement for a similar job in Burnt Fork, Wyoming, and went there. Within eight weeks she had married her employer, a widowed rancher. She also filed a homesteading claim on a 160-acre plot of land right beside his. According to the Homestead Act, Stewart had to live on the land for five years in order to make it legally her own; consequently she built a house that was attached to her husband's and that straddled the border between his property and her claim so she could satisfy the residency requirement.

During her homesteading years, Stewart wrote detailed letters to a former employer describing her life on the ranch, and the friend showed the letters to the editor of *Atlantic Monthly* magazine. He published a series of them beginning in 1913, and they were so popular that the following year they were published as a collection. In 1915 another collection of Stewart's writings, *Letters on an Elk Hunt*, was published.

Letters of a Woman Homesteader includes letters written from 1909 to 1913 at her Burnt Fork homestead. In some of these letters Stewart promotes homesteading as a way for women to better themselves financially. For example, she says:

> I am very enthusiastic about women homesteading. It really requires less strength and labor to raise plenty to satisfy a large family than it does to go out to wash [by taking a job as a laundress], with the added satisfaction of knowing that their job will not be lost to them if they care to keep it. Even if improving the place does go slowly, it is that much done to stay done. Whatever is raised is the homesteader's own, and there is no house-rent to pay. . . . To me, homesteading is the solution of all poverty's problems, but I realize that temperament has much to do with success in

> any undertaking, and persons afraid of coyotes and work and loneliness had better let ranching alone. At the same time, any woman who can stand her own company, can see the beauty of the sunset, loves growing things, and is willing to put in as much time at careful labor as she does over the washtub, will certainly succeed; will have independence, plenty to eat all the time, and a home of her own in the end. . . . I am only thinking of the troops of tired, worried women, sometimes even cold and hungry, scared to death of losing their places to work, who could have plenty to eat, who could have good fires by gathering the wood, and comfortable homes of their own, if they but had the courage and determination to get them. (pp. 214–217)

Stewart repeatedly expresses her enjoyment of homesteading, love of nature, and pride in her accomplishments, and her work encouraged other women to follow in her path. After *Letters* made her a well-known author, Stewart continued to live and work on her ranch, but during the early 1920s she experienced financial trouble and had to lease out her land. She lived in Boulder, Colorado, until 1925, whereupon she returned to the ranch. Two years later she was nearly killed while operating dangerous farming equipment. She remained in poor health following the accident and died in 1933. *See also* HOMESTEADING; NORTH AMERICA.

Further Reading: Stewart, Elinore Pruitt. *Letters of a Woman Homesteader*. 1914. Reprint. Boston: Houghton Mifflin, 1982.

Stinson, Katherine (1892–1977) and Marjorie (1896–1975)

Born in 1892 and 1896, respectively, Katherine and Marjorie Stinson were the first women sworn in by the U.S. Post Office in 1913 and 1914 in Illinois to fly airmail deliveries. Katherine was also the first woman to perform an aerial "loop the loop" in San Antonio, Texas, where the Stinson School of Flying was established, the first person to "loop the loop" at night, and the first person to perform multiple loops. Katherine first decided to learn to fly because she heard that pilots earned $1,000/day in air shows, and she wanted to earn enough money to travel to Europe to study music so she could become a famous pianist. She then encouraged her sister to fly too. Marjorie was the youngest women of her time to earn a pilot's license, a goal she accomplished in 1914 at the age of 18. In 1915, the two women opened the aviation school and trained male cadets to fly during World War I. In 1917, Katherine Stinson broke a nonstop distance flying record set by Ruth Law a year earlier by going 610 miles without refueling. In 1920 Katherine retired from aviation because of poor health; she died in 1977. Meanwhile, Marjorie became a barnstormer, flying in air shows until 1928. She retired from aviation in 1930. In her later years, she studied architecture and became an award-winning home designer. She died in 1975. *See also* AIR TRAVEL; LAW, RUTH.

Further Reading: Rogers, Mary Beth; Sherry A. Smith, and Janelle D. Scott. *We Can Fly, Stories of Katherine Stinson and other Gutsy Texas Women*. Austin: Ellen C. Temple Publishing and Texas Foundation for Women's Resources, 1983.

Strahorn, Carrie Adell (1854–1925)

Born in 1854, American pioneer Carrie Adell Strahorn traveled throughout the United States with her husband Robert, a journalist who was hired by a railroad company to write pamphlets on various parts of the country. Each spring for six months, beginning in 1878, the two set out to visit new places, and each winter they returned to their home in Cheyenne, Wyoming, so Robert could write. In 1884 he took a job managing the construction of railroad lines; this job also required travel, although not as much. Six years later, Robert again changed jobs, this time becoming an investment banker in Boston. He remained in this position for seven years, during which time his work made him wealthy. The Strahorns then retired to Spokane, Washington, and Carrie Strahorn decided to write a book about her life of travel. The book, *Fifteen Thousand Miles by Stage: A Woman's Unique Experience during Thirty Years of Path Finding and Pioneering from the Missouri to the Pacific and from Alaska to Mexico,* was published in 1911. She died in 1925. *See also* NORTH AMERICA; TRANSPORTATION, GROUND.

Further Reading: Strahorn, Carrie Adell. *Fifteen Thousand Miles by Stage*. New York: G. P. Putnam's Sons, 1911.

Strong, Anna Louise (1885–1970)

Born in 1885, American journalist Anna Louise Strong is representative of women who traveled because of political ideology. She received a Ph.D. in philosophy at the University of Chicago in 1908 and shortly thereafter became involved in political causes. In 1915, after settling in Seattle, Washington, she began writing about her views for a local labor newspaper. In 1918 she supported a labor strike; when it failed, she decided to travel to Russia, where socialism was taking hold. There she taught English, opened a trade school, and continued to write and lecture about political issues. She also occasionally traveled to the United States to gain support for various causes and acted as a technical advisor for the 1943 MGM movie *The Song of Russia*.

In 1946 Strong became involved with the Chinese Revolution; when she later promoted this revolution in Russia, she was deported as a Chinese spy. She then went to the United States, where she lived until 1958. That year she returned to China to write about Chinese issues. Her most important work in this area was *Letter from China* (1962), which told about life in Communist China and the Soviet Union. Strong also wrote an autobiography, *I Change Worlds: The Remaking of an American* (1935), as well as several other books about her travels and political views. They include *The Road to the Grey Pamir* (1931), *This Soviet World* (1936), *One-Fifth of Mankind* (1938), *The New Lithuania* (1941), *I Saw the New Poland* (1946), *Dawn over China* (1948), *Tibetan Interviews* (1959), and *When Serfs Stood Up in Tibet* (1960).

In 1965 the leader of China, Mao Tse-tung, made Strong an honorary member of his country's Red Guard in recognition of her promotion of Chinese political causes. When she died in 1970 she was buried in China's National Memorial Cemetery of Revolutionary Martyrs in Beijing.

Further Reading: Strong, Anna Louise. *I Change Worlds: The Remaking of an American*. 1935. Reprint, with an introduction by Barbara Wilson. Seattle, WA: Seal Press, 1979; ———. *I Saw the New Poland*. Boston, MA: Little, Brown, and Company, 1946; ———. *Tibetan Interviews*. Peking: New World Press, 1959; Strong, Tracy B. *Right in Her Soul: The Life of Anna Louise Strong*. New York: Random House, 1983.

Swale, Rosie (1947–)

Born in 1947 in Switzerland, Rosie Swale gained notoriety by hitchhiking 16,000 miles through Europe, Turkey, Iran, India, Scandinavia, and Russia when she was only 17 years old. In the early 1970s, she embarked with her husband on a 30-foot catamaran, sailing from Italy west across the Atlantic and Pacific Oceans to Australia and back again. In 1983, now divorced, Swale sailed in a 17-foot yacht solo across the Atlantic Ocean, and in 1984 she traveled on horseback down the length of Chile to Cape Horn. She wrote several books about her travel experiences, including *Children of Cape Horn* (1974) and *Back to Cape Horn* (1986). In 1997 she jogged across Romania (from Transylvania to the Black Sea) and plans to write about her experiences; she also raised money through her run sponsorships for Romanian orphans. Previous to this, she ran 150 miles across the Saharan desert.

Further Reading: Swale, Rosie. *Back to Cape Horn*. London: Collins, 1986; ———. *Children of Cape Horn*. New York: Walker, 1974.

Sykes, Ella Constance (ca. 1850–1939)

Born in the mid-1800s, Ella Constance Sykes was the first Englishwoman to visit certain parts of Persia (present-day Iran), Russia, and China. She traveled to other parts of the world as well, usually accompanied by her brother Percy, a British diplomat. She wrote several articles and books about her experiences, the latter of which include *Through Persia on a*

Side-Saddle (1898), *Persia and Its People* (1910), and *Through Deserts and Oases of Central Asia* (1920), which she wrote with her brother. In her works, Sykes expresses her pleasure at leaving civilization behind; she found great freedom in travel and sought remote areas. She particularly enjoyed living with her brother at a British outpost in Chinese Turkestan, even though there were few comforts there. According to historian Jane Robinson, Sykes was "a purist for whom real travel had nothing to do with the familiar" (p. 60). Sykes died in 1939.

Further Reading: Robinson, Jane. *Wayward Women: A Guide to Women Travelers.* Oxford: Oxford University Press, 1990; Sykes, Ella C. *Through Persia on a Side-Saddle*. London: A. D. Innes, 1898.

T

Tabei, Junko (1939–)

Born in 1939, Japanese mountaineer Junko Ishibashi Tabei was the first woman to reach the summit of Mt. Everest. In 1975 she was part of a 15-woman team, but she was the only one of her team to reach the summit. Her feat was even more exceptional because during the climb she was injured in an avalanche but insisted on continuing to lead the expedition despite bruises and torn muscles. In addition to Everest, Tabei has climbed 69 major mountains, including Annapurna in the Himalayans, Kilimanjaro in Africa, Aconcagua in Argentina, McKinley in the United States, El'brus in Russia, Vinson Massif in the Antarctica, and Mount Carstensz in Indonesia. With this last ascent, she became the first woman to have climbed the tallest mountain in each of the world's seven major geographical regions. She has also climbed all of Japan's major peaks, and in 1969, she founded the first women's mountaineering club in Japan, the Joshi-Tohan. She continues to climb mountains today. *See also* MOUNTAINEERING.

Junko Tabei. © *AP/Wide World Photos.*

Further Reading: Birkett, Bill, and Bill Peascod. *Women Climbing: 200 Years of Achievement.* Branson, MO: Mountaineer Books, 1990.

Taylor, Annie Royle (1855–ca. 1909)

Born in 1855, British missionary Annie Royle Taylor was the first European woman to visit Tibet. She first went to China in 1884 as part of the China Inland Mission, whose aim was to send women to the interior of China to spread Christianity. In 1892 Taylor set out to reach the Tibetan city of Lhasa, which was closed to foreigners. She began the more-than-1,000-mile journey into the country's interior dressed as a Tibetan nun, accompanied by several Tibetan converts to Christianity. During her seven months of travel, she suffered a variety of mishaps, including being robbed by bandits; three days before she reached Lhasa she was betrayed to authorities by a disgruntled servant. She was arrested and eventually told to leave the country. For approximately 20 years she remained near the Tibetan border doing missionary work. She then returned to England, where she died around 1909. *See also* ASIA; MISSIONARIES.

Further Reading: Carey, William. *Travel and Adventure in Tibet, Including the Diary of Miss Annie R. Taylor's Remarkable Journey from Tau-chau to Ta-chien-Lu through the Heart of the Forbidden Land.* Delhi: Mittal Publications, 1983.

Taylor, Elizabeth (1856–1932)

Born in 1856, Victorian traveler Elizabeth Taylor was particularly attracted to rugged, cold places such as Iceland, Canada, Norway, and the Faroe Islands. She journeyed extensively through these regions, writing essays about the scenery and cultures she observed along the way. A collection of her work, *The Far Islands and Other Cold Places*, was published in 1997.

In her writings Taylor expresses her love of nature—a love that made her appreciate the people of the north. She felt that their closeness to the land as well as their simple lifestyle and their courage in the face of hardships were to be emulated.

Taylor did not travel in order to write; instead she wrote in order to travel. Her travel articles, which were published in various magazines, financed her trips north. She also produced sketches and paintings inspired by her travels. Her favorite subjects were birds and flowers.

Further Reading: Taylor, Elizabeth. *The Far Islands and Other Cold Places: Travel Essays of a Victorian Lady*. Edited by James Taylor Dunn. St. Paul, MN.: Pogo Press, 1997.

Thayer, Helen (1938–)

Born in 1938, American explorer Helen Thayer was the first woman to ski to the magnetic North Pole alone, which she did in 1988. Accompanied only by her dog, she spent 27 days traveling 345 miles across a polar ice cap to reach the site. Along the way, she encountered bad weather, ferocious polar bears, and other dangers. She wrote a book about her adventures, *Polar Dream: The Heroic Saga of the First Solo Journey by a Woman and Her Dog to the Pole* (1993). Later Thayer repeated her trip with her husband; the two camped in the Arctic. On a subsequent trip they hiked through the Sahara Desert together. *See also* POLES, NORTH AND SOUTH.

Further Reading: Thayer, Helen. *Polar Dream: The Heroic Saga of the First Solo Journey by a Woman and Her Dog to the Pole*. New York: Simon & Schuster, 1993.

Thomas, Elizabeth Marshall (1931–)

Born in 1931, American anthropologist Elizabeth Marshall Thomas made new discoveries regarding the Bushman people during her travels to Africa in the 1950s. She first went to Africa in 1951 in the company of her father, a retired corporate executive who wanted to study the Bushmen himself, as well as her mother and younger brother. All of the family members were particularly interested in Bushman language; during their 1951 trip and two subsequent trips in 1952 and 1955, they visited four different linguistic groups: the Naron, the Ko, the Gikwe, and the Kung. By their third trip, Thomas had received her bachelor's degree in anthropology, and, in 1959, she wrote a book about her travels among the Bushmen, *The Harmless People*. This work led to grants and financial support from *New Yorker* magazine that enabled Thomas to embark on a 1961 expedition to Uganda, where she studied a warrior tribe, the Dodoth. Because of the hostility of these people, few foreigners had ever gone among them, yet Thomas never felt threatened. She remained with the Dodoth for six months, writing about her experiences in *Warrior Herdsmen* (1965). Thomas is currently involved in documenting the behavior of wild animals and exploring the ways in which this behavior manifests itself in domestic animals. Her works related to animal studies include *The Tribe of the Tiger: Cats and*

Their Culture (1995) and *The Social Lives of Dogs: The Grace of Canine Company* (2000).

Further Reading: Thomas, Elizabeth Marshall. *The Harmless People*. New York: Knopf, 1959; ———. *The Social Lives of Dogs: The Grace of Canine Company*. New York: Simon and Schuster, 2000; ———. *The Tribe of the Tiger: Cats and Their Culture*. Thorndike, ME: GK Hall, 1995; ———. *Warrior Herdsmen*. 1965. Reprint. New York: Norton, 1981.

Tibet. *See* ASIA; MISSIONARIES.

Tinling, Marion (1904–)

Born in 1904, American author Marion Tinling is an expert on women travelers. She has produced two important books on the subject, *Women into the Unknown: A Sourcebook on Women Explorers and Travelers* (1989) and *With Women's Eyes: Visitors to the New World, 1775–1918* (1993). She is also the author of *Women Remembered: A Guide to Landmarks of Women's History in the United States* (1986).

Women into the Unknown: A Sourcebook on Women Explorers and Travelers is her best-known work. It provides detailed biographical information on 42 of the most notable women explorers and travelers of the nineteenth and twentieth centuries and offers Tinling's views on why women travel. In an introduction she writes:

> What motivates a man or woman to leave the comforts of home to venture into possible danger and certain hardship? The old male reasons for exploration—conquest, acquisition of wealth, and imperialism—are out of favor, and today different reasons are given for travel. Some wish to take the benefits of Christianity or modern technology to other lands. Some want to study plants or wildlife. Others are interested in uncovering the remains of ancient civilizations. Many are devoted to preserving a vanishing culture or to protecting the world's environmental balance. A few confess they want simply the thrill of discovery, or just the fun of being on the open road. The testing of one's strength and ability to survive is a sufficient motive to others. . . . [But many women were motivated by] the need to escape stultifying Victorian social restrictions. . . . Many made their escape only after they were freed from family responsibilities and when they were far from young. . . . [These early women travelers] did not see themselves as pioneers for feminism. They were drawn by curiosity and attracted by the color and excitement of foreign lands. [1989, pp. Xxiii-xxiv]

The book also provides information on other books related to women's exploration and travel. *See also* ADAMS, HARRIET CHALMERS; AKELEY, DELIA DENNING; ATKINSON, LUCY; BAKER, FLORENCE; BATES, DAISY MAY; BELL, GERTRUDE; BISHOP, ISABELLA LUCY BIRD; BLUNT, ANNE; BOYD, LOUISE ARNER; CABLE, MILDRED; CHEESMAN, LUCY EVELYN; CRESSY-MARCKS, VIOLET OLIVIA; DAVID-NEEL, ALEXANDRA; DIXIE, FLORENCE; DODWELL, CHRISTINA; HANBURY-TENISON, MARIKA; JOHNSON, OSA; KINGSLEY, MARY; MAILLART, ELLA; MURPHY, DERVLA; NILES, BLAIR; NORTH, MARIANNE; PFEIFFER, IDA REYER; STARK, FREYA MADELINE; TINNE, "ALEXINE" (ALEXANDRINE); WORKMAN, FANNY BULLOCK.

Further Reading: Tinling, Marion, Compiler and ed. *With Women's Eyes: Visitors to the New World, 1775–1918*. Norman: University of Oklahoma Press, 1999; ———. *Women into the Unknown: A Sourcebook on Women Travelers*. Westport, CT: Greenwood Press, 1989.

Tinne, "Alexine" (Alexandrine) (1835–1869)

Born in 1835, wealthy Dutch explorer "Alexine" Alexandrine Petronella Francisca Tinne was the first white woman to see some parts of Africa as she searched for the source of the Nile River. Her African travels began in 1856, when she, her mother Harriet, and their servants visited Egypt while on a tour of Europe and the Middle East. On this trip they took a boat up the Nile but did not go far. However, when the two women visited Egypt again in 1861, this time accompanied by Harriet's sister, Adriana, Tinne successfully reached the city of Khartoum, at the fork that divides the river into the White Nile and the Blue Nile.

Once in Khartoum, Tinne planned to continue up the Blue Nile, but after hearing about a group of male explorers that was searching along the White Nile to find the river's source, she decided to go that way too. To this end, she equipped several boats, including a steamer, with a large amount of supplies and a great number of servants. The journey was difficult, but the expedition went further up the White Nile than any previous boats had managed. However, after five months on the river, Tinne fell ill, and the group returned to Khartoum. As soon as she recovered, she and Harriet embarked on another journey, this time with several European scientists and soldiers who had arrived in Khartoum intending to explore the African interior. The expedition group included more than 300 men and over 100 camels, but Adriana chose to remain in Khartoum. From the outset the party was plagued by bad weather and illness; Harriet died, as did some of the scientists and servants. A grief-stricken Tinne made it back to Khartoum in March 1864, having been gone almost a year and having collected many valuable botanical specimens. She then parted company with Adriana and spent the next several months sailing around the Mediterranean before settling in Cairo. In 1867 she moved to Algiers, but two years later she set out on another expedition, attempting to be the first European woman to cross the Sahara Desert. When her party was attacked by a hostile desert tribe, Tinne was killed. *See also* AFRICA.

Further Reading: Gladstone, Penelope. *Travels of Alexine: Alexine Tinne, 1835–1869*. London: J. Murray, 1970.

Tours. *See* ADVENTURE TRAVEL; CLASS DISTINCTIONS; ECO-TOURISM; GROUP TOURS.

Traill, Catherine (1802–1899)

Born in 1802, English author and naturalist Catherine Parr Strickland Parr Traill wrote several important books about Canada that promoted travel to the country, including *The Backwoods of Canada: Being Letters from the Wife of an Emigrant Officer, Illustrative of the Domestic Scenery of British America* (1836), *The Canadian Settler's Guide* (1860), and *Afar in the Forest; or, Pictures of Life and Scenery in the Wilds of Canada* (1869). She also wrote an autobiography, *Pearls and Pebbles: Or, Notes of an Old Naturalist* (1894). Traill first went to Canada in 1832, immediately after marrying a Scots army officer who had received a land grant there. By this time, she had already had several stories published, and, in Canada, she turned her attention to writing about her surroundings. Her works were extremely popular and profitable, and they allowed her to support herself after her husband died in 1857. Traill died in 1899. *See also* NORTH AMERICA.

Further Reading: Traill, Catherine Parr. *The Backwoods of Canada: Being Letters from the Wife of an Emigrant Officer, Illustrative of the Economy of British America.* 1836. Reprint. Toronto: Coles Publishing, 1971.

Trains. *See* TRANSPORTATION, GROUND.

Transportation, Ground

Women have played an important role in the history of ground transportation, not only as travelers but as inventors, particularly in the United States. Prior to the nineteenth century most Americans traveled on foot or via horseback, horse and carriage, steamship, or riverboat. But with the advent of the first public railway in 1830, people increasingly relied on train travel; the first railway line to cross the American continent, finished in 1869, was used by thousands of women traveling west to settle the frontier.

Several women made significant contributions in the nineteenth century. One was Rebecca Lukens, who took over the Brandywine Iron Works after her husband's death in 1825. Under her successful management, the mill produced iron for locomotives and rails, as well as for ship boilers; the company still exists today as Lukens, Inc. Other women important to railroad technology during the 1800s were Eliza Murfey,

Mary Walton, Catherine L. Gibbon, and Mary I. Riggin. In 1870, Murfey invented a lubricating system that kept railroad-car axles well oiled, thereby reducing the chance that they would seize and cause a derailment. In the 1880s, Walton invented one system that diverted locomotive smoke emissions into a water tank and another that cushioned the sound made when a train traveled along a railroad track, thereby reducing the noise of New York City's elevated trains. During the same period, Gibbon made several improvements to railroad track construction, and Riggin invented the railway crossing gate. Other women, like Carrie Strahorn, promoted railroad use through advertisements that increased ridership, and, beginning in 1838, women were employed by the railroads to serve passengers, either at train stations or in dining and passenger cars, on certain routes.

Meanwhile, other women were working to improve other types of vehicles. Annie H. Chilton, for example, redesigned the braking system on horse carriages in 1891 so that the driver could both apply a brake to the carriage and also release the bar connecting the horse to the carriage in times of emergency. Martha J. Coston focused her attention on boat travel, inventing a communication system in 1859 whereby ship captains used signal flares to send messages to one another or to ask for help from those on shore in times of distress.

The Car

After automobiles were invented, women worked to improve these vehicles too. One of the first women inventors to concentrate on automobile improvements was Mary Anderson, who developed the first windshield wiper in 1903. Her device was operated manually, using a handle inside the car. By 1923, women had patented more than 175 inventions related to automobiles and traffic signals, along with approximately 170 other transportation-related inventions. These included new types of carburetors, clutches, and engine starters. In the 1930s, Helen Blair Bartlett made improvements in spark plugs that are still valued today.

The first automobiles also attracted adventurous women who wanted to try out the new vehicle for themselves. The first woman to appear in public driving a car was Daisy Post, a relative of the socially prominent Vanderbilt family. Sometime in the late 1890s, she drove a car around the grounds of her estate, went off the road, destroyed some landscaping, and crashed into the steps of her house. In 1899, Mrs. Mary Landon fared much better than Post in demonstrating women's driving skills to the public. A clerk–secretary for the Hayes Apperson Automobile Company—she used an instruction sheet that she herself had written to drive one of her employer's cars through Kokomo, Indiana. In 1900, three women received driver's licenses: Anne Rainsford French of Washington, D.C., Mrs. John Howell Phillips of Chicago, Illinois, and Jeannette Lindstrom, who was only 13 years old.

As the Victorian era ended and the women's rights movement gained strength during the early twentieth century, even more women decided to become motorists. They were aided by changes in clothing that made it easier for them to handle the dust generated by their open cars, as well as by the establishment of automobile clubs that supported women's driving activities. One such club, the Women's Automobile Club of New York, was involved in a major milestone in automotive history. In 1909 the club president, Alice Heyler Ramsey, became the first woman to drive across the United States, accompanied by three other club members. The women traveled from New York to San Francisco in a Maxwell touring car whose fuel tank had been enlarged to hold 22 gallons of gas; their journey took them 3,800 miles in 41 days, during which they changed tires and navigated unmarked roads without assistance from any men. However, the Maxwell Company, which had arranged the trip to promote its cars, did hire publicity expert John D. Murphy to plan the trip and arrange for hotels and refueling along the way. The following year, Blanche Scott became the first woman to drive an automobile across the United States without such help and is listed in record books as being a solo driver, although she was accompanied by a journalist who wrote about her exploits along the way. Like Ramsey, Scott traveled from New York to San Francisco.

Around the same time, automobile races were becoming popular among women. Thirteen women drivers participated in a long-distance race from New York to Philadelphia in 1909. The winner was Mrs. J. Newton Cuneo, who drove a Lancer automobile; four cars failed to complete the race. Cuneo had become famous just four years earlier, in 1905, for being the only woman out of 77 contestants to compete in the first Glidden Tour, a 26-day automobile race from New York City to St. Louis, Missouri. However, she failed to complete the race; when a car stalled in front of her, she slammed into it and flipped her own vehicle into a creek. Another competitive motorist during this time was mountaineer Elizabeth Le Blond, who participated in a popular automotive sport that involved driving cars up mountainsides.

Sightseeing and Longer Journeys

Automobiles were also used for sightseeing, and many women travelers of the early twentieth century went on pleasure tours by car. One of them, Mildred Mary Bruce, drove a car into remote regions of Lapland and the Sahara, where no one else had ever driven. She wrote a book about her experiences, *Nine Thousand Miles in Eight Weeks: Being An Account of an Epic Journey by Motor-Car Through Eleven Countries and Two Continents* (1927). Another unusual driving experience was a feat accomplished by Adeline and Augusta Van Buren, who in 1916 became the first women to travel across the United States on motorcycles. As part of that trip, which took them, in 60 days, from New York City to San Francisco, California, they also became the first women to drive up Pikes Peak. Accomplishments like theirs proved that women could partcipate in adventures commonly associated with men and consequently furthered the early feminist movement.

Books about the adventures of women motorists also encouraged other women to drive vast distances. For example, one popular work of 1923, *Through Algeria & Tunisia on a Motor-Bicycle* by Englishwoman Lady Warren, reports on a 1,707-mile trip that the author and her husband made in a motorcycle and sidecar through Algeria and Tunisia in 1921. The book describes scenery and customs as well as various problems the couple experienced with their motorcycle, which included a broken connection between the bike and the sidecar that caused the latter to swerve and somersault. But despite such mishaps, the couple made it through their trip safely, averaging 99 miles a day at 19 miles per hour and spending an average of just over six hours on the road every day. The author suggested that other women make equally adventurous journeys. Similarly, Englishwoman Rose Macaulay wrote about driving alone through Spain in *The Fabled Shore* (1949), and other women subsequently followed her route for their own solo driving adventures.

Today women continue to embark on such journeys. For example, in 1996 Australian Kay Forwood set off with her husband Peter to travel around the world on a Harley Davidson motorcycle and, as of December 1999, was still on the road. Support for today's motorists is offered by the modern version of the auto club, which emphasizes roadside assistance. Britain's Royal Automobile Club (RAC) provided the first telephone box for roadside assistance in 1919, and the Automobile Club of Southern California began installing road signs with directions and mileage information in California in 1905. The American Automobile Association (AAA) (www.aaa.com), formed in 1902, provides towing and roadside aid. Although women and men use such services equally, women arguably gain more psychological comfort from their availability, and as a result the existence of automobile clubs has helped encourage ordinary women—not just to adventurers like Ramsey and Scott—to travel cross-country.

In fact, by 1915 automobile travel had become so commonplace and so affordable that many women began using their cars for everyday outings, such as driving to work or to shops. After World War II, insurance companies, oil companies, and national retailers formed their own automobile clubs to encourage both men and women throughout the world to start driving. These businesses benefited economically from the resulting increase in motorists, but women have also benefited, because automobiles give them increased mobility and freedom.

Employment

Women have found many job opportunities within the modern transportation industry. For example, in 1922 Helen Schulz established the first woman-owned busline, Red Ball Transportation Company. She was denounced by railroad executives who said it was improper for a woman to drive a bus, when in actuality they were upset that Schulz was cutting into their business. Schulz's bus line, which initially traveled between two cities in Iowa, was extremely popular with women passengers.

During World War II, the number of women working in transportation increased dramatically—from 12 million to 18 million. Women could be seen operating all kinds of vehicles, from street cars and taxicabs to cranes and tractors. In modern times, women have continued to work in the transportation industry, not only as vehicle operators but also as decision makers. In 1977 Joan Claybrook became the first woman to head the National Highway Traffic Safety Administration, and in 1983 Elizabeth Dole became the first female U.S. Secretary of Transportation. Also in 1983, Carmen Turner became the first African American woman to head a major transportation agency, the Washington Metropolitan Area Transit Authority in Washington, D.C.

Modern women also continue to invent new transportation technologies. In 1938, for example, Katherine Burr Blodgett developed the first nonreflecting glass, which is used in car windows and windshields. American chemist Edith Flanigen has spent over 40 years working for Union Carbide, producing inventions related to fuel efficiency, gasoline cleanliness, and lubricating oil effectiveness. Many other women currently work to improve vehicle efficiency, roadway conditions and design, and passenger comfort and safety. *See also* Bruce, Mildred Mary; Le Blond, Elizabeth; Macaulay, Rose; Ramsey, Alice Heyer; Scott, Blanche; Strahorn, Carrie Adell.

Further Reading: Mathison, Richard R. *Three Cars in Every Garage*. New York: Doubleday, 1968; Warren, Lady. *Through Algeria & Tunisia on a Motor-bicycle*. Boston and New York: Houghton Mifflin, 1923; Withey, Lynne. *Grand Tours and Cook's Tours: A History of Leisure Travel, 1750–1915*. New York: W. Morrow, 1997.

Travel Writers

Up until the mid-twentieth century, it was customary for the most adventurous women travelers to write travel books about their experiences. This had more to do with social conventions than with the travelers' desire to write. Prior to modern times, travel for sport was considered a male activity; women's travel was acceptable only if it was justified by some worthwhile purpose, such as learning about something and then sharing that knowledge. As scholars I. S. MacLaren and Lisa N. LaFramboise explain in *The Ladies, the Gwich'in, and the Rat: Travels on the Athabasca, Mackenzie, Rat, Porcupine, and Yukon Rivers in 1926* (1998), British women earned the right to travel

> by invoking the Victorian ethics of self-improvement and community service. In a society that identified travel outside the domestic sphere with male activity, women who wished to travel often offered socially useful excuses. Famous traveller Isabella Bird Bishop . . . justified her Asian travels with the excuse of missionary work; Kate Marsden succoured lepers in Siberia; and Marianne North produced paintings of tropical flowers. . . . Botanizing, sketching in watercolours, or recording ladylike "impressions" of the exotic spaces of the world could justify an otherwise inappropriate delight in leaving home to wander abroad. (p. xxxvii)

Collecting plant specimens, painting travel scenes, and gathering material for travel books were all acceptable reasons for women to visit exotic places, and many women engaged in one or all of these activities. For example, British artist Gwendolen Dorrien Smith painted and sold watercolors based on her travels and also collected plant specimens for botanical gardens. However, most women found that writing books had a benefit other pursuits did not: it paid well. For example, during the early nineteenth century, Englishwoman Louisa Stuart Costello, who is considered by many historians to be the first woman to make a living writing travel books, intended to support her family with her paintings but found authorship more lucrative, as did English travel writer Constance Cumming, who visited such exotic places as India, Fiji, Ceylon, Tonga, Samoa, Tahiti, Japan, and China, and illustrated her books with watercolors she had painted during her travels. English travel writer Mary Edith Durham also illustrated her own books; she had been trained as an artist at the Royal Academy of Arts in London, and her naturalist paintings earned her some acclaim during her lifetime.

Some women travel writers had an even more socially acceptable reason to travel: they were accompanying their husbands. For example, Englishwoman Maria Graham, who wrote *Journal of a Residence in India* (1812) and *Journal of a Voyage to Brazil and Residence There During Part of the Years 1821, 1822, 1823* (1824), was married to a sea captain. Englishwoman Mrs. F. D. Bridges and her husband traveled throughout the world from 1877 to 1880, and when they returned from their world tour she published her diary writings as *Journal of a Lady's Travels Round the World* (1883).

In addition to needing to justify their voyages, another feature that distinguished women travelers from men was what they sought from their travel experiences. During the nineteenth and early twentieth centuries, men often traveled to prove something to themselves by mastering the wilderness, while women primarily sought a chance to get out of the house. This difference between men's and women's motivations for travel resulted in differences in their travel literature. Men's travel books of the period offer many exciting and dangerous adventures, while women's travel books primarily concern themselves with observations of scenery, people, and cultures. The works of Austrian Ida Pfeiffer, American Ann Wharton, and Englishwomen Isabella Lucy Bird Bishop, Isabella Christie, Mabel Crawford, Frances Trollope, and Ethel Tweedie are examples of women's travel writing that places the author in the role of observer. Some of these writers also offered travel advice, and a few women, most notably American Anne Royall and Englishwoman Mariana Stark, wrote guidebooks that told readers which places to visit and which to avoid. As the twentieth century approached, however, women travel writers began to discuss the political situations in the places they visited. Englishwoman Florence Dixie, for example, discussed not only her travel experiences but also British policies in South Africa in her books *A Defence of Zululand and Its King: Echoes from the Blue Books* (1882) and *In the Land of Misfortune* (1882).

The practice of including political commentary in travel writing continued as the century progressed. For example, Irish travel writer Dervla Murphy mentioned the political situation in Tibet in her travel books of the 1960s, which include *Full Tilt: Ireland to India with a Bicycle* (1965), *Tibetan Foothold* (1966), and *The Waiting Land: A Spell in Nepal* (1967). Similarly, Mary Russell comments on the political situation in Russia while describing her travel experiences in *Please Don't Call It Soviet Georgia: A Journey Through a Troubled Paradise* (1991). British travel writer Christina Dodwell also mentions various political situations in her work, although politics are not the main focus of her writings. Her books include *Travels with Fortune: An African Adventure* (1979), *In Papau New Guinea* (1983), *A Traveller in China* (1985), *A Traveller on Horseback: In Eastern Turkey and Iran*

(1987), *Travels with Pegasus: A Microlight Journey Across West Africa* (1990), *Beyond Siberia* (1993), and *Madagascar Travels* (1995).

Another trend in modern women's travel writing is to include intensely personal comments and memories along with observations about the place being visited. For example, although *A Journey with Elsa Cloud* by American Leila Hadley (1997) describes the author's trip to India, its focus is her attempts to reconnect with the estranged daughter she was visiting. Similarly, Barbara Grizzuti Harrison's 1989 book *Italian Days* not only describes her travels through Italy but also her attempts to learn about her heritage as the daughter of Italian-American parents.

In setting out on journeys for such personal reasons, these travel writers are far different from such earlier travel writers as Australian Mary Gaunt, Englishwoman Dorothy Mills, and Englishwoman Freya Stark. Gaunt and Mills were African explorers; Gaunt's *Alone in West Africa* (1912) describes two expeditions to Africa in 1908 and 1910 while Mills—the first white woman to visit Timbuktu (in present-day Mali) and parts of West Africa and South America—is the author of *The Road to Timbuktu* (1924), *Through Liberia* (1926), and *The Country of the Orinoco* (1931). Freya Stark was an archaeologist who wrote about her travel adventures in such books as *The Valleys of the Assassins and Other Persian Travels* (1934), *Baghdad Sketches* (1937), *Letters from Syria* (1940), *A Winter in Arabia* (1940), *Ionia: A Quest* (1954), *Riding to the Tigris* (1959), and *The Minaret of Djam: An Excursion in Afghanistan* (1970). These women were interested in finding new geographical landscapes, while modern travel writers—living in a time when there are few new places to explore—concentrate on spiritual landscapes instead. *See also* Bishop, Isabella Lucy Bird; Costello, Louisa Stuart; Crawford, Mabel Sharman; Cumming, Constance; Dixie, Florence; Dodwell, Christina; Durham, Mary Edith; Gaunt, Mary; Graham, Maria; Hadley, Leila; Harrison, Barbara Grizzuti; Marsden, Kate; Mills, Dorothy; Murphy, Dervla; North, Marianne; Pfeiffer, Ida Reyer; Royall, Anne Newport; Smith, Gwendolyn Dorrien; Stark, Freya Madeline; Stark, Mariana; Trollope, Frances; Tweedje, Ethel Brilliana.

Further Reading: Bridges, Mrs. F. D. *Journal of a Lady's Travels Round the World*. London: J. Murray, 1883; Dixie, Florence. *In the Land of Misfortune*. London: R. Bentley and Son, 1882; Dodwell, Christina. *Beyond Siberia*. London: Hodder and Stoughton, 1993; ———. *In Papau New Guinea*. Somerset, England: Oxford Illustrated Press, 1983; ———. *Madagascar Travels*. London: Hodder and Stoughton, 1995; ———. *A Traveller on Horseback: In Eastern Turkey and Iran*. New York: Walker, 1989; ———. *Travels with Fortune: An African Adventure*. London: W. H. Allen, 1979; ———. *Travels with Pegasus: A Microlight Journey Across West Africa*. New York: Walker, 1990; Gaunt, Mary. *Alone in West Africa*. London: W. T. Laurie, 1912; Graham, Maria. *Journal of a Residence in India*. Edinburgh: A. Constable, 1813; ———. *Journal of a Voyage to Brazil and Residence there During Part of the Years 1821, 1822, 1823*. 1824. Reprint. New York: Praeger, 1969; Hadley, Leila. *A Journey with Elsa Cloud*. New York: Penguin Books, 1997; Harrison, Barbara Grizzuti. *Italian Days*. New York: Weidenfeld and Nicolson, 1989; MacLaren, I. S., and Lisa N. LaFramboise, eds. *The Ladies, the Gwich'in, and the Rat: Travels on the Athabasca, Mackenzie, Rat, Porcupine, and Yukon Rivers in 1926 by C. C. Vyvyan*. Edmonton: The University of Alberta Press, 1998; Mills, Dorothy. *The Country of the Orinoco*. London: Hutchinson, 1931; ———. *The Road to Timbuktu*. London: Duckworth, 1924; ——— *Through Liberia*. London: Duckworth, 1926; Murphy, Dervla. *Full Tilt: Ireland to India with a Bicycle*. 1965. Reprint. Woodstock, NY: Overlook Press, 1987; ———. *Tibetan Foothold*. London: Murray, 1966; ———. *The Waiting Land: A Spell in Nepal*. Woodstock, NY: Overlook Press, 1987; Russell, Mary. *Please Don't Call It Soviet Georgia: A Journey through a Troubled Paradise*. London: Serpent's Tail, 1991; Stark, Freya. *Ionia: A Quest*. London: J. Murray, 1954; ———. *Letters from Syria*. London: J. Murray, 1942; ———. *The Minaret of Djam: An Excursion in Afghanistan*. London: J. Murray, 1970; ———. *Riding to the Tigris*. London: J. Murray, 1959; ———. *The Valleys of the Assassins and Other Persian Travels*. 1934. Reprint. Los Angeles: J. P. Tarcher; Boston: Houghton Mifflin, 1983; ———. *A Winter in Arabia*. 1940. Reprint. Woodstock, NY: Overlook Press, 1987.

Travelers' Tales Guides

Published by Travelers' Tales, Inc., the Travelers' Tales guides are a series of award-winning books that offer first-person essays and tips from travelers. Within this series are subdivisions; some focus on specific countries, some on cities, and some on particular subjects such as food or women's travel.

Books in the women's travel subdivision include *A Woman's World* (1997), *A Mother's World* (1998), and *Women in the Wild* (1998). Edited by Marybeth Bond, *A Woman's World* offers essays by women travelers from throughout the world. Bond reports:

> In this anthology women relate their most intimate travel experiences, from the absurd to the sublime. . . . The authors come from many walks of life. They are writers, doctors, teachers, nurses, athletes, young and old, mothers and grandmothers, novice and well-seasoned travelers, women traveling alone, in a couple, in a group. But whatever their background or mode of travel, in each story a female voice resonates. . . . (p. xvi)

The book is organized into five parts. The first, "Essence of Travel," contains essays that address the meaning of travel from a women's perspective. The second, "Some Things to Do," focuses on enjoyable activities for women travelers. The third, "Going Your Own Way," offers unusual travel experiences. The fourth, "In the Shadows," presents stories that show the dangers of travel for women. The fifth, "The Last Word," has just one essay, "The Next Destination" by Ann Jones, which concerns a woman's desire to visit new places. Other material in *A Woman's World* include an excerpt from Robyn Davidson's book *Tracks* (1980), about her experiences crossing the Australian Outback, and an essay by environmentalist and nature writer Gretel Ehrlich about significance of wilderness travel.

A Mother's World is more tightly focused, concentrating on the stories of women traveling with children. Editors Marybeth Bond and Pamela Michael describe the essay collection by saying:

> These "journeys of the heart" illuminate, celebrate, lampoon, and examine a rich panoply of human experience, often exploring terrain not usually covered in travel writing—a joint decision made in Italy to conceive or not to conceive, camping alone with a one-year-old on the rugged Northern California coast, realizing an aging mother's mortality while trying to navigate a Cairo marketplace. . . . Mother Nature (and mother nature) is revealed again and again in these stories—as goddess, healer, teacher. And finally, mothers speak of the pain of letting go, separation, the empty nest, and the "last goodbye." The combined voices in *A Mother's World* remind us that traveling—like motherhood—is not always easy, but certain to be an endless fascinating and challenging experience. (p. xvii)

Women in the Wild is also tightly focused, but in this case the essays revolve around wilderness experiences. Editor Lucy McCauley explains that the authors in her collection use their experiences to "affirm [their] self-sufficiency and creative force" by getting in touch with their own instincts. (p. xiv) She adds:

> They offer stories of high adventure in the wild—rafting a river in Borneo, diving in Mexican cenotes, climbing Mt. Everest. They share with us their journeys into the natural world: a lone hiker trekking the Appalachian Trail, a wildlife worker who is attacked by a hyena in Israel, a traveler who pulls off a gutsy rescue of endangered animals in Vietnam. They give us stories of women exploring the wilderness of their own natures—the affinity a mother-to-be feels with skunks, bears, and mice, a woman who confronts her hunter's instincts while fishing for mackerel in Ireland. . . . These tales transform possibility into reality. They remind us that there is a seasonal longing in our natures . . . to foray into the wild, both within and without. (p. xiv)

Other titles in the Travelers' Tales series that feature women's essays include *Gutsy Women: Travel Tips and Wisdom for the Road* (1996) and *Gutsy Mamas: Travel Tips and Wisdom for Mothers on the Road* (1997), both edited by Marybeth Bond.

Further Reading: Bond, Marybeth, ed. *A Woman's World*. San Francisco: Travelers' Tales, 1997; Bond, Marybeth, and Pamela Michael, eds. *A Mother's World: Journeys of the Heart*. San Francisco: Travelers' Tales, 1998; McCauley, Lucy, ed. *Women in the Wild*. San Francisco: Travelers' Tales, 1998.

Tristan, Flora (1803–1844)

Born in 1803, French political and social reformer Flora Tristan is representative of women who traveled for a cause. During the 1830s she journeyed throughout her native country promoting workers' and women's rights, keeping a journal of her experiences that was published posthumously as *Tour de France* (1925).

Tristan's interest in human rights was spurred during a trip to Peru in 1833. While there she noted injustices in the treatment of different classes of people. After returning to Paris in 1834, she wrote a book about her travels, *Peregrinations of a Pariah, 1833–34*, which was published in in 1838. By this time, she had spent more than 10 years engaged in a child custody battle with her ex-husband, who became so upset over the struggle for their two children that he tried to kill Tristan and was sent to prison. Tristan lived the last years of her life in relative obscurity and died in 1844.

Further Reading: Strumingher, Laura. *The Odyssey of Flora Tristan*. New York: P. Lang, 1988; Tristan, Flora. *Flora Tristan, Utopian Feminist: Her Travel Diaries and Personal Crusade*. Selected, translated, and with an introduction to her life by Doris and Paul Beik. Bloomington: Indiana University Press, 1993.

Trollope, Frances (1780–1863)

Born in 1780, English author Frances Trollope wrote a series of travel books as well as 34 novels. Her first work was *Domestic Manners of the Americans* (1832), which was based on her experiences visiting United States from 1827 to 1830. She went there to found a department store in Cincinnati, Ohio, called the Bazaar. Slightly more than three years later, the Bazaar had failed; all of Trollope's money was gone and her furniture had been seized. Destitute and dependent upon the charity of friends, she returned to England. By this time, however, she had already begun to write *Domestic Manners of the Americans*, and its publication rescued her financially.

About the success of the work, historian Donald Smalley writes in a 1949 introduction to the book:

> Britons . . . read her book and relished it for the racy genre pictures of American life with which its pages abound. Before the end of the year, *Domestic Manners* had run through four editions, "Yankeeisms" were a conversational fad, and Mrs. Trollope was a literary lion, well launched at the age of fifty-two upon her remarkable career as a writer of novels and travel books. . . . In the United States an outraged citizenry read *Domestic Manners* as they had read no travel book before it. . . . Newspapers in every section of the country made a pastime of reviling Mrs. Trollope, though they also quoted long sections from her book. Even the polished quarterlies denounced her "coarse exaggeration" and "bitter cari-

Frances Trollope in an 1852 engraving. *CORBIS*.

cature"; A Western editor indexed his review under "Lies of an English lady." (pp. viii-ix)

Smalley quotes a writer of the period, British Lieutenant E.T. Coke, who was in New York when the book appeared there as saying:

> At every table . . . , on board of every steamboat, in every stage-coach, and in all societies, the first question was "Have you read Mrs. Trollope?" And one half of the people would be seen with a red or blue half-bound volume in their hand, which you might vouch for being the odious work; and the more it was abused the more rapidly did the printers issue new editions. (p. ix)

In fact, Trollope was so reviled in America that cartoons and satirical works poked fun at her, and the word "Trollope" was often used to deride someone for behaving improperly. Americans who had met with Trollope in person described her as "rough-cast and misshapen," "of coarse and vulgar expression," "singularly unladylike," and "robust and masculine," with a shrill voice and sarcastic tone. (p. xxxix) However, Smalley suggests that criticisms of Trollope were unfair, because she did not intend to insult Americans. Instead she merely gave her honest impressions of Americans, particularly of those in Cincinnati. Because her experiences with the Bazaar were so bad, they affected her view of the people she encountered. Moreover, some Americans of the period, such as author Washington Irving, accepted Trollope's criticisms as at least somewhat valid, and most modern historians believe that she offers one of the most honest portraits of ordinary life in the United States. Consequently the book was successful among people in Europe who had never visited America and was translated into French and Spanish.

Trollope's second travel book, *Belgium and Western Germany in 1833* (1839)was also successful, and she went on to write many more travel books as well as novels. In all, she published 113 volumes. In her later years she lived in Italy, where she died in 1863. *See also* NORTH AMERICA; TRAVEL WRITERS.

Further Reading: Heineman, Helen. *Frances Trollope*. Boston: Twayne, 1984; Trollope, Frances. *Domestic Manners of the Americans*. 1832. Reprint. New York: Alfred A. Knopf, 1949.

Tubman, Harriet (ca. 1820–1913)

Born in approximately 1820, Harriet Tubman is significant in terms of travel because she helped many African American make their way from slave plantations in the South to freedom in the North. She herself was an African American slave who escaped from a Southern plantation in 1849 and went on to become a leading abolitionist in the North. She initially reached the North via the Underground Railroad, a secret system for guiding slaves along the route to freedom. This system was not, as some people mistakenly believe, a series of underground tunnels, nor was it an actual railroad; the word "underground" refers to its secret nature, and "railroad" refers to the fact that its users all employed railroad terms as code words for various activities. For example, slaves were called packages, while hiding places en route were called stations, and the people who helped guide the slaves were called conductors.

Once she became a free woman, Tubman helped many other slaves make the same journey along the Underground Railroad, personally conducting more than 300 escapees north, and several prominent Southerners offered rewards for her capture. During the Civil War, Tubman worked as a spy for the United States government along the coast of South Carolina, pretending to be a laundress. After the war she fell into poverty, although shortly before her death she was granted a government pension because of her war work. She died in 1913.

Further Reading: Blockson, Charles L. *The Underground Railroad*. New York: Prentice-Hall, 1987; Janney, Rebecca Price. *Harriet Tubman*. Minneapolis, MN: Bethany House Publishers, 1999.

Tullis, Julie (1939–1986)

Born in 1939, English mountaineer Julie Tullis was the first woman to climb Broad Peak on the Pakistan-China border. At 26,470 feet, it is one of the tallest mountains in the world. No woman prior to Tullis had climbed as high, an experience she wrote about in a book, *Clouds from Both Sides* (1986). Tullis specialized in climbing at extremely high altitudes, scaling peaks not only in Asia but also in the Alps, the Andes, and the Himalayas. Tullis was killed in 1986 while descending from K2 in central Asia; at 28,253 feet it is the second-highest mountain in the world. *See also* MOUNTAINEERING.

Further Reading: Tullis, Julie. *Clouds From Both Sides*. London: Grafton, 1986.

Tweedie, Ethel Brilliana (? –1940)

Englishwoman Ethel Brilliana Tweedie (later Alec-Tweedie) was one of the most prolific female travel writers of the late nineteenth and early twentieth centuries. She wrote many articles and books, the latter of which include *Through Finland in Carts* (1897), *Mexico As I Saw It* (1901), *America As I Saw It* (1913), and *An Adventurous Journey (Russia-Siberia-China)* (1926), before her death in 1940.

An Adventurous Journey is perhaps her best known work and includes Tweedie's own watercolor sketches and photographs. The text describes her travels in Russia and China from April 1925 to January 1926. Her intent was to reach Peking, China, as quickly as possible, so that she could accomplish two tasks: paint scenes there for a June 1926 art exhibition in Paris and finish a book she had been writing about China. However, en route through Russia, she became interested in the political ties between the two countries. She explains:

> I had no intention of writing anything about Russia when I went there—I was on my way to China, to both paint and write. But Russia so moved me, and so amazed me, that write I had to. Circumstances forced me. And week by week the veil dropped from my eyes, and I saw China was Russia, or rather Russia was China. The two could not be divided. (p. 27)

Consequently *An Adventurous Journey* not only describes Tweedie's travel adventures but also criticizes the political system of the two countries, blaming these systems for widespread social problems. For example, in describing a scene of street vendors, Tweedie takes the opportunity to condemn child labor, saying:

> A boy of about ten pulling a rickshaw is a sad sight, and not so uncommon as it should be, for youth begins to work far too early. Even toddling children thread beads or carve peach stones in the native-owned shops—sweated labour with a vengeance on every side. (p. 368)

Tweedie also compares China and Russia to Japan, which she believes is the superior Asian country. She states: "Japan . . . is the land of progression—China the land of retrogression. Russia is the land of destruction." (p. 368) Tweedie predicts that Japan will be at war with the United States before 1930 and argues that Japan will be victorious.

Tweedie concludes by describing her departure from China. Returning home, she reflects on the price she has paid for her years of travel, saying:

> These seven years have produced two books and several picture exhibitions, and a trunk has become a nightmare. A passport a nuisance. Strange beds a worry, and tips to servants a curse. Twice I have circled this globe—have side-tracked north and south, east and west, have endured every climate, and tackled several tongues. . . . Seven years are a lifetime with friends in the passing. And the future? . . . Luckily, no man knoweth his future. And kills time till time graciously kills him. (pp. 395-396)

Yet Tweedie also expresses a reluctance to leave her friends in China, and her book ends with her emotions in conflict. In this regard, the book is typical of of Tweedie's works; she is often more concerned with emotions that with basic descriptions. *See also* ASIA.

Further Reading: Alec-Tweedie, Ethel. *An Adventurous Journey (Russia-Siberia-China), with Four Water Colour Sketches by the Author, Two Maps and Sixty-Six Other Illustrations*. London: Hutchinson, 1926; ———. *America as I Saw It; Or, America Revisited*. New York: Macmillan, 1913.

V

Victorian Era

The Victorian era was a period of dramatic increase in women's travel and exploration. Named for Queen Victoria (1819–1901), it was a time when Great Britain aggressively promoted exploration and colonization, and many men who came home from adventures abroad were rewarded with fame, honors, and sometimes wealth. Consequently, women came to view travel as something desirable. As historian Dea Birkett explains in *Spinsters Abroad: Victorian Lady Explorers* (1989):

> Raised in a Britain of confidence and Empire, expansion and conquest, [the foremost women travelers of the period] shared common perceptions of a foreign and, as yet, untouchable world. Many of their families were connected, directly or indirectly, with British imperial expansion. . . . Women prepared themselves for their journeys by drawing upon the plethora of visual and written images of exotic places and peoples offered in Victorian Britain. These provided the ingredients for the imaginary arenas in which they could act out their most daring dreams and adventures, and forge a picture of themselves as travellers. (pp. 19–20)

Birkett also points out that the rigid social codes that prevailed under Queen Victoria, who believed that duty came before pleasure, dictated that women bear heavy responsibilities—responsibilities that could only be escaped honorably by travel. She says:

> While women travellers were able to paint canvasses on which to exercise their dreams of active and fulfilling lives, theirs was an uneasy inclusion in the myth of exploration and discovery. They were continally [sic] torn between the two conflicting landscapes of self-fulfillment and duty. These clashing voices would never be resolved, and their passage from their rooms to the world outside was a troubled one. . . . Dutiful daughters, loving sisters and maiden aunts would be haunted by a responsibility so heavily felt as young women and demanded of them by the prevailing though often unspoken demands on women of the time. They could only shirk this early load by geographically removing themselves from the society in which it operated. But in shedding the load they acquired the heavy burden of the guilt of its shedding. (p. 27)

To assuage this guilt, Victorian women travelers typically justified their travels by performing scientific research while abroad or by writing about their experiences with the expressed intent of educating others. By having a noble purpose, they felt better about escaping their duties. Consequently many of these women collected plant, insect, and animal specimens for museums while traveling, or returned home to write didactic travel books, often illustrated by their own artwork—another way of showing that they had indeed been involved in productive activities while abroad.

Representative of these Victorian travelers are Englishwomen Gertrude Bell, Isabella Bird Bishop, Amelia Edwards, Mary Kingsley, and Marianne North and Frenchwoman Alexandra David-Neel. Each one had a seemingly purposeful reason to travel, but in actuality all really sought the pleasure of travel as well as an escape from responsibility. Bell was an archaeologist, Bishop a travel writer, David-Neel a missionary, and Kingsley an explorer, while Edwards and North produced artwork based on their travels. But once their work was done, they continued to seek out travel opportunities. In their later years, when they were no longer young enough or healthy enough to travel, all of these women continued to chafe at the restrictions imposed upon them by Victorian society. Birkett reports:

> Crippling disease was especially painful for women to whom strength and movement had meant mental as well as physical freedom. Many shared a horror of being buried on land, locked in the dismal grey vaults of Victorian Britain, decorated with posies and poems, symbols of the lives they fled. While Mary Kingsley asked to be buried . . . [at sea], Alexandra David-Neel, at the age of 101, asked her devoted secretary to scatter her Buddhist ashes into the flow of the Ganges [River]. Even in death, they wanted to be travellers, on the move. (p. 275)

After the Victorian era, women were much freer to travel than before. Yet ironically, the coming of the modern age also brought a dwindling supply of unexplored regions, making it more difficult to women to find adventures to match those of their Victorian predecessors. *See also* BELL, GERTRUDE; BISHOP, ISABELLA LUCY BIRD; DAVID-NEEL, ALEXANDRA; EDWARDS, AMELIA; GREAT BRITAIN; KINGSLEY, MARY; NORTH, MARIANNE; QUEENS.

Further reading: Birkett, Dea. *Spinsters Abroad: Victorian Lady Explorers.* Oxford: Basil Blackwell, 1989.

Vikings (Alfhild)

Alfhild was perhaps the most prominent of several female Vikings who roved the seas plundering villages and engaging in trade and/or battles. Also known as Alvild or Alwilda, she was a ninth century Danish princess who, when asked to marry a Danish prince, rebelled by donning men's clothing and becoming a warrior of the sea. Moreover, she gathered together enough other female Vikings to form a complete all-woman crew. About these women's new lifestyle, Joan Druett, in her book *She Captains: Heroines and Hellions of the Sea* (2000), says:

> Obviously, in opting to abandon a soft . . . life at the palace to take on this kind of existence, Alfhild and her companions had accepted quite a challenge. The Norsemen were consummate seamen, navigating by the sun, the stars, the tides, the ocean currents, and the migratory patterns of birds and whales, so the women had a great deal to learn. Viking rovers were hardy, too, sleeping in leather sleeping bags with their weapons close to their hands. This was usually on some deserted beach, after their ships had been drawn up on the sand and lashed together for safety, because longships [Viking boats] were not well designed for stretching out full-length. It was very difficult to cook in longships, too, so . . . raw meat . . . was the rule. . . . Somehow, Alfhild managed. For weapons she would have had swords, axes, bows and arrows, and spears, and in battle she would have worn a horned helmet and perhaps an iron breastplate. She must have had a feasting hall somewhere, even if it was some humble and secluded hut made of mud and wattle, for she and her companion valkyria [women warriors] to recruit their strength, bury their treasure, and brag about their deeds. . . . Without a doubt, Alfhild and her force would have created terror and havoc wherever they landed, and the villagers and monks who fled from their ravening screams and slashing weapons would have had no idea they were women. (pp. 37–38)

Alfhild's exploits as a warrior ended when the man she had been supposed to marry, Prince Alf, attacked and boarded her ship. He forced himself on her, claimed her as his wife, and took her back to Denmark, where she soon gave birth to a daughter. Many of her crew met a worse fate; they were killed by Alf's boarding party, which mistook the women for men.

Further Reading: Druett, Joan. *She Captains: Heroines and Hellions of the Sea*. New York: Simon & Schuster, 2000.

Visser-Hooft, Jenny (1888–1939)

Born in 1888, Dutch mountaineer Jenny Visser-Hooft explored, mapped, and researched the flora and fauna of the Karakorum Glaciers in Pakistan and India. She and her husband, a physician and diplomat, made four expeditions to the region during the 1920s, writing about some of their experiences in *Among the Kara-Korum Glaciers in 1925* (1926). Visser-Hooft also climbed other mountains in central Asia, as well as in the Alps and the Caucasus. She died in 1939. *See also* MOUNTAINEERING.

Further Reading: Robinson, Jane. *Wayward Women: A Guide to Women Travellers*. Oxford: Oxford University Press, 1990.

Vyvyan, C. C. (1885–1976)

Born in 1885 as Clara Coleman Rogers, Lady C. C. Vyvyan was an adventurous traveler who wrote several books about her experiences. She published her first book in 1924; *Cornish Silhouettes* offered sketches of Cornwall in England. Although she was born in Australia, Vyvyan was reared in England and lived there her entire life. Consequently most of her writings concern the British Isles. However, she sometimes toured other countries, and these tours produced such works as *Temples and Flowers: A Journey to Greece* (1955), *Down the Rhone on Foot* (1955), and *Arctic Adventure* (1961; later retitled *The Ladies, the Gwich'in, and the Rat*). The last was by far Vyvyan's most adventurous and extensive journey.

Arctic Adventure describes a historic trip that Rogers took into Canada with friend Gwendolen Dorrien Smith in 1926. The two traveled from Cornwall, England, to Edmonton, Alberta, then into northern Alberta, the Northwest Territories, the Yukon, and across the Canadian north to Alaska. Once on the coast, they returned home by ship.

En route across Canada, Rogers and Smith went by train, boat, and canoe down the Athabasca, Mackenzie, Rat, Porcupine, and Yukon Rivers. Most of their trip was taken in the company of a series of guides they met at prearranged places along their route. However, for three days they traveled alone down part of the Porcupine River, an opportunity they welcomed. In *Arctic Adventure* Vyvyan writes:

> We were absolutely alone now in this wilderness. If we wanted help, there would be none forthcoming except from our own resourcefulness. If we met with an accident, no one would be any the wiser. If we should encounter rapids, snags or swift water, we should have to call on our own reserves of initiative and skill. I do not think that all these things were clearly in our minds as our guides walked away, but we did feel, as we looked about us, a queer depth and breadth of solitude that we had never known before. However, instead of dwelling on unknown difficulties ahead, we decided to look upon our solitude as complete freedom and to celebrate this freedom we undressed completely, stepped into the water, each with a cake of soap, and washed thoroughly from head to toe. (MacLaren and LaFramboise, p. 142)

But while their time on the Porcupine River was unique for its solitude, their time on the Rat River had more historical significance, according to scholars I. S. MacLaren and Lisa N. LaFramboise. In an introduction to Vyvyan's book, they say:

> While far from the first Europeans to travel on the Rat, Vyvyan and Dorrien Smith were two of the earliest *recreational* travellers on the transmontane route, and information concerning it was still not readily obtainable. So a keen sense of adventure still prevailed. (p. xxx)

MacLaren and LaFramboise also criticize Vyvyan for omitting Canadian history—particularly history related to women—from *Arctic Adventure*. They say:

> From the beginning of her book, and despite millennia of indigenous habitation (not to mention more than a century of European presence), Vyvyan presents her

> readers with a land where people leave no trace on the imagination: "Among all the people that we met I cannot remember any faces . . . yet certain rivers, mountains, forests, glaciers, birds and rapids, remain vivid. . . ." . . . Vyvyan's disinclination to learn more about the North before publishing *Arctic Adventure* in 1961 compounds this sense of vacancy. In combination with her disposition as a character sketcher to write about the adventure of men, this disinclination leaves the representation of women almost entirely out of her picture. In particular, women travellers, either before or after 1926, are all but absent from the book. Instead, Vyvyan leaves the impression that white women, if they were there at all, did not constitute a significant presence in the North. (pp. xlvi–xlvii)

MacLaren and LaFramboise present the full text of *Arctic Adventure,* along with detailed notes and commentary on the work, retitled as *The Ladies, the Gwich'in, and the Rat: Travels on the Athabasca, Mackenzie, Rat, Porcupine, and Yukon Rivers in 1926* (1998). They also report that after returning from Canada, Vyvyan wrote several articles about her experience and made plans to travel to other wilderness regions, including Labrador. In 1929, however, she met Sir Courtenay Bouchier Vyvyan and abandoned her travel plans to marry him. Until his death in 1941, she remained close to home and did not resume traveling until her later years. She made the walking tour described in *Down the Rhone on Foot*, for example, at age 67. Vyvyan died in 1976 at the age of 91.

Further Reading: MacLaren, I. S., and Lisa N. LaFramboise, eds. *The Ladies, the Gwich'in, and the Rat: Travels on the Athabasca, Mackenzie, Rat, Porcupine, and Yukon Rivers in 1926 by C. C. Vyvyan*. Edmonton: The University of Alberta Press, 1998.

W

Wall, Rachel (?–1789)

During the eighteenth century, American pirate Rachel Wall sailed along the eastern seaboard with her husband George, who attacked ships using Wall as a lure. After a storm, Wall's band of pirates would pretend to be ordinary seamen whose ship had been damaged; Wall was their distraught passenger. When a trading vessel came alongside to offer help, the pirates would attack and kill everyone on board—unless the crew on the trading vessel seemed too difficult to overcome, in which case the Walls pretended that they merely needed help fixing a leak. On one occasion, however, the pirates' ship really was damaged in a storm, and Wall's husband was swept overboard and drowned. Without him Wall lost interest in the sea, but back on land she remained a thief. She was soon arrested for stealing money from docked ships. Wall was hanged for her crimes in 1789. *See also* PIRATES.

Further Reading: Druett, Joan. *She Captains: Heroines and Hellions of the Sea*. New York: Simon & Schuster, 2000; Stanley, Jo, ed. *Bold in Her Breeches: Women Pirates Across the Ages*. London and San Francisco: Pandora, 1996; Wall, Rachel. *Life, Last Words, and Dying Confession of Rachel Wall: who, with William Smith and William Dunogan, were executed at Boston, on Thursday, October 8, 1789, for high-way robbery*. Boston, 1789.

Watson, Patty Jo (1932–)

Born in 1932, American anthropologist and archaeologist Patty Jo Watson is representative of women who traveled to further knowledge about ancient people. During the 1960s and 1970s, she went to excavations in Turkey, Iran, and Iraq to study Near Eastern prehistory, and, in subsequent years, she focused on exploring the Mammoth Cave system in Kentucky, seeking evidence related to the growth of maize and other crops in prehistoric times. She continues her research into this subject today, in an attempt to determine how ancient people cultivated maize and how this pursuit affected their culture.

Further Reading: Watson, Patty Jo. *Archaeological Ethnography in Western Iran*. Tucson: University of Arizona Press, 1979; ———, ed. *Archaeology of the Mammoth Cave Area*. St. Louis, MO : Cave Books, 1997.

Wayward Women

Published in 1990, *Wayward Women: A Guide to Woman Travellers* by Jane Robinson offers brief biographical information about women who have written about their own travel experiences. Approximately 400 women are mentioned; however, the book emphasizes women from Great Britain. In addition, travel writers are grouped according to their reason for traveling. For example, the chapter "Quite Safe Here With Jesus" lists women missionaries, and the chapter "In

Camp and Cantonment" lists women who traveled because of military campaigns. In discussing her book's organization, the author says:

> I rashly resolved to label these unruly individuals according to the sort of traveller they were on setting forth . . . But even these labels, it must be said, have a tendency to flutter away in the melee. One traveller frequently encroaches upon the territory of another, so that missionaries who write about their travels . . . might easily become confused with explorers who discover God on a mountain summit. Governesses and nurses fall prey to wanderlust on their way to an appointment in St. Petersburg or the Crimea, and a female foreign correspondent turns anthropologist with as much ease as an army officer's wife turns professional tourist. In fact one of the few things these women have in common is their very originality. (p. ix)

See also ANTHROPOLOGISTS AND ARCHAEOLOGISTS; EMIGRATION SOCIETIES (BRITISH); MISSIONARIES.

Further Reading: Robinson, Jane. *Wayward Women: A Guide to Women Travellers*. Oxford: Oxford University Press, 1990.

Wells, Fay Gillis (1908–)

Born in 1908, American aviator Fay Gillis Wells was the first airplane saleswoman; she traveled across the country in the late 1920s and early 1930s demonstrating aircraft. In the 1930s she went to Russia, where she became the first woman to fly a Soviet civil aircraft and worked as a journalist. In 1935 she married another American pilot and returned to the United States, where she covered presidential news. In 1972 she went with President Richard M. Nixon on an historic trip to China. Meanwhile she continued to fly as a hobby. *See also* AIR TRAVEL.

Further Reading: The Ninety-Nines: International Organization of Women Pilots. http://www.ninety-nines.org (May 2000); Welch, Rosanne. *Encyclopedia of Women in Aviation and Space*. Santa Barbara, CA: ABC-Clio, 1998.

West, Rebecca (1892–1983)

Born in 1892, Rebecca West is the pseudonym of Cicily Isabel Fairfield Andrews, who wrote about her experiences traveling with her husband through Yugoslavia in the late 1930s in *Black Lamb and Grey Falcon: A Journey Through Yugoslavia*. Published in two volumes in 1941, this work first appeared in *The Atlantic Monthly* magazine in 1940 and immediately increased interest in Yugoslavia among her peers.

Black Lamb and Grey Falcon not only recounts West's travel adventures but also offers an immense amount of detail about the scenery, people, and political conflicts at the time of her visit. In fact, it was West's interest in these political conflicts that guided her during the writing of her book. In its epilogue she writes:

> This experience made me say to myself, "If a Roman woman had, some years before the sack of Rome, realized why it was going to be sacked and what motives inspired the barbarians and what the Romans, and had written down all she knew and felt about it, the record would have been of value to historians. My situation, though probably not so fatal, is as interesting." Without doubt it was my duty to

Rebecca West in 1956. *Library of Congress.*

> keep a record of it . . . to put on paper what a typical Englishwoman felt and thought in the late nineteen-thirties when, already convinced of the inevitability of the second Anglo-German war, she had been able to follow the dark waters of that event back to its source. That committed me to . . . write a long and complicated history, and to swell that with an account of myself and the people who went with me on my travels, since it was my aim to show the past side by side with the present it created. (pp. 1088–1089)

The remainder of the book's epilogue discusses the problems facing Yugoslavia. It also underscores the theme presented in the book's title, presenting an untitled poem in which a grey falcon warns the leader of the country that a war will destroy his people, which West later refers to as black lambs sacrificed for religious reasons. But although West addresses contemporary issues, her book is still a work of travel literature, and as such it includes descriptions of scenery, maps of West's travel route, and the author's black-and-white photographs of various sites.

West also wrote nonfiction books on other subjects, as well as novels. Raised in Edinburgh, Scotland, she lived most of her life in England, where she died in 1983.

Further Reading: West, Rebecca. *Black Lamb and Grey Falcon: A Journey through Yugoslavia.* New York: Viking Press, 1941.

Wharton, Edith (1862–1937)

Born in 1862, American author Edith Wharton is best known for her novels, but she also wrote several travel books, including *In Morocco*. First published in 1920, this work was the first guidebook for Europeans traveling to Morocco. Wharton initially visited the country in 1917, when it was inhospitable to tourists. In writing a new edition of her guidebook in 1927, she reported that Morocco had gone through dramatic changes, saying:

> It is hard to believe that in the interval since my visit this guide-book-less and almost roadless empire has become one of the most popular and customary scenes of winter travel—travel by rail and motor; still more difficult to conceive that, in spite of its accessibility and its conveniences, it has kept nearly all of the magic and mystery of forbidden days. So skilfully, in fact, have the 5,000 kilometers of rail and road been insinuated into the folds of the brown hills, so tastefully and tactfully have crumbling Moorish palaces been transformed into luxurious modern hotels, that, from the vantage-ground of the new Morocco, the tourist may still peep down at ease into the old. (p. 15)

Both editions of *In Morocco* offer detailed descriptions of Moroccan sights, people, customs, and the author's reaction to them. For example, in describing her visit to Fez, the oldest city in Morocco, Wharton says:

> We visited old palaces and new, inhabited and abandoned, and over all lay the same fine dust of oblivion, like the silvery mould on an overripe fruit. Overripeness is indeed the characteristic of this rich and stagnant civilization. Buildings, people, customs, seem all about to crumble and fall of their own weight: the present is a perpetually prolonged past. To touch the past with one's hands is realized only in dreams; and in Morocco the dream-feeling envelops one at every step. (pp. 76–77)

Wharton also offers separate chapters on Moroccan architecture and history. In addition, she provides a list of books related to these two subjects, although most of these are in French.

A member of upper-class society, Wharton first traveled to Europe as a girl and continued to travel as an adult, primarily to study architectural and landscaping styles. From 1907 until her death in 1937, she lived in France, where she was one of the first women to go on a driving tour. She wrote a book about her automotive adventure, *A Motor-Flight Through France* (1908).

Further Reading: Auchincloss, Louis. *Edith Wharton: A Woman in Her Time.* New York: Viking Press, 1971; Joslin, Katherine, and Alan Price, eds. *Wretched Exotic: Essays on Edith Wharton in Europe.* New York: P. Lang, 1993; Wharton, Edith. *A Motor-Flight Through France.* DeKalb: Northern Illinois University Press, 1991; ———. *In Morocco.* London: Jonathan Cape, 1927.

Wheeler, Sara

Journalist Sara Wheeler writes travel articles and books. The latter include *Travels in a Thin Country: A Journey Through Chile* (1995), which concerns her solo journey through Chile; *An Island Apart: Travels in Evia* (1995), which describes her travels in Greece; and *Terra Incognita: Travels Through Antarctica* (1996), which is considered a classic of polar literature.

Terra Incognita came about because Wheeler decided to visit Antarctica while doing a book on Chile, and, once in the polar region, she had a new sense of attachment to a place. In *Terra Incognita* she writes:

> Standing on the edge of the ice field in a wind strong enough to lean on, squinting in the buttery light, it was as if I were seeing the earth for the very first time. I felt less homeless than I have ever felt anywhere, and I knew immediately that I had to return. (p. xiv)

Consequently Wheeler joined the American National Science Foundation's Antarctic Artists' and Writers' Program and went to live with a polar scientific team for seven months as the only woman at their research center. Wheeler talks about the scarcity of women at the poles in *Terra Incognita,* suggesting that this scarcity is attributable to chauvinism. She writes:

> Men had always wanted to keep Antarctica for themselves. . . . Sir Vivien Fuchs wrote in his book *Of Ice and Men,* published in 1982, "Problems will arise should it ever happen that women are admitted to base complement [a British research facility]," and when I sat drinking tea with him . . . before I left England he suggested that the answer was to put women in all-female camps. . . . Rear Admiral George Dufek, an early commander of U.S. operations on the ice, summed it up when he said: "I think the presence of women would wreck the illusion of the frontiersman—the illusion of being a hero." (p. 221)

Wheeler describes the difficulties she encountered as one of few women to have visited the Antarctic. But despite these difficulties, she found her Antarctic experience restorative. She concludes *Terra Incognita* by saying: "It was the greatest thrill of my life. . . . It had allowed me to believe in paradise, and that, surely, is a gift without price." (p. 334). *See also* POLES, NORTH AND SOUTH.

Further Reading: Wheeler, Sara. *Terra Incognita: Travels in Antarctica.* New York: Modern Library, 1999.

Winternitz, Helen (1951–)

Born in 1951, American reporter Helen Winternitz is best known for two travel books, *East Along the Equator: A Journey Up the Congo and Into Zaire* (1987) and *A Season of Stones: Living in a Palestinian Village* (1991). Both discuss regional politics as well as daily life. For the first, Winternitz traveled along the Congo River; for the second, she lived from 1988 to 1990 in Nahalin, a farming village near Bethlehem, Israel. In both cases, she lived among the indigenous people and learned not only their languages but also their concerns. In Nahalin, she discovered that almost every villager belonged to the Palestine Liberation Organization (PLO), and she reports on problems between them and the Israeli police. This type of information is juxtaposed with comments on local embroidery techniques and other everyday activities.

Further Reading: Winternitz, Helen. *East Along the Equator: A Journey Up the Congo and Into Zaire.* London: Bodley Head, 1987; ———. *A Season of Stones: Living in a Palestinian Village.* New York: Atlantic Monthly Press, 1991.

Wollstonecraft, Mary (1759–1797)

Born in 1759, Mary Wollstonecraft was an English feminist who went to Scandinavia in 1795 to conduct business for her first husband, Gilbert Imlay. The following year she published a book about her experiences, *Letters Written During a Short Residence in Sweden, Norway, and Denmark* (1796), which was popular in England. The work not only describes Wollstonecraft's travel adventures but also comments on the differences between Scandinavia and her native England, as well as between one Scandinavian country and another. For example, in writing about Sweden and Norway, she says:

> I could not avoid feeling surprise at observing the difference in the manners of the inhabitants . . . for every thing shows that the Norwegians are more industrious and more opulent. The Swedes, for neighbours are seldom the best friends, accuse the Norwegians of knavery, and they retaliate by bringing a charge of hypocrisy against the Swedes. Local circumstances probably render both unjust, speaking from their feelings, rather than reason: and is this astonishing when we consider that most writers of travels have done the same, whose works have served as materials for universal histories. (Hamalian, p. 24)

Wollstonecraft also visited France in 1792 to study women's rights there and became famous for a feminist book published the same year, *A Vindication of the Rights of Women*. She is also notable for her connection to two prominent authors: William Godwin was her second husband and Mary Shelley was her daughter. Wollstonecraft died in 1797 during childbirth.

Further Reading: Bouten, Jacob. *Mary Wollstonecraft and the Beginnings of Female Emancipation in France and England.* Philadelphia: Porcupine Press, 1975; Hamalian, Leo, ed. *Ladies on the Loose: Women Travellers of the 18th and 19th Centuries.* New York: Dodd, Mead, 1981.

Workman, Fanny Bullock (1859–1925)

Born in 1859, Fanny Bullock Workman is one of the foremost American women explorers. She came from a wealthy political family, and in 1881 married a Massachusetts physician. When he retired in 1889, the couple moved to Europe and became passionate travelers. Their favorite mode of travel was the bicycle, and they wrote several books together about their cycling tours through Spain, Italy, Algeria, and India. Their first book was *Algerian Memories: A Bicycle Tour over the Atlas to the Sahara* (1895), which describes a trip over the Atlas Mountains to the Sahara Desert. Subsequent works include *Sketches Awheel in Modern Iberia* (1897) and *Through Town and Jungle: Fourteen Thousand Miles A-wheel among the Temples and People of the Indian Plain* (1904).

The latter is one of the Workmans' most popular books. It describes the couple's bicycle trip through India to study the ruins of ancient temples, palaces, mosques, and tombs during 1901–1903. They also spent three months in the winter of 1903–1904 reshooting photographs that had been damaged during a flood. The book includes more than 200 of the couple's black-and-white photographs.

In all, the Workmans traveled more than 1,400 miles by bicycle, as well as by foot, train, steamboat, cart, and a variety of other conveyances. However, it was the cycling portion of their journey that was the most difficult. They explain:

Fanny Workman in the Karakoram with climbing equipment. *Library of Congress.*

> A cycle tour in India is quite a different thing from what it is in the countries of Europe, in Algeria, or even in Ceylon and Java. In all of these countries what may be called European conditions exist, i.e., one can always find shelter at night or something that passes under the name of an inn or hotel, where one's most pressing necessities are provided for. . . . In India hotels . . . are found only in the larger cities, which are comparatively few in number and scattered over a wide area. The cyclist has to find shelter on the greater part of his route in dark bungalows, the only places accessible to the public that represent an inn, which are by no means always to be found in localities convenient to him, or in inspection or engineer bungalows built at certain places for the use of Government officials when on duty, which can only be occupied by permission of the Executive Engineer or some other officer of the district, who usually lives too far away from the bungalow to admit of the required permission being readily obtained. . . . Failing to meet with a bungalow the cyclist may occasionally find a refuge in the waiting-room of a small railway station, containing two wooden chairs and a wooden bench, or he may be obliged to occupy the porch of some native building. On rare occasions a missionary or planter may take pity on him and lodge him, which hospitality he regards as a godsend and duly appreciates. (pp xi–xii)

Through Town and Jungle offers many details about the Workmans' difficulties, as well as information about Indian artifacts and culture. However, the authors specifically omitted criticisms of the Indian government from their work, arguing that "no traveller, unless he has lived a long time in a country and had special opportunities of studying its institutions, is in a position to understand fully the problems that confront its Government, much less to criticise the administration of its affairs." (p. xi) Even without political commentary, however, *Through Town and Jungle* is a lengthy work.

In addition to bicycling trips like the one described in *Through Town and Jungle,* the Workmans enjoyed mountain climbing, and they wrote books about these experiences too. These books include *Ice-Bound Heights of the Mustagh: An Account of Two Seasons of Pioneer Exploration and High Climbing in the Baltistan Himalaya* (1908), *Peaks and Glaciers of the Nun Kun: A Record of Pioneer Exploration and Mountaineering in the Punjab Himálaya* (1909), and *Two Summers in the Ice-Wilds of Eastern Karakoram: The Exploration of Nineteen Hundred Square Miles of Mountains and Glaciers* (1917).

The Workmans' climbing expeditions, which took place from 1898 to 1906, made Fanny Workman the first woman to explore several peaks and glaciers. Her ascent of Mount Koser Gunge in the Karakoram mountain range in 1899 gave her the record for the highest climb by a woman, as did her ascent of Nun Kun in 1906. She and her husband also first mapped some of the peaks they reached. In 1912 they embarked on a major expedition to the Siachen Glacier. They received some loaned surveying instruments from the Royal Geographic Society, but the Workmans financed the expedition themselves.

Unable to travel during World War I, the Workmans planned to continue their explorations once the conflict was ended, but in 1917 Fanny developed an illness that prevented her from further travel. She settled in France, where she died in 1925. *See also* BICYCLES; EXPLORERS.

Further Reading: Workman, William Hunter, and Fanny Workman. *Algerian Memories: A Bicycle Tour over the Atlas to the Sahara*. London: T. F. Unwin, 1895; ———. *Ice-Bound Heights of the Mustagh: An Account of Two Seasons of Pioneer Exploration and High-Climbing in the Baltistan Himálaya.* New York: C. Scribner's Sons, 1908; ———. *Sketches Awheel in Modern Iberia.* New York and London: G. P. Putnam's Sons, 1897; ———. *Through Town and Jungle: Fourteen Thousand Miles A-Wheel among the Temples and People of the Indian Plain*. London: T. Fisher Unwin, 1904; ———. *Two Summers in the Ice-Wilds of Eastern Karakoram: The Exploration of Nineteen Hundred Square Miles of Mountains and Glacier*. New York: E. P. Dutton, 1917.

Wortley, Emmeline Stuart (1806–1855) and Victoria (1837–1912)

Lady Emmeline Stuart Wortley (1806–1855) and her daughter Victoria (1837–1912) were members of British nobility who traveled throughout the world, even in regions not normally visited by Englishwomen. They went to North and South America, Spain, and Egypt as well as Africa. Emmeline wrote several books about their adventures, including *Travels in the United States &c. during 1849 and 1850* (1851) and *A Visit to Portugal and Madeira* (1854). On a trip to the Middle East in 1855, the women grew sick from heat, exhaustion, and bad food. Emmeline died from her illness, and Victoria never traveled again. *See also* AFRICA; ASIA; NORTH AMERICA.

Further Reading: Cust, Mrs. Henry. *Wanderers: Episodes from the Travels of Lady Emmeline Stuart-Wortley and Her Daughter Victoria, 1849–1855.* Reprint. London: J. Cape, 1928; Stuart-Wortley, Emmeline. *Travels in the United States, etc., during 1849 and 1850.* 3 vols. New York: Harper & Brothers, 1851.

Wright, Frances (1795–1852)

Born in 1795, nineteenth century British social reformer Frances Wright is representative of women who traveled to educate themselves about politics. Wright traveled from England to America in 1818 to study democratic government, touring America and Canada for two years before returning to London. In 1821 she published *Views of Society and Manners in America* to express her opinions on what she had seen; the book praised Americans and criticized the British. In 1824 Wright returned to the United States to become an abolitionist, lecturing in favor of freeing slaves. She also promoted women's rights. Wright died in 1852.

Further Reading: Wright, Frances. *Views of Society and Manners in America.* 1821. Reprint, edited by Paul R. Baker. Cambridge: Belknap Press of Harvard University Press, 1963.

Wright, Irene (1879– 1972)

Born in 1879, American historian Irene Aloha Wright left her Colorado home at the age of 16 to travel through Mexico, using $300 her mother had given her. Once she was out of money, she took a job as a governess so she could stay in the country. Three years later she returned to the United States to attend school, and, after graduating in 1904, she took a job as a New York newspaper reporter on assignment in Cuba. She spent 12 years reporting on Cuban events, first for the New York paper and later for her own publications, a newspaper and a magazine for Americans living in Cuba. During this period, she decided to write about Cuban history and began studying historical documents in both Cuba and Spain, where the Archives of the Indies had been established to house early writings on South and Central America. She used this information to write several books on the region, including *Cuba* (1912), *The Early History of Cuba, 1492–1586* (1916), *Spanish Documents Concerning English Voyages to the Caribbean, 1527–1568* (1929), and *Documents Concerning English Voyages to the Spanish Main, 1569–1580* (1932). In later years, Wright worked in the United States as a specialist on Cuba and the Caribbean for the National Archives and, later, the State Department. She also served as president of the Society of Women Geographers from 1953–1956. *See also* ORGANIZATIONS AND ASSOCIATIONS.

Further Reading: Wright, Irene Aloha. *The Early History of Cuba, 1492–1586.* 1916. Reprint. New York: Octagon Books, 1970.

Y

Yeager, Jeana (1952–)

Born in 1952, American aviator Jeana Yeager was the first woman to circumnavigate the globe nonstop in an airplane without refueling. She accomplished this feat over nine days in 1986, taking off from Edwards Air Force Base in California with fellow aviators Burt and Dick Rutan. Their aircraft, the *Voyager*, which can fly over 28,000 miles without refueling, weighed 9,000 pounds, of which 7,000 pounds was fuel. This craft also has a unique wing design that makes it more fuel-efficient, a development that has already begun to influence designs for future aircraft. *See also* AIR TRAVEL; CIRCUMNAVIGATORS AND ROUND-THE-WORLD TRAVELERS.

Further Reading: Yeager, Jeana, and Dick Rutan with Phil Patton. *Voyager*. Boston: G.K. Hall, 1989.

Young Women's Christian Association (YWCA)

The establishment of an organization known as the Young Women's Christian Association (YWCA) helped encourage women's travel by providing safe places for young women to live while far from home. The YWCA began as a movement rather than as an official organization. In 1855 in England and in 1858 in the United States, charity and religious groups began turning their attention to the plight of women factory workers who, because of the Industrial Revolution, were leaving their homes in large numbers to labor in industrial centers. Working conditions were often unhealthy, work hours were always long, and the women—some as young as seven—had little opportunity to socialize and no families to offer them guidance. Therefore, associations formed throughout New England to establish boarding houses for these women and to provide them safe places to meet. Some of these associations were secular but most were religious; in 1859, an association in Boston became the first to use the name Young Women's Christian Association. Nonetheless, an association formed a year earlier in New York is generally considered the oldest YWCA in existence.

In 1867, YWCAs sprung up in Hartford, Connecticut; Pittsburgh, Pennsylvania; Cleveland, Ohio; and, Cincinnati, Ohio. The following year St. Louis, Missouri, had a YWCA, and in 1870 they appeared in Dayton, Ohio; Washington, D.C.; Buffalo, New York; and, Philadelphia, Pennsylvania. Within five years, the United States had 28 YWCAs, and the number continued to grow. In 1890 the first YWCA on an Indian reservation appeared in Oklahoma, and by the early 1900s there were YWCAs in primarily African American areas of the South. As YWCA facilities proliferated, travelers moving across the country in search of work were encouraged to stay at YWCA boarding houses at each of their stops.

In 1907 the YWCA officially became a national organization rather than a loose association of independent, local groups. Its headquarters was established in New York City and remains there today. In 1911, the YWCA held its first national conference and declared its commitment to promoting a minimum wage law for women. Since that time, the organization has been involved with issues related to social justice as well as women's health and well-being. The YWCA currently supports a variety of social and health programs and has established shelters for homeless and/or battered women. The YWCA maintains approximately 326 locations throughout the United States that offer help to approximately 2 million women and their families. *See also* EMIGRATION SOCIETIES (BRITISH); INDUSTRIAL REVOLUTION.

Further Reading: Mjagkij, Nina, and Margaret Spratt. *Men and Women Adrift: The YMCA and YWCA in the City*. New York: New York University Press, 1997.

Z

Zwinger, Ann

American nature writer Ann Zwinger has traveled extensively in pursuit of wilderness experiences and in writing about them has encouraged other women to do the same. One of her most popular books—involving one of her most extensive travel experiences—is *Downcanyon: A Naturalist Explores the Colorado River Through the Grand Canyon*. Published in 1995, this book tells of Zwinger's adventures and observations rafting down the Colorado River once per month during a full year. In discussing her decision to embark on such journeys, Zwinger suggests that the only way to truly understand a particular wilderness area is to visit it in all seasons. She says that by viewing the river throughout the year she gained "a different perspective on the river than that of the summertime visitor with only a week to shoot the rapids. The river is not folded up like a neoprene raft and stored away between summers, but is an ever-flowing, energetic, whooping and hollering, galloping presence, whether there's ice along the edge or a flotilla of yellow willow leaves floating in a back-eddy or summer sunshine glinting off a cross chop." (p. 6)

Zwinger's other books include *Run, River, Run: A Naturalist's Journey Down One of the Great Rivers of the West* (1975), which tells of Zwinger's trip down the Green River in Wyoming and Utah; *Wind in the Rock: The Canyonlands of Southeastern Utah* (1978), which describes her visits to canyon country wilderness areas of southeast Utah; and *The Mysterious Lands* (1989), which describes Zwinger's experiences studying bighorn sheep in desert wildlife areas. *Run, River, Run* won the John Burroughs Medal for excellence in nature writing.

Further Reading: Zwinger, Ann. *Downcanyon: A Naturalist Explores the Colorado River Through the Grand Canyon*. Tucson: University of Arizona Press, 1995; ———. *Run, River, Run: A Naturalist's Journey Down One of the Great Rivers of the West.* New York: Harper & Row, 1975; ———. *Wind in the Rock: The Canyonlands of Southeastern Utah*. New York: Harper & Row, 1978.

BIBLIOGRAPHY

Works Consulted

Adams, W. H. Davenport. *Celebrated Women Travellers of the Nineteenth Century*. London: W. S. Sonnenschein & Co., 1883.

Aebi, Tania, with Bernadette Brennan. *Maiden Voyage*. New York: Ballantine, 1989.

Alec-Tweedie, Ethel. *An Adventurous Journey (Russia-Siberia-China)*. London: Hutchinson, 1926.

Alexander, Joan. *Voices and Echoes. Tales from Colonial Women*. New York: Quartet books, 1983.

Allen, Alexander. *Travelling Ladies*. London: Jupiter, 1980.

Allison, Stacy, with Peter Carlin. *Beyond the Limits: A Woman's Triumph on Everest*. New York: Delta (Dell), 1996.

Ashby, Ruth, and Deborah Gore Ohrn, eds. *Herstory: Women Who Changed the World*. New York, Viking, 1995.

Berger, Karen. *Hiking and Backpacking: A Complete Guide*. New York: W. W. Norton, 1995.

Berlage, Gai Ingham. *Women in Baseball: The Forgotten History*. Westport, CT: Praeger Publishing, 1994.

Bird, Isabella L. *A Lady's Life in the Rocky Mountains*. 1879. Reprint, with an introduction by Daniel J. Boorstin. Norman: University of Oklahoma Press, 1960.

———. *Unbeaten Tracks in Japan*. 1880. Reprint, with an introduction by Pat Barr. London: Virago Press, 1984.

Birkett, Dea. *Spinsters Abroad: Victorian Lady Explorers*. Oxford: Basil Blackwell, 1989.

Blum, Arlene. *Annapurna: A Woman's Place*. San Francisco: Sierra Club Books, 1980.

Bond, Marybeth, ed. *A Woman's World*. San Francisco: Travelers' Tales, 1997.

Bond, Marybeth, and Pamela Michael, eds. *A Mother's World: Journeys of the Heart*. San Francisco: Travelers Tales, 1998.

Clappe, Louise Amelia Knapp Smith. *The Shirley Letters: From the California Mines, 1851–1852*. 1933. Reprint, edited with an introduction by Marlene Smith-Baranzini. Berkeley: Heyday Books, 1998.

Clarke, Patricia. "Private Lives Revealed." http://www.nla.gov.au/events/ private/ clarke.html (1999).

Conlon, Faith, Ingrid Emerick, and Jennie Cook, eds. *Gifts of the Wild: A Woman's Book of Adventure*. Seattle, WA: Seal Press, 1998.

Conway, Jill Ker, ed. *Written by Herself: Autobiographies of American Women: An Anthology*. New York: Vintage Books, 1992.

David-Neel, Alexandra. *My Journey to Lhasa*. 1927. Reprint, with an introduction by Peter Hopkirk. Boston: Beacon Press, 1986.

Davidson, Robyn. *Tracks*. New York: Pantheon Books, 1980.

Davis, Gwenn. *Personal Writings by Women to 1900: A Bibliography of American and*

British Writers. Norman: University of Oklahoma Press, 1989.

De Pauw, Linda Grant. *Seafaring Women*. Boston: Houghton Mifflin, 1982.

De Watteville, Vivienne. *Speak to the Earth: Wanderings and Reflections Among Elephants and Mountains*. New York: Harrison Smith and Robert Haas, 1935.

Dinesen, Isak. *Out of Africa*. New York: Random House, 1952.

Donelson, Londa. *Out of Isak Dinesen in Africa: Karen Blixen's Untold Story*. Iowa City, IA: Coulsong List, 1998.

Druett, Joan. *She Captains: Heroines and Hellions of the Sea*. New York: Simon and Schuster, 2000.

The Editors of Aventura Books. *Gifts of the Wild: A Woman's Book of Adventure*. Seattle, WA: Seal Press, 1998.

Farnham, Eliza W. *Life in Prairie Land*. New York: Harper & Brothers, 1846.

Foster, Shirley. *Across New Worlds: Nineteenth-Century Women Explorers and Their Writings*. New York: Harvester Wheat–sheaf, 1990.

Gendergap. "American Women in the Military." http://www.gendergap.com.

Golde, Peggy, ed. *Women in the Field: Anthropological Experiences*. Berkeley: University of California Press, 1986.

Griffin, Lynne, and Kelly McCann. *The Book of Women*. Holbrook, MA: Bob Adams, 1992.

Hadley, Leila. *A Journey with Elsa Cloud*. New York: Penguin Books, 1998.

Hamalian, Leo. *Ladies on the Loose: Women Travellers of the 18th and 19th Centuries*. New York: Dodd, Mead, 1981.

Harding, Les. *The Journeys of Remarkable Women: Their Travels on the Canadian Frontier*. Waterloo, Ontario: Escart Press, 1994.

Harrison, Barbara Grizzuti. *Italian Days*. New York: Weidenfeld & Nicholson, 1989.

Hobson, Sarah. *Masquerade: An Adventure in Iran*. Chicago: Academy Chicago Ltd., 1979.

Jansz, Natania, and Miranda Davies. *More Women Travel: Adventures and Advice from More than 60 Countries*. London: Rough Guides, 1995.

Kaufman, Polly Welts. *National Parks and the Woman's Voice*. Albuquerque: University of New Mexico Press, 1996.

Kingley, Mary H. *Travels in West Africa*. 1897. With an introduction by Elizabeth Claridge. London: Virago Press, 1982.

Leonowens, Anna. *The English Governess at the Siamese Court*. New York: Roy Publishers, 1870.

McCarry, Charles, ed. *From the Field: The Best of National Geographic*. Washington, D.C.: National Geographic Society, 1997.

McCauley, Lucy, ed. *Women in the Wild*. San Francisco, CA: Travelers' Tales, 1998.

Macaulay, Rose. *Fabled Shore: From the Pyrenees to Portugal*. 1949. Oxford: Oxford University Press, 1986.

MacLaren, I. S., and Lisa N. LaFramboise, eds. *The Ladies, the Gwich'in, and the Rat: Travels on the Athabasca, Mackenzie, Rat, Porcupine, and Yukon Rivers in 1926 by C. C. Vyvyan*. Edmonton: The University of Alberta Press, 1998.

Maillart, Ella. *The Land of the Sherpas*. London: Hodder and Stoughton, 1955.

Mayes, Frances. *Under the Tuscan Sun: At Home in Italy*. San Francisco: Chronicle Books, 1996.

Mayotte, Judy. *Disposable People: The Plight of Refugees*. Maryknoll, NY: Orbis Books, 1992.

Mead, Margaret, and Rhoda Metraux. *A Way of Seeing*. New York: The McCall Publishing Company, 1970.

Middleton, Dorothy. *Victorian Lady Travellers*. London: Routledge & Kegan Paul, 1965.

Miller, Luree. *On Top of the World: Five Women Explorers in Tibet*. London: Pad–dington Press, 1976.

Milnes Walker, Nicolette. *When I Put Out to Sea*. Collins: London, 1972.

Morris, Mary. *Maiden Voyages: Writings of Women Travelers*. New York: Vintage Books, 1993.

———. *Nothing to Declare: Memoirs of a Woman Traveling Alone*. Boston: Houghton Mifflin, 1988.

Murphy, Dervla. *The Waiting Land: A Spell in Nepal*. London: John Murray, 1967.

O'Brien, Kate. *Farewell Spain.* London: Virago Press, 1985.

Oliver, Caroline. *Western Women in Colonial Africa.* Westport, CT: Greenwood Press, 1982.

Outside magazine. "Howl: What Goodall and Fossey Did for Primates, a Lone Biologist Has Done for *Canis lupus.*" *Outside,* August 1999, 57.

Outside magazine, "Letting It Be." *Outside,* August 1999, 58.

Peace Corps. "Celebrating Women's History Month." http://www.peacecorps.gov/essays/women/index.html (June 2000).

Pfeiffer, Ida. *A Lady's Second Voyage Round the World.* New York: Harper & Brothers, 1856.

Rau, Santha Rama. *Gifts of Passage.* New York: Harper Brothers, 1961.

Robinson, Jane. *Unsuitable for Ladies: An Anthology of Women Travellers.* Oxford: Oxford University Press, 1995.

———. *Wayward Women: A Guide to Women Travellers.* Oxford: Oxford University Press, 1990.

Russell, Mary. *The Blessings of the Good Thick Skirt.* London: Flamingo, 1994.

Saxby, Maurice, and Robert Ingpen. *The Great Deeds of Heroic Women.* New York: Peter Bedrick Books, 1990.

Scribner, Mary Suzanne, ed. *Telling Travels: Selected Writings by Nineteenth Century American Women Abroad.* DeKalb: Northern Illinois University Press, 1995.

Shaw, David. *Flying Cloud: The True Story of America's Most Famous Clipper Ship and the Woman Who Guided Her.* New York: William Morrow, 2000.

Stark, Freya. *A Winter in Arabia.* Woodstock, NY: The Overlook Press, 1987.

Stefoff, Rebecca. *Women Pioneers.* New York: Facts On File, 1995.

Stewart, Elinore Pruitt. *Letters of a Woman Homesteader.* 1914. Reprint. Boston: Houghton Mifflin, 1982.

Thurman, Judith. *Isak Dinesen: The Life of a Storyteller.* New York: Picador USA, 1995. (Originally published in 1982.)

Tinling, Marion. *Women Into the Unknown: A Sourcebook on Women Travelers.* Westport, CT: Greenwood Press, 1989.

———, ed. *With Women's Eyes: Visitors to the New World, 1775–1918.* Hamden, CT: Archon Books, 1993.

Trollope, Frances. *Domestic Manners of the Americans.* 1932. Reprint. New York: Alfred A. Knopf, 1949.

Truman, Margaret. *Women of Courage.* New York: William Morrow, 1976.

Warren, Lady. *Through Algeria & Tunisia on a Motor-bicycle.* Boston & New York: Houghton Mifflin, 1923.

West, Rebecca. *Black Lamb and Grey Falcon.* New York: The Viking Press, 1941.

Wharton, Edith. *In Morocco.* London: Jonathan Cape, 1927.

Wheeler, Sara. *Terra Incognita: Travels in Antarctica.* New York: Modern Library, 1999.

Withey, Lynne. *Grand Tours and Cook's Tours: A History of Leisure Travel, 1750–1915.* New York: W. Morrow, 1997.

Women's Voices on Africa: A Century of Travel Writings. New York: M. Wiener, 1992.

Workman, William Hunter, and Fanny Bullock Workman. *Through Town and Jungle: Fourteen Thousand Miles A-Wheel Among the Temples and People of the Indian Plain.* London: T. Fisher Unwin, 1904.

Internet Web Sites

Academy of Achievement. http://www.achievement.org (May 2000).

AdventureWomen: Adventures Worldwide for Women Over 30. http://www.rainbowadventures.com (May 2000).

Age of Reason: Information for Senior Travelers. http://www.ageofreason.com (May 2000).

The Alpine Club. http://www.alpine-club.org.uk/acg/acg.htm (May 2000).

American Association of American Geographers. http://www.aag.org (May 2000).

ASTA (American Association of Travel Agents): "Becoming a Travel Agent." http://www.astanet.com/www/asta/pub/car/becomingagent.htmlx

Bibliography

The Automobile Club of Southern California. http://www.aaa-calif.com (May 2000).

Clarke, Patricia. "Private Lives Revealed: Letters, Diaries, History." National Library of Australia. http://www.nla.gov.au/events/private/clarke.html (May 2000).

Comet. Issue 3, June 1998. http://www.lonelyplanet.com/comet/issue3.htm

Damron Travel Guide Information. http://www.damron.com/catdamron.html

Department of Foreign Affairs and International Trade. "Her Own Way: Advice for the Woman Traveller," http://www.dfait-maeci.gc.ca/travel/consular/16009-e.htm

Desert USA. "Nellie Cashman: The 'Angel' of Tombstone." http://www.desertusa.com/mag98/may/papr/du_cashman.html (May 2000).

Ecosource Network. http://www.ecosourcenetwork.com (May 2000).

Fox, Barbara Radin, and Larry Fox. "Why We Cruise." http://www.romanticgetaways.com/why.html (May 2000).

Exploratorium's Science of Baseball "The Bloomer Girls." http://www.exploratorium.edu/baseball/girls_2.html (May 2000).

The Explorers Club: World Center for Exploration. http://www.explorers.org.

Federal Highway Administration. "Women in Transportation." http://www.fhwa.dot.gov/wit/admin.htm (May 2000).

Girl Scouts USA. http://www.gsusa.org (May 2000).

History of Women in Sports Timeline. http://www.northnet.org/stlawrenceaauw/timeline.htm (May 2000)

Journeywoman: The Premier Travel Resource for Women. "A Journeywoman's Beat-the-Heat Fact Sheet." http://www.journeywoman.com/journeydoctor.beatheheat.html.

Long Island History. "Deborah Moody." http://www.lihistory.com/3/hs304a.htm (May 2000).

Lustig, Lawrence. "Bargaining Tips for Travellers." http://www.travel-library.com/general/bargaining.html (May 2000).

MCW HealthLink. Medical College of Wisconsin Physicians & Clinics. "Travel Medicine." http://healthlink.mcw.edu/travel-medicine/ (June 2000).

The Ninety-Nines: International Organization of Women Pilots. http://www.ninety-nines.org (May 2000).

Notable Women Ancestors. "Maud Griffin, AKA 'Tugboat Annie.'" http://www.rootsweb.com/~nwa/tugboat.html (May 2000).

Outward Bound USA. http://www.outwardbound.com (May 2000).

Peace Corps History. http://www.peacecorps.gov/about/history/index.html (May 2000).

Red Cross Society. http://www.redcross.org (May 2000).

The Rosie Swale Website. www.homeusers.prestel.co.uk/davidadcooke/swale/contact.html.

Royal Geographical Society. http://www.rgs.org (May 2000).

Society of Women Geographers Fact Sheet. http://www.pleiades-net.com/org/SWG.1.html.

The Subculture Pages: Isabelle Eberhardt Biographical Notes. http://www.fringewear.com/subcult/Isabelle-Eberhardt.html

Travel Document Systems Passport/Visa Information. http://www.traveldocs.com (May 2000).

University of Minnesota Library, biographical information. http://www.lib.umn.edu/etrc/travbio.htm (May 2000).

Women's History in America, Women's International Center. http://www.wic.org/misc/history.htm (May 2000).

Women's Internet Information Network. http://www.undelete.org.

Women's Travel Club. http://www.womenstravelclub.com (May 2000).

WorldWide Classroom. Dr. Lalervo Oberg, "Culture Shock and the Problem of Adjusting to New Cultural Environments." http://www.worldwide.edu/planning_guide/Culture_Re-entry_Shock.htm.

YWCA. http://www.ywca.org (May 2000).

INDEX

by Debbie Lindblom

Page numbers in bold type refer to *Encyclopedia* entries. Page numbers in italic type refer to illustrations.